Recent Advances in Biomedical Imaging

Recent Advances in Biomedical Imaging

Editor

Cuneyt M. Alper

MDPI

Basel • Beijing • Wuhan • Barcelona • Belgrade • Novi Sad • Cluj • Manchester

Editor
Cuneyt M. Alper
University of Pittsburgh School of Medicine
Pittsburgh
USA

Editorial Office
MDPI
St. Alban-Anlage 66
4052 Basel, Switzerland

This is a reprint of articles from the Special Issue published online in the open access journal *Bioengineering* (ISSN 2306-5354) (available at: https://www.mdpi.com/journal/bioengineering/special_issues/24F01PJ541).

For citation purposes, cite each article independently as indicated on the article page online and as indicated below:

Lastname, A.A.; Lastname, B.B. Article Title. *Journal Name* **Year**, *Volume Number*, Page Range.

ISBN 978-3-7258-0131-2 (Hbk)
ISBN 978-3-7258-0132-9 (PDF)
doi.org/10.3390/books978-3-7258-0132-9

Contents

About the Editor

Cuneyt M. Alper

Dr. Cuneyt M. Alper is a Professor of Otolaryngology and Clinical and Translational Research. He served as Vice Chair of Research in Pediatric Otolaryngology at the University of Pittsburgh, School of Medicine. He is an Attending Physician at Children's Hospital of Pittsburgh (CHP) of University of Pittsburgh Medical Center (UPMC). Dr. Alper obtained his M.D. degree at Hacettepe University Medical Faculty, Ankara, Turkey. Dr. Alper had his otolaryngology residency training in Turkey and came to Pittsburgh in 1992 to continue his research on otitis media. Dr. Alper completed a pediatric otolaryngology fellowship at CHP in 1996, and joined its faculty at the Division of Pediatric Otolaryngology. Dr. Alper has published more than 100 peer-reviewed articles and 25 book chapters. He has edited four textbooks, including the fourth edition of the most comprehensive textbook in pediatric otolaryngology, "Bluestone and Stool's Pediatric Otolaryngology". Dr. Alper was the first editor of the books "Decision Making in Ear Nose Throat Disorders" and "Advanced Therapy in Otitis Media"—which is one of the most comprehensive and contemporary books on middle ear infections. The second edition of "Decision Making in Otolaryngology" was also published. Dr. Alper's research interest on otitis media, the Eustachian tube, and medical imaging have an extensive track record of NIH as well as industry funding for about three decades, including a number of grants of which he has been either the principal or co-investigator.

bioengineering

MDPI

Article

Fabrication of a Reflective Optical Imaging Device for Early Detection of Breast Cancer

Huu Thuan Mai [1,*], Duc Quan Ngo [1], Hong Phuong Thi Nguyen [2] and Duong Duc La [3,*]

1 School of Engineering Physics, Hanoi University of Science and Technology, No. 1 Dai Co Viet, Hanoi 100000, Vietnam; quan.ngoduc@hust.edu.vn
2 School of Chemical Engineering, Hanoi University of Science and Technology, No. 1 Dai Co Viet, Hanoi 100000, Vietnam; phuong.nguyenthihong@hust.edu.vn
3 Institute of Chemistry and Materials, 17 Hoang Sam, Hoang Quoc Viet, Cau Giay, Hanoi 100000, Vietnam
* Correspondence: thuan.maihuu@hust.edu.vn (H.T.M.); duc.duong.la@gmail.com (D.D.L.); Tel.: +84-966185368 (D.D.L.)

Abstract: This work presented the design and fabrication of a blood vessel and breast tumor detection device (BKA-06) based on optical energy spectroscopy. The BKA-06 device uses red-to-near-infrared light-emitting diodes that allow physicians or physicians to visualize blood vessels and surface structures such as breast tumors with the naked eye. The device consists of a built-in current control circuit to have the appropriate brightness (maximum illuminance of 98,592 lux) for the examination of superficial tumors deep under the skin, with a scan time of 3–5 min. The device BKA-06 can facilely observe each layer of blood vessels at the depth of the skin. For breast tumors, the location, size, and invasive areas around the tumor can also be visualized with the naked eye using the BKA-06 sensor. The results show that the BKA-06 sensor can provide clear breast tumor and vascular images, with a penetration of up to 15 cm in the skin and tissue layers of the breast. The breast tumor scanning tests with the BKA-06 sensor gave patients quick results and compared them through cell biopsy and MRI, respectively. The device has the advantages of being simple and easy to use, providing potential practical applications in the medical field and reducing costs for patients when taking MRI or CT scans. Therefore, the BKA-06 device is expected to help doctors and medical staff overcome difficulties in infusion, as well as identify breast tumors to support early breast cancer diagnosis and treatment.

Keywords: BKA-06; blood vessels detection device; red LED lights; medicine

Citation: Mai, H.T.; Ngo, D.Q.; Nguyen, H.P.T.; La, D.D. Fabrication of a Reflective Optical Imaging Device for Early Detection of Breast Cancer. *Bioengineering* **2023**, *10*, 1272. https://doi.org/10.3390/bioengineering10111272

Academic Editors: Cuneyt M. Alper and Andrea Cataldo

Received: 13 August 2023
Revised: 14 October 2023
Accepted: 24 October 2023
Published: 1 November 2023

1. Introduction

In medicine, the skill of identifying veins in infusions and taking blood is a skill that not all doctors can perform well. Finding a vein is difficult (especially for children, women, and obesity) [1]. It is not uncommon for cases of vein deviation or the rupture of vessels when injected or drawn blood that has adversely affected the patient. The vein retriever process requires the precise operation of people with experience that are not allowed to make mistakes because, if performed many times, will cause pain and discomfort to the patient [2].

On the other hand, the problem of breast cancer shows that the incidence of breast cancer is increasing in many parts of the world, especially in transition countries [3]. Sadly, an estimated 685,000 women lost their lives to breast cancer in 2020, making up 16% of all cancer deaths in women or 1 in every 6 deaths. To address this concerning trend, the World Health Organization (WHO) recently initiated the Global Breast Cancer Initiative [4]. In 2018, Vietnam ranked first in the world in terms of breast cancer incidence, with 15,229 new cases and 6103 deaths. In 2020, there were over 21,555 new cases, including more than 9345 deaths because of breast cancer [5]. Currently, Vietnam annually records about 183,000 new cases of cancer, of which breast cancer accounts for 25.8% of cancers in women, with nearly 22,000 new cases and more than 9000 deaths [6]. Notably, often, the younger breast cancer patients are, the worse the prognosis and the lower the cure rate compared to the elderly. Epidemiological studies indicated that the most effective method to increase the cure rate of breast cancer, especially in young people, is screening examination to detect the disease early. While the over 5-year survival rate for the very early stage reaches 98%, for the late stage, this rate is only about 10%. This collaborative effort aims to reduce breast cancer mortality by promoting prompt diagnosis, effective treatment, and comprehensive patient management. Currently, there are many advanced diagnostic techniques, such as the BRCA1/2 gene test, breast ultrasound elastography, 3D mammogram, magnetic resonance imaging (MRI), breast tumor biopsy, ultrasound, computed tomography (CT), etc.

Past studies have indicated that the timely detection and appropriate treatment of breast cancer could significantly decrease mortality rates in the long run [7]. At present, there is a wide array of clinical approaches employed for the screening and diagnosis of breast cancer. Although mammography is currently the standard screening method for breast cancer, it has limitations. It is less effective for individuals under 40 years old and with dense breasts, and it is not sensitive enough to detect small tumors (less than 1 mm, approximately 100,000 cells). Furthermore, it does not provide any information on the potential outcome of the disease [8,9]. Contrast-enhanced (CE) digital mammography, on the other hand, offers a more precise diagnosis than mammography and ultrasound for individuals with dense breasts. However, this technique is not widely accessible due to its cost and the high levels of radiation involved [10]. Ultrasound has been utilized as an additional tool for screening. Magnetic resonance imaging (MRI) has the capability to identify small lesions that may not be detectable through mammography, but it is also costly and has low specificity, potentially leading to overdiagnosis [9,10]. Positron emission tomography (PET) is recognized as the most accurate method for visualizing tumor spread or monitoring their response to treatment [11].

However, the aforementioned methods face challenges when it comes to widespread implementation in early-stage cancer screening due to their limited availability in local medical facilities, high costs, lengthy result turnaround times, and complex processes [12]. On the other hand, thermal imaging has emerged as an effective method for the early detection and diagnosis of breast cancer. Reflectance optical imaging operates by quantifying both transmitted and reflected photons and utilizing this data alongside tomography reconstruction algorithms to create images of spatial absorption and/or scattering patterns [13,14]. These optical imaging systems have the capacity to generate three-dimensional (3D) images [15]. In recent decades, efforts have been made to advance these techniques for clinical research [16]. However, the complexity and high expenses associated with time domain (TD) systems have impeded their widespread adoption [17]. On the other hand, frequency domain (FD) systems function by modulating the light source's amplitude at a high frequency (>50 MHz), with detectors measuring the reduction in amplitude and phase shift of the transmitted signal [18]. Conversely, continuous wave (CW) systems emit near-infrared (NIR) light at a consistent intensity or with low-frequency modulation (a few kHz) to enhance the signal-to-noise ratio. In CW systems, detectors gauge the reduction in amplitude of the transmitted signals [19]. CW systems have been enhanced with multiwavelength laser diodes and dense arrays of sources and detectors, making them valuable in clinical research [20,21]. CW systems only require photo detectors with a slow response rate and low-frequency circuits. Consequently, CW systems are significantly more cost-effective and less complex than FD systems, allowing for the integration of more sources and detectors [22]. Nevertheless, relying solely on amplitude data from CW systems presents limitations in quantitative optical reconstruction, and clinical applications have predominantly focused on monitoring dynamic changes in breast abnormalities [23]. To address this limitation and strike a balance between cost and complexity, FD and CW systems have been combined [24].

In this study, we have focused on the development and testing of the CW light-based reflectance device (named BKA-06 device), which simulates the energy absorption image of cancer (particularly breast cancer and superficial cancers). BK06 can provide valuable contribution to the field by addressing the limitations of the optical reflectance imaging method, such as enhanced accuracy, providing real-time results, cost-efficient, and less harmful to tissues. Through testing on both volunteers and patients, this device has demonstrated outstanding advantages compared to similar devices, such as (1) cheap price (250–300 USD/device); (2) to give fast results (in 3–5 min); (3) no side effects; (4) to produce a clear and accurate image and location of the tumor within a range of less than 15 cm (compared with MRI). Additionally, the device also aids in identifying blood vessels during patient treatment, which is crucial for procedures such as blood collection and injection/infusion. It is worth noting that not all medical professionals possess the expertise to accurately identify blood vessels, and misidentification can have detrimental effects on patients, especially children.

2. Materials and Methods

2.1. Research Methods

BKA-06 is based on the biomedical physics principles. Analyze and select electronic elements and suitable materials for fabrication equipment.

2.2. Physical Basis and Operating Principle of BKA-06

In the composition of red blood cells, there is a heme pigment, the porphyrin type that may combine with metal atoms. Heme in humans is protoporphyrin IX in combination with Fe. Heme has 4 pyrol nuclei linked together by the menten bridge (-CH=). Porphyrin rings are attached to the groups as metyl (-CH_3) at the sites 1, 3, 5, and 8; vinyl (-CH=CH_2) at the sites 2 and 4; and propionyl (-CH_2-CH_2-C00H) at the sites 6 and 7.

Within heme, the iron atom is attached to the pyrrole nucleus by two symmetrical and two coordinated links, as well as to the protein globin through a histidine residue at the base. These bonds are crucial for the functioning of hemoglobin (Hb) (Figure 1), as they help to stabilize the structure and allow for the reversible binding of oxygen.

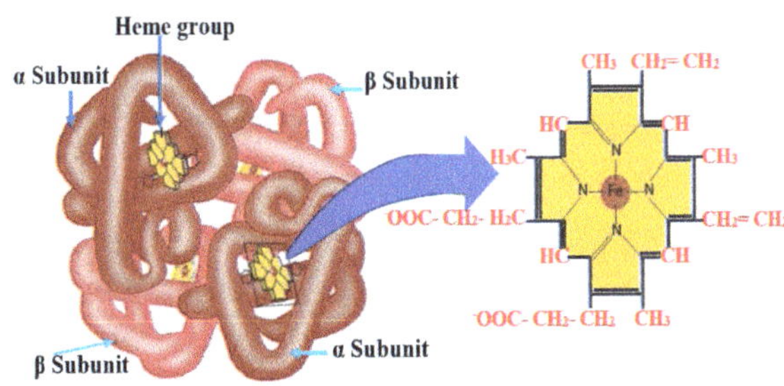

Figure 1. Chemical formular and structure of isoleucine and hemoglobin—the nucleus of red blood cells.

Each Hb molecule consists of four hemes, with one heme containing 1 Fe^{2+} Spectrometry analysis of blood with the maximum absorption spectrum within the 540–700 nm region, with the maximum absorbance of the SHb component at 622.8 nm [11,25]. Human blood with a hematocrit of 10% and oxygen saturation of 98% displays optical properties of 0.210 ± 0.002 mm^{-1} for μa, 77.3 ± 0.5 mm^{-1} for μs, and 0.994 ± 0.001 for the g factor [26]. As hematocrit increases up to 50%, there is a linear increase in absorption and a decrease in scattering. Variations in osmosis and the wall cutting speed affect all three parameters, while oxygen saturation has a significant impact solely on the absorption coefficient. Measurements of oxidized and deoxidized blood absorption spectra in the 400–2500 nm wavelength range demonstrate that blood absorption follows that of hemoglobin and water.

The scattering factor for λ = 500 nm reduces to approximately 1.7, and the g coefficient is 0.9 higher for the entire wavelength range [17].

2.3. Techniques

Analysis of the circuit components and electronic devices.

Determine the physical parameters of the device by modern measuring devices and reliable devices such as: Kyoritsu 1052 (Kyoritsu, Tokyo, Japan), Lux Meter Testo 0500 (Testo, PA, USA), and Testo 608-H2-US electronic thermometer (Testo, PA, USA).

2.4. Fabrication Design

Component selection: RED LIGHT LED Chip, IC Stable Pressure, and Line Stabilization.

Line stabilization circuit design for LED as shown in Figure 2.

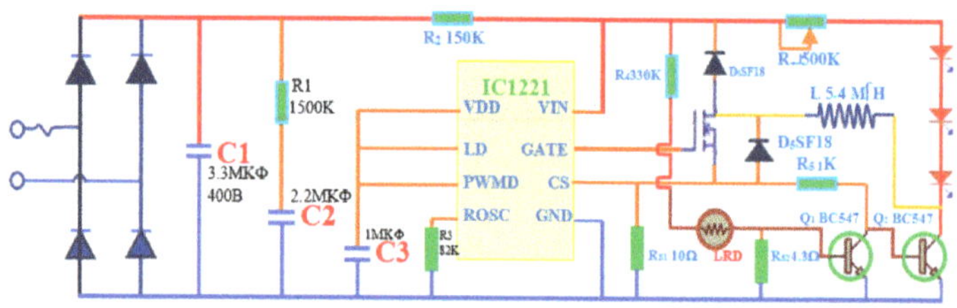

Figure 2. Current stabilizer circuit for LED and sensor (the red line is the positive wire, the blue line is the ground wire, the orange line is the neutral wire; D1 to D4 are the 1N4001 diodes, C1 to C3 are the capacitors; R refers to the resistor; Q refers to the transistors; L refers to the coils; D5 to D7 are the red light-emitting diodes).

The equipment shell of the device is exhibited in Figure 3.

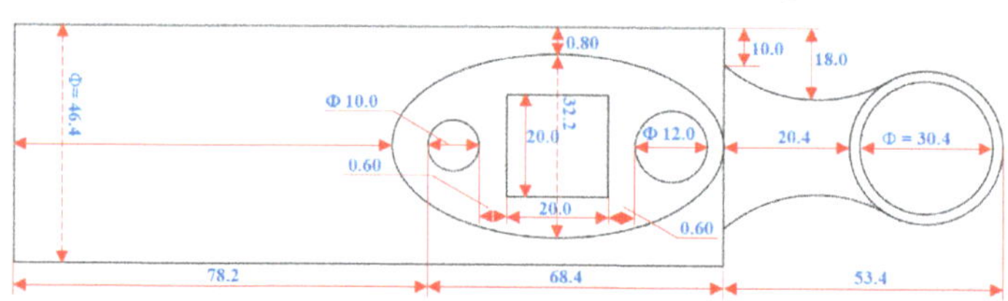

Figure 3. Device shell scheme. Device case: aluminum. Projector material: hard plastic. Projector diameter: Φ = 32.90 mm. Projector length: 34.50 mm. Device size: 200.00 × 38.20 × 50.06 mm.

2.5. Measurement Process

Experiment 1: After designing and testing the elements of the device, the equipment is assembled according to the design scheme, and then, the basic electrical parameters of the BKA-06 (voltage and current intensity when operating) are measured using a universal meter (Kyoritsu 1052—Japan) and some other indicators.

Experiment 2: Survey the parameters of BKA-06: measure the brightness intensity using the three modes of the device (start, shallow test, and deep test) and check the heat increase of the device in the screening area for a period of 1 to 25 min.

Experiment 3: Screening of BKA-06 on the surface of the hands of adults and children.

Experiment 4: Use BKA-06 to identify the position on volunteers and breast cancer patients and then test with MRI.

The results obtained from both patients and volunteers have been officially certified with ethics certificate number V11.CN6.3000.23 by the Vietnam Metrology Institute.

3. Results

3.1. Studies of Basic Parameters of BKA-06

3.1.1. Device Specifications

Figure 4 shows the printed circuit scheme, BKA-06 device, and schematic layout of the device with following specifications:

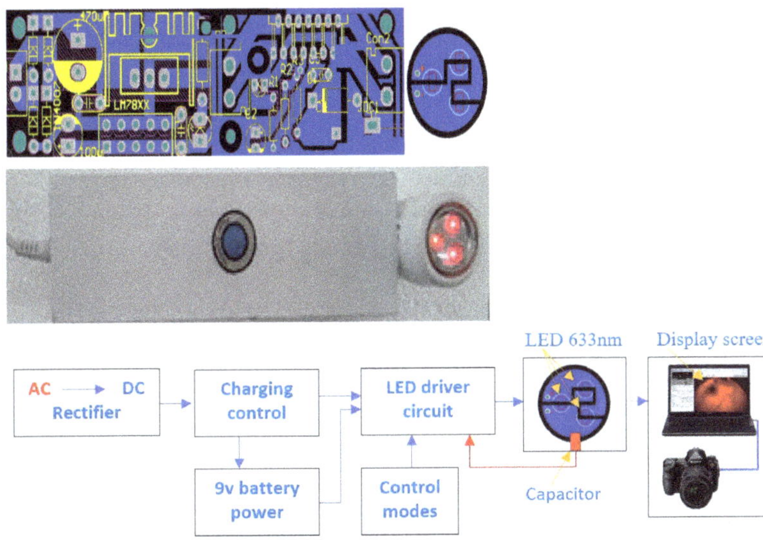

Figure 4. Printed circuit scheme, BKA-06 device, and schematic layout of the device.

Changing power source: 220 V~50 Hz, power supply: lithium–ion battery of 9 V
Power Consumption: 12.19 W
Dimensions: 210.08 × 41.22 × 50.08 mm; Φ 32.92 ± 0.02 mm (use caliper, micrometer)
Weight: 892 g
The ability to penetrate the layers of tissue in the breast is 15 cm.

Device BKA-06 (early vein detection and breast cancer detection device by the energy spectrum) has been licensed by the Metrology Institute—Vietnam Academy (attached to certificate no.: V11.CN6.300.23).

3.1.2. Measure Brightness

Using the lux meter (Testo 0500, Testo, PA, USA) and multimeter (Kyoritsu 1062, Kyoritsu, Tokyo, Japan), place the sensor head of the lux meter close to the header of the BKA-06 equipment and measure the brightness intensity in different changing modes. The result is shown in Table 1 and Figure 5.

The survey results show that the amperage reaches 0.75 to 2.80 A, the maximum power consumption is 12.19 W, and the illuminance increases from 5.03 to 143.93 W/m^2. With an illuminance from 5.03 to 21.73 W/m^2, which is bright enough to examine capillaries close to the skin, this mode is suitable for superficial tumor surfacing or vascular examination with the BKA-06 device (Figure 6b,c). With an illuminance from 28.08 to 91.31 W/m^2, suitable for the vascular examination of tumors and superficial subcutaneous lymph nodes (neck, nasopharynx, and breast), this mode has a penetration depth of about 3 ÷ 7 cm (Figure 6a). The illuminance from 91.31 to 143.93 W/m^2 is suitable for deep lymph nodes under the skin of the breast 7 ÷ 15 cm for breast cancer diagnosis (using the Sony Alpha ILCE-6400L/A6400 Kit and 16–50 mm F3.5–5.6 OSS Camera).

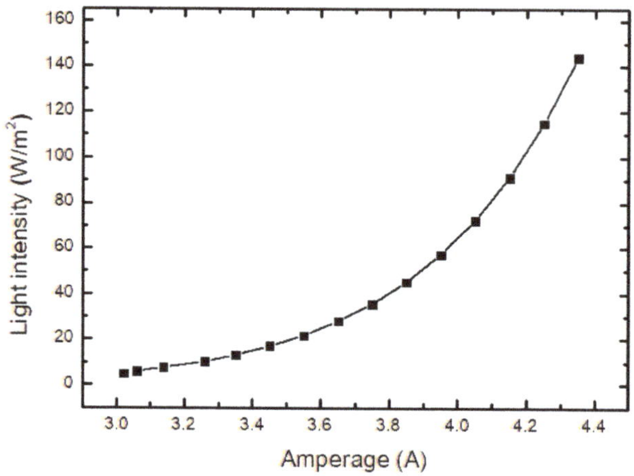

Figure 5. Relationship between amperage and illuminance.

Table 1. Results of measuring the device power and illuminance.

Amperage I (A)	Voltage U (V)	Power P (W)	Light Intensity (W/m^2)
3.02	0.75	2.27	5.03
3.06	0.82	2.50	6.07
3.14	0.90	2.82	7.70
3.26	1.00	3.25	10.24
3.35	1.10	3.68	13.16
3.45	1.21	4.17	17.07
3.55	1.33	4.73	21.73
3.65	1.46	5.35	28.08
3.75	1.61	6.04	35.59
3.85	1.77	6.81	45.22
3.95	1.94	7.67	57.33
4.05	2.13	8.63	72.49
4.15	2.33	9.69	91.31
4.25	2.56	10.87	114.99
4.35	2.80	12.19	143.93

3.2. Vascular Results in Adults and Children

An angiogram of the blood vessels in the arm and hands (6 adults) is shown in Figure 6b,c.

The obtained results show that the images of the blood vessels under the skin are very clear, and their positions are easily recognized with the naked eye.

The results obtained show that the images of blood vessels under the skin are very clear and easily identifiable with the naked eye when using the BKA-06 device. Because the BKA-06 device heats the area for imaging, the thermal camera also provides clear images of the blood vessel locations. Additionally, for blood vessels located deeper at 2 cm to 3 cm, heating with the BKA-06 also yields images with clearer contrast of the blood vessels on the thermal camera. The maximum temperature increase of the device during scanning is ΔT °C < 1 °C; this temperature increase is very small compared to the change in body temperature according to environmental temperature. The comparison of the images clearly indicates that the contrast of blood vessel images by the BKA-06 is more distinct.

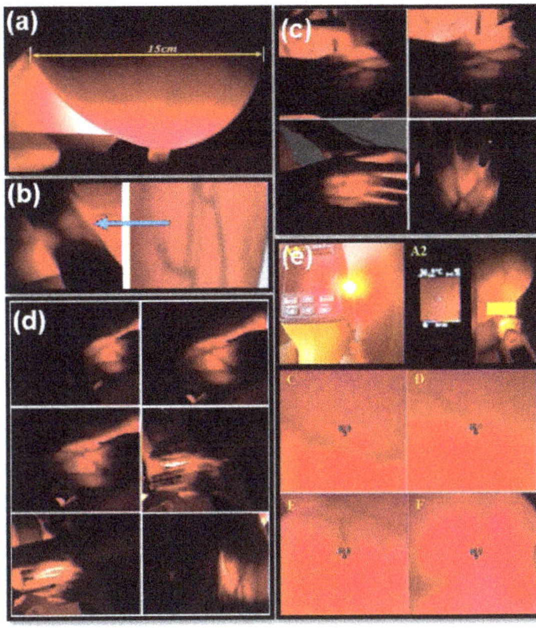

Figure 6. (**a**) BKA-06 equipment image that scans through layers of adult breast tissue. (**b**) Vascular screening on the surface of an adult's arm using BKA-06 equipment. (**c**) Vascular screening on an adult's hand using BKA-06 equipment. (**d**) Vascular screening on a child's hands using BKA-06 equipment. (**e**) The thermal images of blood vessels illuminated by BKA-06 on (A1, C, and D) a child's forearm and (A2, E, and F) adult's breast.

3.3. Breast Test Results on Volunteers and Patients

3.3.1. Volunteers

Volunteers were randomly selected between the ages of 18 and 60 (never had breast cancer tested with any medical devices, no suspicious or conclusive signs of breast cancer). The initial scan results are shown in Figures 7 and 8 and Table 2.

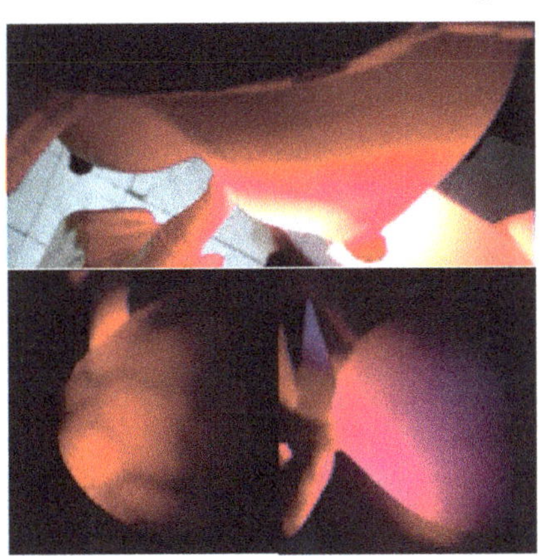

Figure 7. Breast test photos of volunteer no. 9.

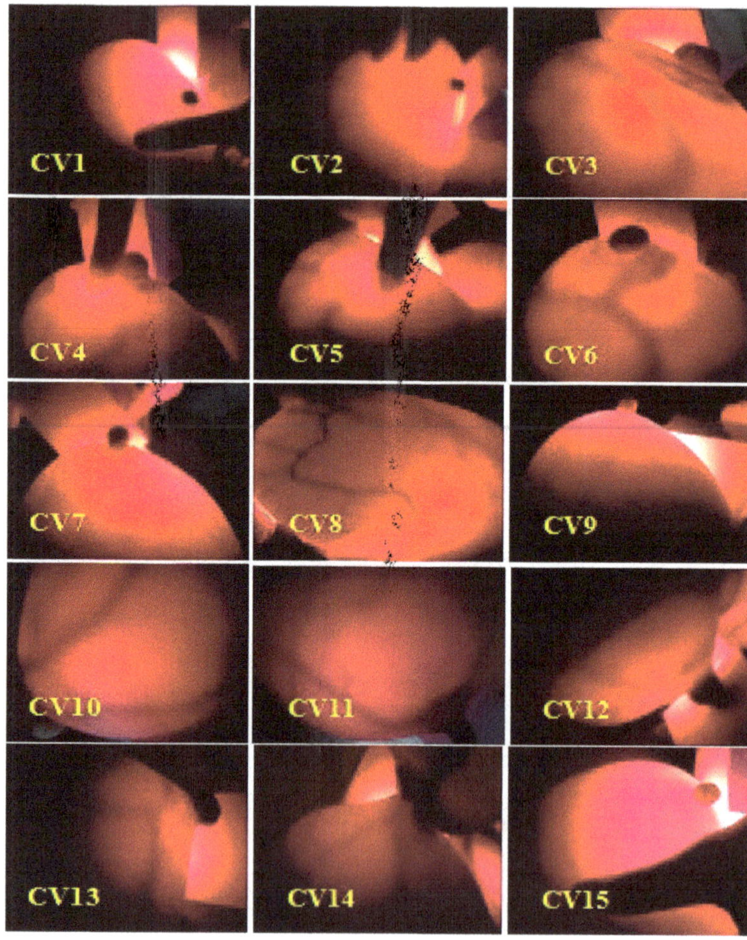

Figure 8. Summary of breast cancer test results using the BKA-06 device.

Table 2. List of volunteers for breast cancer screening (3 April–28 May 2022).

No.	Full Name	Age	Code
1	Volunteer 1	43	CV1
2	Volunteer 2	31	CV2
3	Volunteer 3	28	CV3
4	Volunteer 4	41	CV4
5	Volunteer 5	55	CV5
6	Volunteer 6	30	CV6
7	Volunteer 7	60	CV7
8	Volunteer 8	40	CV8
9	Volunteer 9	65	CV9
10	Volunteer 10	40	CV10
11	Volunteer 11	18	CV11

Table 2. *Cont.*

No.	Full Name	Age	Code
12	Volunteer 12	42	CV12
13	Volunteer 13	36	CV13
14	Volunteer 14	25	CV14
15	Volunteer 15	58	CV15

(In this table, some volunteers requested not to publish their phone number, the last two numbers of the symbol XX, and the image of their breasts is committed to each individual to be allowed to publish the images and the information in scientific journals. They volunteered to participate for scientific purposes, not for any other purposes).

Figure 9 clearly shows the non-screening area and the scanning area. At the illuminated area, we observe with the eye the mammary gland blood vessels with a dark red color, and with the appropriate brightness intensity, we can observe the layers of breast tissue. Therefore, there is no need for an infrared camera auxiliary device to connect a computer to simulate the image, as we can also see the inside image of the breast when using the equipment.

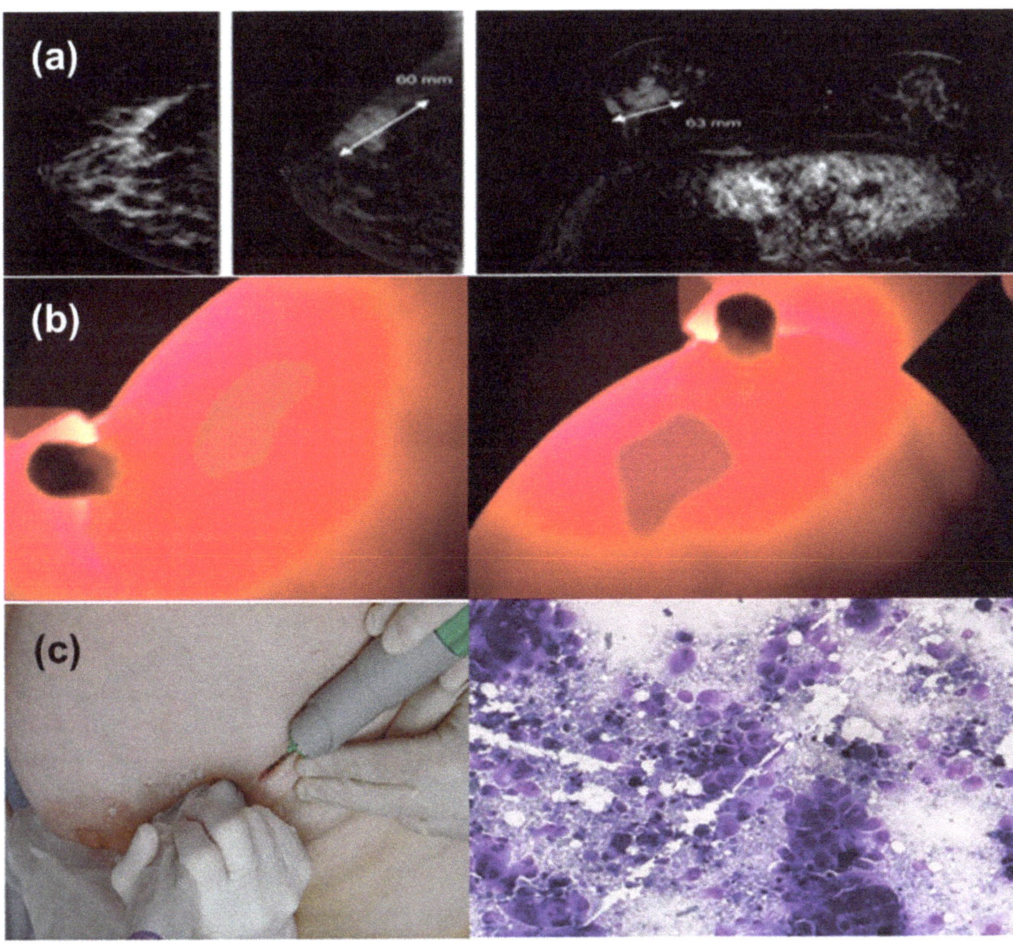

Figure 9. Images of (**a**) MRI scans, (**b**) BKA-06, and (**c**) cell biopsies of breast cancer patient no. 6.

The results of breast cancer screening with the BKA-06 device on 15 volunteers showed quality images, observed with the naked eye; the mammary glands and blood vessels were very clear, with no abnormal organization and no lymphoma.

3.3.2. Patient Participation

Twelve (12) patients who took part in the scan had abnormal breast abnormalities. Two MRI patients were found to have tumors, then had breast cancer scans using BKA-06 devices; the rest were clinically examined and suspected of a tumor from a MRI scan, so conducted a BKA-06 scan, MRI scan, and cell biopsies.

Results:

+ Patient profile: Patient No. 6, age of 58. Profile number: 7673/20. Diagnose: Mammary tumor (T), size 60 × 63 mm, middle tumor in the breast (T), non-mobile soft tumor (Δ(t): left mammary K (T) $T_4N_0M_0$. GTB UT GDII). The MRI image shows the tumor, location, and size of 60 × 63 mm of these images taken after the photoreceptor injection and through a computer simulation system. Observing two perpendicular images from the BKA-06 device, we also see that the tumor area is darker than the area around the tumor; the border area is especially very clear, so it is easy to recognize the shape of the tumor's position and size. Because the device has not been integrated with a computer, the tumor size on the image of BKA-06 is not quantized, but through the naked eye, we also estimate the size of the tumor. The results are shown in Figure 9.

+ Patient profile: Patient No. 7. Profile number: 7692/20. Diagnose: Mammary tumor (T), size 16 × 29 mm, inner protrusive bottom (T), I am probably not mobile (Δ(t): left mammary gland (T) T4N0M0. GTB UT GDII). Results of MRI scans, BKA-06, and cell biopsies are shown in Figure 10.

Similar to patient 6, the images taken of the disease give clearer and darker images (it may be because one patient has a different type with solid lymph nodes, according to the results of the clinical examination) so we can observe the depth of the tumor; the area around the tumor (yellow vinculum) develops many capillary organizations, so it is darker than the middle area tumor center (blue vinculum in Figure 11). The cell biopsy results show melanoma. The results on the MRI show the location and size of the tumor (16 × 29 mm).

The results of the BKA-06, MRI, and cell biopsies are summarized in Table 3. Observations from the images taken with BKA-06 shows that the location and size of the tumor is very clear and with a shallow depth or same cell density.

Table 3. List of breast cancer test patients (April–December 2022).

Number	Full Name	Age	Encrypted Image
1	Patient 1	25	UHV1–MRI1
2	Patient 2	67	UHV1–MRI2
3	Patient 3	45	UHV3–MRI3
4	Patient 4	65	UHV4–MRI4
5	Patient 5	45	UHV5–MRI5
6	Patient 6	58	UHV6–MRI6
7	Patient 7	64	UHV7–MRI7
8	Patient 8	52	UHV8–MRI8
9	Patient 9	46	UHV9–MRI9
10	Patient 10	43	UHV10–MRI10
11	Patient 11	59	UHV11–MRI11
12	Patient 12	58	UHV12–MRI12

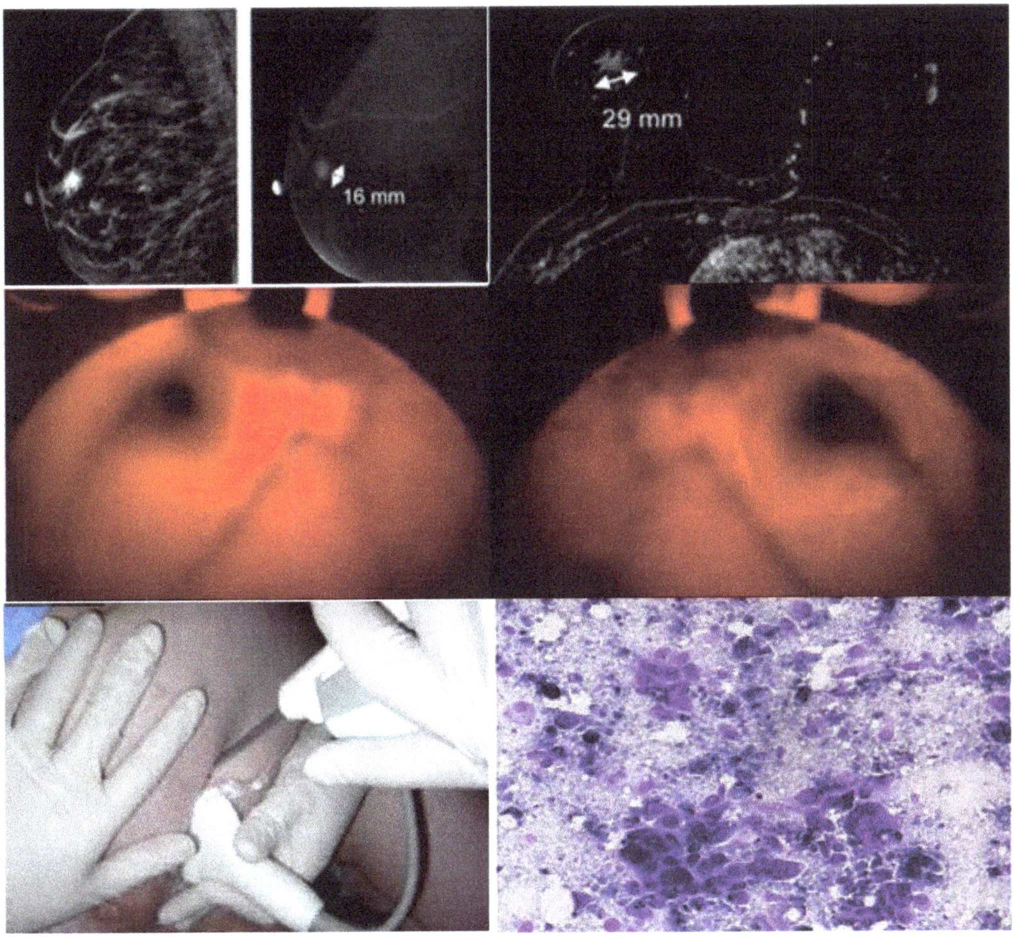

Figure 10. Images of MRI scans, BKA-06, and cell biopsies of breast cancer patient no. 7.

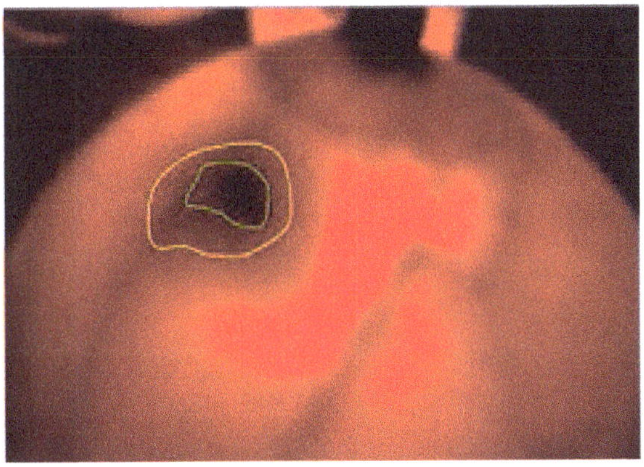

Figure 11. BKA-06 image of patient no. 7.

The results of the MRI scans, BKA-06, and cell biopsies are exhibited in Figure 12. Observations from the images taken using the BKA-06 show that the location and size of the tumor are very clear, no matter the depth or density of the cells and the internal environment in and outside the typical tumors in patients UHV1, UHV7, UHV8, and UHV9. On the other hand, the upper breast MRI images in Figure 12 from the patients also show the location and size of the tumor with the same results as BKA-06. The advantage is that the penetrating image compared to the black and white image on the MRI is recommended to observe the tumor boundary area.

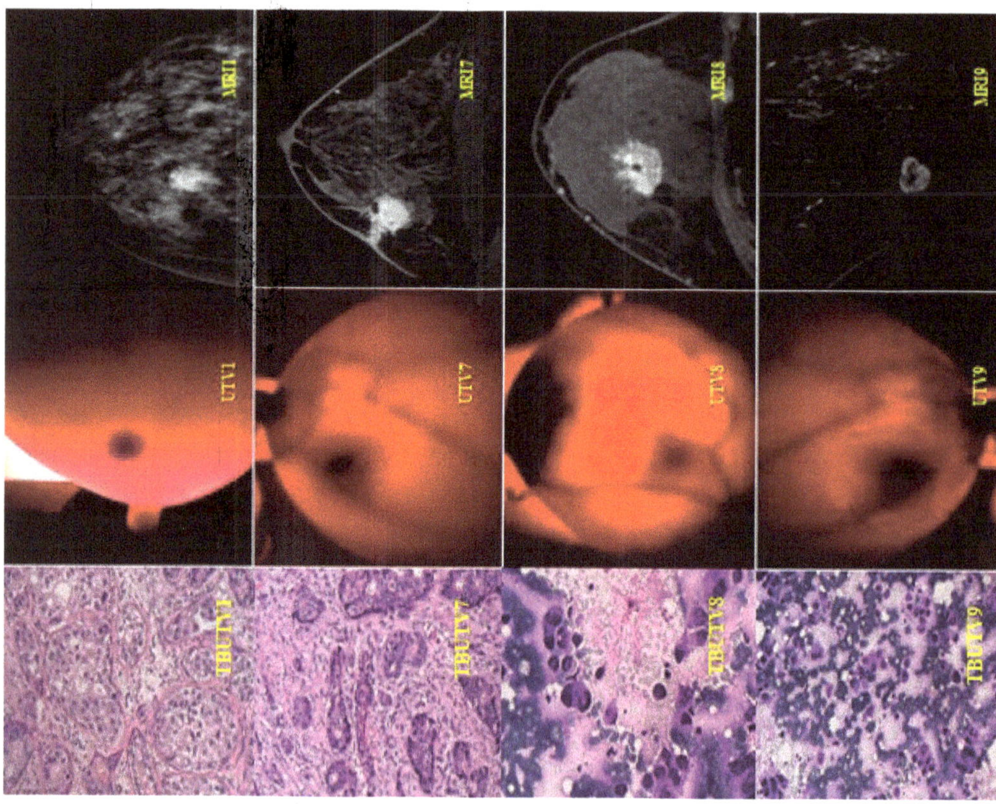

Figure 12. Images of MRI scans, BKA-06, and cell biopsies of breast cancer patients.

4. Discussion

The BKA-06 device obtains 2D images of the screening area on the body. Visually observing the images with the naked eye, we can see some of the structures inside the skin that are layer-by-layer vascular systems in depth [27–29]. For breast tumors, it is also close to the location, size, and invasive areas around the tumor.

The device has the advantages of a simple structure, easy to use, and clear images that can be observed with the naked eye. The area of light emitted in the red area of the spectrum is not harmful to the eyes or skin, because infrared lights are used in the treatment of skin diseases in the cosmetic industry and other areas in health.

The initial results of the successful manufacture of BKA-06 and some potential results can be upgraded to create new essential products for the medical industry, improving the quality of infusions and blood draws.

BKA-06 is specialized for a blood angiogram of shallow tumors. There are outstanding advantages when it comes to breast scans compared to CT scan or MRI methods. Without photovoltaic injections, the 3 ÷ 5 min scan time (CT scan, MRI 25 ÷ 30 min) costs an estimated 7 ÷ 9 million USD, so if the cost of a test is negligible compared to the MRI (2.3 ÷ 2.8 million USD/1 degree), the patient can take a selfie and easily see their results, making it easy to use and with a low cost about 250–300 USD/1 BKA-06 device.

BKA-06 has not been able to capture internal organs, deeper locations, and bone organizations that can be taken by MRI or CT scans, and deep positions on breasts larger than 15 cm cannot be observed.

The theory successfully helped design BKA-06 and initially achieved some good results that can be upgraded to create new essential products for the medical industry and improve diagnostic methods for medical facilities on the domestic and international levels.

The imaging method based on the principle of energy absorption of the mass is a new method in diagnostic imaging, both in the world and in Vietnam, so this study could help to access and update research and applications in other industries. The BKA-06 device is also expected to have superior applications with other diagnostic methods. The BKA-06 device is also expected to have applications that are superior to reality.

The BKA-06 has many advantages, such as technological advancements: The device is capable of providing high-resolution, real-time images of breast tissue, allowing for better visualization of potential abnormalities; non-invasiveness: Unlike traditional mammography, it does not involve ionizing radiation, making it a safer option for regular breast cancer screening; cost-effectiveness: Reflectance optical imaging has the potential to be a cost-effective breast cancer detection method, especially in comparison to more expensive imaging modalities like MRI; and patient comfort: Patients may find reflectance optical imaging more comfortable and less intimidating than traditional mammography, which could encourage more regular screenings.

However, for clinical application, it might also pose several challenges, such as clinical validation (large-scale clinical trials are needed to establish its accuracy and reliability), limited depth penetration (this can affect its sensitivity in detecting tumors located deeper within the breast), operator training (effective use of these devices requires specialized training for healthcare providers), false positives and negatives, integration into the clinical workflow, and reimbursement and regulations (regulatory approval and reimbursement policies need to align with the adoption of this technology. It may take time to navigate these aspects.). Addressing the challenges related to clinical validation, operator training, and integration into clinical practice will be critical for its successful implementation as a routine screening tool for breast cancer. Ongoing research and collaboration between healthcare providers, researchers, and regulatory agencies are essential to overcome these challenges.

5. Conclusions

In short, we successfully designed and fabricated equipment (BKA-06) with basic specifications: power consumption of 12.19 W, size of 210.08 × 41.22 × 50.08 mm, Φ 32.92 ± 0.02 mm, and weight of 892 g. The device is capable of penetrating the layers of tissue on the breast up to 15 cm. BKA-06 does not burn during screening, emits red light (the main wavelength of 633 nm is healthy radiation and does not cause side effects), there are three screening modes, the maximum brightness intensity is 98,238 lux, inspection time is 3 ÷ 5 min, and the image is visible to the naked eye. We have reviewed the safety specifications when using the device in various modes (starting, shallow test, and deep test mode) during venipuncture, as well as breast scan examinations. We have successfully tested the device on volunteers and patients, with very good initial results. The XCS observations identified blood vessels, tumors in the breast, and surrounding organizations very clearly to determine the location, size of the tumor, and the area around the tumor directly without surgery or other diagnostic methods. A parallel examination using MRI and cell biopsies were performed, with reliable results. In future work, we will continue the testing of patients on a larger scale to evaluate the reliability of the device, as well as optimizing the processing factors of the device, to obtain reliable results for breast cancer detection.

Author Contributions: Conceptualization, H.T.M. and D.D.L.; methodology H.T.M.; software, D.Q.N.; validation, H.T.M. and H.P.T.N.; formal analysis H.P.T.N.; investigation, H.P.T.N.; resources, D.Q.N.; data curation, D.Q.N.; writing—original draft preparation H.T.M. and D.D.L.; writing—review and editing, H.T.M. and D.D.L.; visualization, H.T.M. and D.D.L.; supervision, H.T.M. and D.D.L.; project administration, H.T.M. and D.D.L.; and funding acquisition, H.T.M. All authors have read and agreed to the published version of the manuscript.

Funding: This research was funded by the Vietnam Ministry of Education and Training, grant number: B2021-BKA-06.

Institutional Review Board Statement: The study was conducted in accordance with the Declaration of Helsinki, and approved by the Medical Ethics Council of Hospitals under the Vietnam Ministry of Health (certificate no.: V11.CN6.300.23).

Informed Consent Statement: All patients agreed to participate in the testing of the device. Informed consent was obtained from all subjects involved in the study.

Data Availability Statement: Data will be provided upon requested from the authors.

Conflicts of Interest: The authors declare no conflict of interest.

References

1. Duong Thi Tuan, A. *Assessment of the Actual Situation of Performing Intravenous Injection Procedures by Nurses in Construction Hospitals*; KHKT Publisher: Hanoi, Vietnam, 2011; pp. 6–11,13–20.
2. Dang Thi Thanh, T. *Knowledge, Skills to Practice Safe Injection and Some Related Factors of Students at Kon Tum Medical Secondary School in 2016*; KHKT Publisher: Hanoi, Vietnam, 2016; pp. 12–14.
3. Heer, E.; Harper, A.; Escandor, N.; Sung, H.; McCormack, V.; Fidler-Benaoudia, M.M. Global burden and trends in premenopausal and postmenopausal breast cancer: A population-based study. *Lancet Glob. Health* **2020**, *8*, e1027–e1037. [CrossRef] [PubMed]
4. Anderson, B.O.; Ilbawi, A.M.; Fidarova, E.; Weiderpass, E.; Stevens, L.; Abdel-Wahab, M.; Mikkelsen, B. The Global Breast Cancer Initiative: A strategic collaboration to strengthen health care for non-communicable diseases. *Lancet. Oncol.* **2021**, *22*, 578–581. [CrossRef] [PubMed]
5. Bray, F.; Ferlay, J.; Soerjomataram, I.; Siegel, R.L.; Torre, L.A.; Jemal, A. Global cancer statistics 2018: GLOBOCAN estimates of incidence and mortality worldwide for 36 cancers in 185 countries. *CA A Cancer J. Clin.* **2018**, *68*, 394–424. [CrossRef]
6. Migowski, A. Early detection of breast cancer and the interpretation of results of survival studies. *Cienc. Saude Coletiva* **2015**, *20*, 1309. [CrossRef]
7. Breast Cancer. *Early Detection and Prompt Treatment are Critical, Mayo Clinic Health Letter, English ed.*; Mayo Clinic: Rochester, MN, USA, 2013; pp. 1–8.
8. Onega, T.; Goldman, L.E.; Walker, R.L.; Miglioretti, D.L.; Buist, D.S.; Taplin, S.; Geller, B.M.; Hill, D.A.; Smith-Bindman, R. Facility Mammography Volume in Relation to Breast Cancer Screening Outcomes. *J. Med. Screen.* **2016**, *23*, 31–37. [CrossRef] [PubMed]
9. Lewis, T.C.; Pizzitola, V.J.; Giurescu, M.E.; Eversman, W.G.; Lorans, R.; Robinson, K.A.; Patel, B.K. Contrast-enhanced Digital Mammography: A Single-Institution Experience of the First 208 Cases. *Breast J.* **2017**, *23*, 67–76. [CrossRef] [PubMed]
10. Hassan, A.M.; El-Shenawee, M. Review of electromagnetic techniques for breast cancer detection. *IEEE Rev. Biomed. Eng.* **2011**, *4*, 103–118. [CrossRef]
11. Xu, P.; Peng, Y.; Sun, M.; Yang, X. SU-E-I-81: Targeting of HER2-Expressing Tumors with Dual PET-MR Imaging Probes. *Med. Phys.* **2015**, *42*, 3260. [CrossRef]
12. Poneros, J.M.; Brand, S.; Bouma, B.E.; Tearney, G.J.; Compton, C.C.; Nishioka, N.S. Diagnosis of specialized intestinal metaplasia by optical coherence tomography. *Gastroenterology* **2001**, *120*, 7–12. [CrossRef]
13. Arridge, S.R. Philosophical Transactions of the Royal Society A: Mathematical. *Phys. Eng. Sci.* **2011**, *369*, 4558–4576.
14. Ren, K.; Bal, G.; Hielscher, A.H. Diffuse optical tomography using the one-way radiative transfer equation. *J. Sci. Comput.* **2006**, *28*, 1463–1489.
15. Cerussi, A.E.; Berger, A.J.; Bevilacqua, F.; Shah, N.; Jakubowski, D.; Butler, J.; Holcombe, R.F.; Tromberg, B.J. Sources of absorption and scattering contrast for near-infrared optical mammography. *Acad. Radiol.* **2001**, *8*, 211–218. [CrossRef] [PubMed]
16. Pifferi, A.; Swartling, J.; Chikoidze, E.; Torricelli, A.; Taroni, P.; Bassi, A.; Andersson-Engels, S.; Cubeddu, R. Spectroscopic time-resolved diffuse reflectance and transmittance measurements of the female breast at different interfiber distances. *J. Biomed. Opt.* **2004**, *9*, 1143–1151. [CrossRef] [PubMed]
17. Grosenick, D.; Rinneberg, H.; Cubeddu, R.; Taroni, P. Review of optical breast imaging and spectroscopy. *J. Biomed. Opt.* **2016**, *21*, 091311. [CrossRef] [PubMed]
18. Pifferi, A.; Contini, D.; Mora, A.D.; Farina, A.; Spinelli, L.; Torricelli, A. Palatino Linotype. New frontier in time-domain diffuse optics, a review. *J. Biomed. Opt.* **2016**, *21*, 091310. [CrossRef]
19. Pogue, B.W.; Testorf, M.; McBride, T.; Osterberg, U.; Paulsen, K. Instrumentation and design of a frequency-domain diffuse optical tomography imager for breast cancer detection. *Opt. Express* **1997**, *1*, 391–403. [CrossRef]

20. Siegel, A.; Marota, J.; Boas, D.A. Design and development of a continuous-wave diffuse optical tomography system. *Opt. Express* **1999**, *4*, 287–298. [CrossRef]
21. Flexman, M.L.; Kim, H.K.; Gunther, J.E.; Lim, E.A.; Alvarez, M.C.; Desperito, E.; Kalinsky, K.; Hershman, D.L.; Hielscher, A.H. Optical biomarkers for breast cancer derived from dynamic diffuse optical tomography. *J. Biomed. Opt.* **2013**, *18*, 096012. [CrossRef]
22. Li, C.; Zhao, H.; Anderson, B.; Jiang, H. Multispectral breast imaging using a ten-wavelength, source/detector channels silicon photodiode-based diffuse optical tomography system. *Med. Phys.* **2006**, *33*, 627–636. [CrossRef]
23. Schmitz, C.H.; Klemer, D.P.; Hardin, R.; Katz, M.S.; Pei, Y.; Graber, H.L.; Levin, M.B.; Levina, R.D.; Franco, N.A.; Solomon, W.B. Design and implementation of dynamic near-infrared optical tomographic imaging instrumentation for simultaneous dual-breast measurements. *Appl. Opt.* **2005**, *41*, 2140–2153. [CrossRef]
24. El-Ghussein, F.; Mastanduno, M.A.; Jiang, S.; Pogue, B.W.; Paulsen, K.D. Medical hyperspectral imaging. *J. Biomed. Opt.* **2014**, *19*, 10901.
25. Higgins, C. Hemoglobin and its measurement. Hemoglobin **2005**. Available online: https://acutecaretesting.org/en/articles/hemoglobin-and-its-measurement (accessed on 13 August 2023).
26. Chandra, F.; Wahyudianto, A.; Yasin, M. Design of vein finder with multi tuning wavelength using RGB LED. *J. Phys. Conf. Ser.* **2017**, *853*, 012019. [CrossRef]
27. Delvo, E.D.; Guin, P. *Implementa, on of Near—Infrared Technology (AccuVein AV—400®) To Facilitate Successful PIV Cannula*; UFHealth: Leesburg, FL, USA, 2012; Available online: https://www.accuvein.com/articles/implementation-of-near-infrared-technology-accuvein-av-400-to-facilitate-successful-piv-cannulation/ (accessed on 13 August 2023).
28. Wang, Y.; Bower, B.A.; Izatt, J.A.; Tan, O.; Huang, D. Retinal blood flow measurement by circumpapillary Fourier domain Doppler optical coherence tomography. *J. Biomed. Opt.* **2008**, *13*, 202–208. [CrossRef] [PubMed]
29. Chiao, F.B.; Resta-Flarer, F.; Lesser, J.; Ng, J.; Ganz, A.; Pino-Luey, D.; Bennett, H.; Perkins, C., Jr.; Witek, B. Vein visualization: Patient characteristic factors and efficacy of a new infrared vein finder technology. *Br. J. Anaesth.* **2013**, *110*, 966–971. [CrossRef] [PubMed]

bioengineering

MDPI

Article

Cost-Effective Full-Color 3D Dental Imaging Based on Close-Range Photogrammetry

Bin Yang *, Jennifer Schinke, Amir Rastegar, Melikhan Tanyeri and John A. Viator

Department of Biomedical Engineering, Duquesne University, Pittsburgh, PA 15282, USA; schinkej@duq.edu (J.S.); rastegara@duq.edu (A.R.); tanyerim@duq.edu (M.T.); viatorj@duq.edu (J.A.V.)
* Correspondence: yangb1@duq.edu

Abstract: Dental imaging plays a crucial role in clinical dental practice. Conventional 2D dental imaging serves general-purpose tasks, such as patient documentation, while high-precision 3D dental scanning is tailored for specialized procedures, such as orthodontics and implant surgeries. In this study, we aimed to develop a cost-effective 3D imaging technique that could bridge the gap between conventional dental photography and high-precision 3D dental scanning, with the goal of improving patient dental care. We developed a 3D imaging technique based on close-range photogrammetry and termed it close-range photogrammetry-based dental imaging (CPDI). We evaluated this technique on both in vitro dental models and in vivo teeth. For dental models, we conducted a parametric study to examine the effects of the depth of field and specular reflection on reconstruction quality. We showed that the optimal results were achieved with an f/5.6 lens and without a circular polarizer for reflection suppression. This configuration generated 3D scans with 57.7 ± 3.2% and 82.4 ± 2.7% of reconstructed points falling within ±0.1 mm and ±0.2 mm error margins, respectively. With such accuracy, these 3D dental models can faithfully represent dental morphology and features. During in vivo imaging, we were able to reconstruct high-quality 3D models of the anterior arch, further demonstrating its clinical relevance. The reconstructed models carry both 3D shapes and detail full-color surface textures, which positions CPDI as a versatile imaging tool in different areas of clinical dental care.

Keywords: photogrammetry; dental imaging; 3D dental reconstruction; intraoral scanner

Citation: Yang, B.; Schinke, J.; Rastegar, A.; Tanyeri, M.; Viator, J.A. Cost-Effective Full-Color 3D Dental Imaging Based on Close-Range Photogrammetry. *Bioengineering* **2023**, *10*, 1268. https://doi.org/10.3390/bioengineering10111268

Academic Editor: Cuneyt M. Alper

Received: 27 August 2023
Revised: 16 October 2023
Accepted: 27 October 2023
Published: 31 October 2023

1. Introduction

Dental photography serves as an effective way for documentation, patient communication and education, and treatment planning in modern dental practice [1–4]. While conventional dental photography captures images in high spatial resolution and in color, it is limited to providing 2D information. In certain cases, without a comprehensive 3D representation, specific dental conditions may necessitate the capture of multiple photos from different angles, thus complicating diagnosis and treatment. In addition to dental photography, 3D dental imaging is also available [5,6]. Applications with 3D dental imaging tend to be more specialized areas, such as orthodontics and dental implants. Within orthodontics, 3D scanning has become a standard procedure for assessment, treatment planning, and the fabrication of dental aligners [7,8]. In dental implant surgeries, the use of 3D imaging improves surgical outcomes by reducing potential complications and enabling more precise implant placement [9–11].

Several imaging techniques have been developed for dental imaging, including structured light and stereo vision [6,12]. Structured light uses active illumination to project patterns, such as coded stripes and speckles, onto the surface [13,14]. Due to the surface elevation (3D shape), these patterns are deformed, and this deformation is analyzed to reconstruct the 3D shape of the scene. On the other hand, stereo vision utilizes two cameras to image the scene. Due to parallax, identical features are captured at different locations on the two sensors [6,15]. This positional difference on the sensor is known as disparity, which is utilized to determine the depth [16,17]. While these imaging techniques provide excellent results for orthodontics and implant surgeries, the cost and complexity of the technology probably do not justify adopting such methods for other general imaging needs in dental care.

In recent years, close-range photogrammetry has emerged as an appealing alternative to the existing 3D imaging technologies [18–20]. Close-range photogrammetry does not require active illumination and only needs a single camera. It can generate full-color 3D models from a set of images or a video. Photogrammetry estimates the 3D positions of points by triangulating common features found in multiple images [21]. Due to its straightforward system setup and ease of use, close-range photogrammetry-based dental imaging (CPDI) has shown great potential in dental practice and has undergone extensive exploration [22–24]. Notably, the majority of CPDI has been performed on plaster cast dental models typically with a professional digital camera. Data on in vivo CDPI are limited, possibly due to the lack of appropriate intraoral imaging devices. In this study, we focused on (1) demonstrating and evaluating the capability of CDPI to generate 3D dental models in an in vivo setting by (2) developing and testing a compact and cost-effective intraoral camera. An affordable camera would lower the barrier for adopting this technology, potentially improving the quality of patient care.

2. Materials and Methods

2.1. Imaging Device Development

We developed two low-cost imaging setups based on a Raspberry Pi 4 (RPi) single-board computer (1) to conduct parametric studies of the effects of depth-of-field and surface reflection on reconstruction and (2) to evaluate the performance of in vivo teeth imaging. A typical configuration of the imaging system is shown in Figure 1A. For parametric studies, we adopted an RPi-compatible camera module featuring an M12 mount (UR-261, Arducam, Nanjing, China). This camera is later referred to as the M12 camera. Although this camera type is typically large and not ideal for in vivo imaging, its standard M12 mount allows for the use of a variety of M12 camera lenses. Illumination was provided by two 1 W white LEDs positioned on both sides of the camera. To reduce stray light, two flexible light-blocking sheets were placed between the LED and the camera. A more compact setup was developed for in vivo imaging, which comprised a compact Rpi imaging module (Module V2, Raspberry Pi, Cambridge, UK) with a fixed lens and a single 1 W white LED for illumination. The imaging setup was assembled into a toothbrush-like structure with a compact imaging head ($2 \times 1 \times 1$ cm, $L \times W \times D$). This camera is later referred to as the intraoral camera. Both imaging setups utilized the same 8 MP imaging sensor (IMX 219, SONY, Tokyo, Japan). Circular polarizers could be added to both imaging modules to reduce specular reflection, given that both illumination and imaging share a common optical path [25]. Figure 1B illustrates the principle of reflection suppression with a circular polarizer. A circular polarizer consists of a linear polarizer and a quarter-wave plate. The quarter-wave plate introduces a 90 degree phase shift to the polarized light [26]. In the incident direction, the 90 degree phase shift turns the linear polarized light into circular polarized light. Upon reflection from the teeth surface, the handedness of the circular polarized light is reversed. The 90 degree phase shift from the quarter-wave plate converts the reflected light into linear polarized light with the polarization direction 90 degrees apart from the original one, which is blocked by the linear polarizer. A custom Python script was developed to adjust imaging parameters and capture images. To ensure high-quality and consistent images, we used a fixed white balance setting and an ISO value of 200. Throughout imaging, an auto-exposure setting was used with a fixed exposure compensation value to reduce the risk of overexposure.

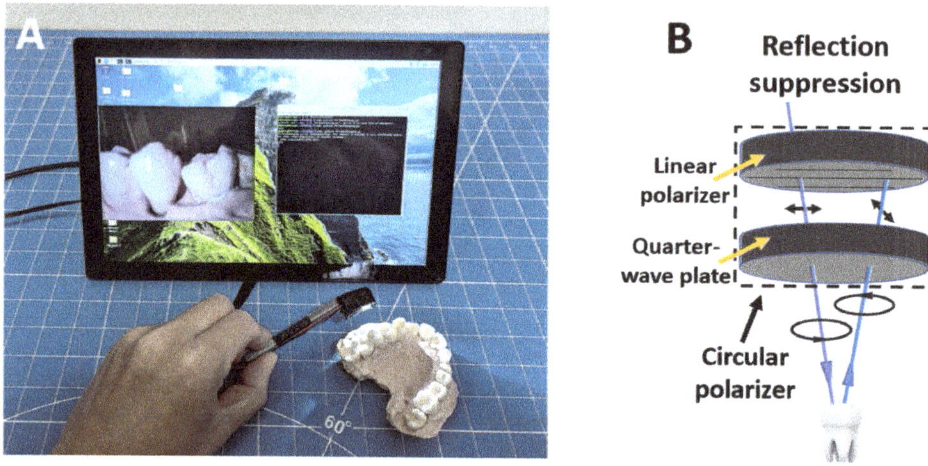

Figure 1. (**A**) Raspberry-Pi-based dental imaging system. (**B**) Working principle of reflection suppression using a circular polarizer.

2.2. Parametric Studies of the Effects of Depth-of-Field and Surface Reflection on Reconstruction Quality

Intraoral imaging poses several challenges, such as a short working distance and surface reflection, that can potentially impact the quality of the 3D reconstruction. The parametric study was designed to understand these challenges and optimize imaging parameters. The working distance of intraoral imaging needs to be short to accommodate the limited space in the oral cavity. A shorter working distance will result in a shallower depth-of-field (DoF), which means a smaller portion of the object will remain in focus. A lens with a smaller aperture can be used to alleviate this issue at the expense of lower light collection efficiency, potentially affecting the image quality. To understand how DoF affects the reconstruction quality, we imaged a dental model using the M12 camera with two lenses sharing the same focal length but differing in aperture sizes. The F/2.8 lens (89342, Edmund Optics, Barrington, NJ, USA) features a larger aperture, producing a shallower DoF, while the F/5.6 lens (89343, Edmund Optics, Barrington, NJ, USA) has a smaller aperture, resulting in a more moderate DoF. The enamel reflects light strongly under illumination, often leading to image saturation and subsequent loss of information. To study whether the suppression of reflected light can improve the reconstruction quality, we imaged the same dental model both with and without a circular polarizer (88-085, Edmund Optics, Barrington, NJ, USA). A circular polarizer effectively reduces reflection when the illumination and imaging share a common optical path. For each testing condition, the dental model was imaged three times.

2.3. In Vivo Teeth Imaging

The anterior sections of both the upper and lower arches of a volunteer were imaged using an intraoral camera. Prior to imaging, the volunteer brushed the teeth, and excess fluids on the teeth were wiped with gauze. The intraoral camera was positioned approximately 2 cm in front of the teeth and followed the curvature of the arch during image acquisition. Each portion of the arch was imaged from multiple angles with a good focus, which was visually confirmed before taking the picture. It is worth noting that the imaging was performed by the same volunteer without the need for additional assistance, further demonstrating the user-friendly nature of this imaging technique. Due to the specular reflections from the moist teeth surface, a circular polarizer (88-085, Edmund Optics, Barrington, NJ, USA) was used throughout the in vivo teeth imaging to mitigate these reflections. Images were subsequently reconstructed for visualization.

2.4. Photogrammetry-Based 3D Reconstruction

Photogrammetry, as a versatile 3D reconstruction technique, has been widely used in remote sensing and aerial applications [27–29], archaeology [30], and augmented/virtual reality applications [31]. Detailed descriptions of the working principles of photogrammetry are beyond the scope of this study and have been well documented elsewhere [20,30]. Briefly, photogrammetry involves capturing a set of images of the subject from various directions. Common features of the subject within a subset of the images are detected and extracted using certain techniques, such as the scale-invariant feature transformation (SIFT) method [32]. These detected features are then used to estimate the 3D coordinates of the features based on triangulation [21]. The point cloud of the subject is generated after the 3D coordinates of all features have been estimated. This point cloud is further processed, meshed, and textured to create a faithful 3D representation of the subject.

In practice, photogrammetry requires that images have sufficient overlap to facilitate feature detection, registration, and reconstruction [20]. In both parametric studies and in vivo imaging, we maintained approximately 50% image overlap for all acquired images. As illustrated in Figure 2A, we followed specific imaging paths to ensure comprehensive coverage from various angles. The camera was positioned sequentially at approximately 45, 90, and 135 degrees towards the dental model (indicated by blue arrows in Figure 2A). At each angle, images were taken while following the curvature of the arch, as depicted by the red arrow. Figure 2B visualizes the camera positions and angles during a CPDI session. For the parametric study, 30–40 images were acquired to reconstruct the full arch, while in in vivo imaging, 20–30 images were acquired.

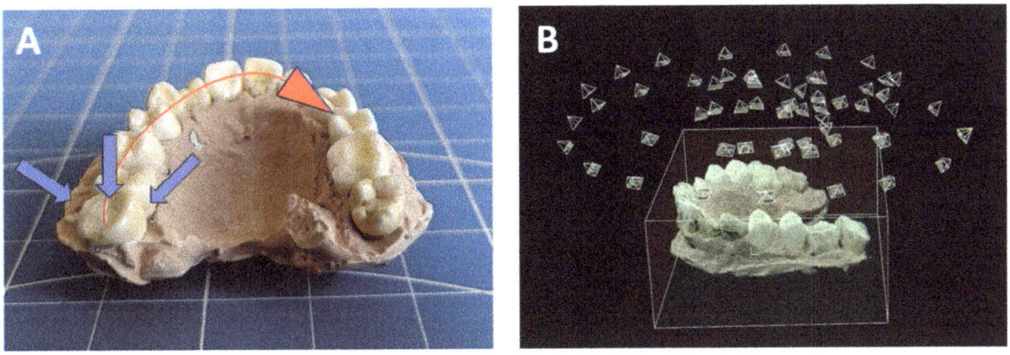

Figure 2. (**A**) Image acquisition paths following the curvature of the arch (red arrow) at different angles (blue arrows) for complete coverage. (**B**) Visualization of camera positions and angles.

2.5. Reconstruction and Trueness Analysis

Following image acquisition, we reconstructed 3D models using PhotoCatch (EOS Innovations LLC) running on a M1 Mac Mini. The reconstructed 3D model was exported in .obj format, which contains both the 3D structure mesh and color texture of the dental model. The 3D model was imported into CloudCompare (V 2.12.4), an open-source point processing software, for visualization and trueness analysis.

To facilitate the trueness analysis, the teeth portion of the reconstructed dental model was manually segmented out in CloudCompare. Due to the lack of a physical dimension of the 3D model generated with the photogrammetry technique, the initial alignment between the 3D teeth model and the ground truth model was performed manually. Specifically, the 3D teeth model was manually scaled to the approximate size of the ground truth mode. Following scaling, the orientation of the 3D teeth model was adjusted to match the orientation of the ground truth model. Once a good manual alignment between these two models was achieved, we utilized the registration function provided in CloudCompare for a more precise registration. During the registration process, we designated the ground truth as the target and allowed the 3D teeth model to be further scaled to minimize the overall error between the two models. Unlike CPDI, the ground truth model was generated on an absolute scale using a commercial intraoral scanner. Local deviations, measured in millimeters, between the ground truth model and the reconstructed model were quantified to indicate the reconstruction errors. As the ground truth for the in vivo dental model was not available, trueness analysis was not conducted for in vivo imaging.

3. Results

3.1. Imaging Device Development

Both the M12 camera for parametric studies (Figure 3A) and the in vivo intraoral camera (Figure 3B) share a similar configuration; however, the intraoral camera is noticeably smaller. The lens on both devices was adjusted to achieve a short working distance of approximately 1.5 to 2 cm. The horizontal field-of-view (FoV) is about 70 degrees for both cameras.

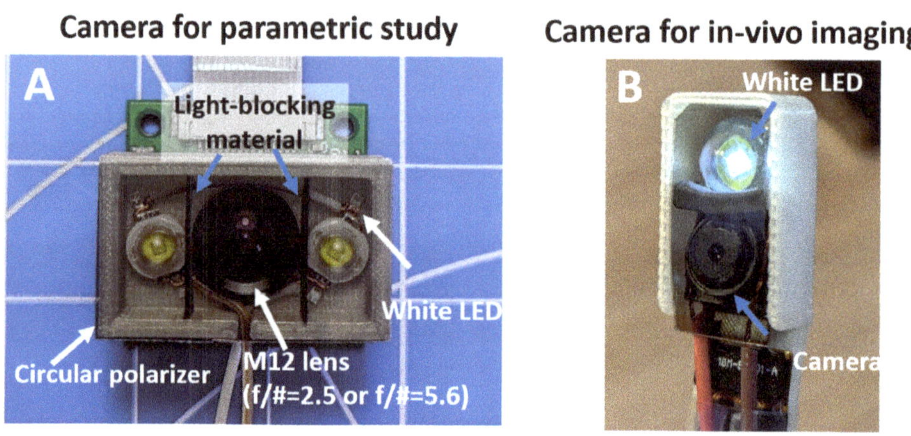

Figure 3. (**A**) Camera setup for parametric study with interchangeable M12 lens. (**B**) Toothbrush-like compact camera for in vivo imaging. Both cameras can be equipped with a circular polarizer to suppress surface reflection.

3.2. Parametric Studies of the Effects of Depth-of-Field and Surface Reflection on Reconstruction Quality

The reconstructed 3D dental models under three camera configurations (Figure 4A–C) exhibited an excellent and detailed visual representation of the physical dental model (Figure 4D). The colors of the teeth and the plaster base were faithfully reproduced along with their surface textures. The morphology of the incisor, pre-molar, and molar teeth are distinctly shown in the reconstructed model. The models generated with f/2.5 (Figure 4A) and f/5.6 (Figure 4B) lenses, both without a polarizer, are visually comparable. However, the model obtained with a polarizer in combination with an f/5.6 lens (Figure 4C) showed noticeable reconstruction errors, as indicated by the red arrows, unlike the models obtained without a polarizer (Figure 4A,B). This result was surprising, given that the polarizer typically improves the image quality by reducing surface glare. Comparing images acquired with and without a polarizer, we noticed that the image acquired with a polarizer showed reduced contrast in certain regions, as indicated by the orange arrows (Figure 5). During the interaction between the polarized light and the teeth, reflection occurs both at the tooth surface and within the deeper layers (diffuse reflection). The surface reflection carries crucial surface contrast information necessary for the reconstruction process. However, the circular polarized light reflected from the surface exhibited a reversed handedness, which was suppressed by the circular polarizer in its return path. As a result, the surface contrast information was reduced, which may have adversely affected the reconstruction accuracy.

Comparing the reconstructed 3D dental models (Figure 4E–G) to the ground true model (Figure 4H) in wireframe mode effectively eliminates potential interference from color texture. It is evident that the ground truth model was reproduced with a higher accuracy, exhibiting a higher level of detail with sharper edges and smoother surfaces. For instance, while the top surfaces of one incisor (indicated with a white arrow) are small, they are clearly visible in the ground truth model. The three reconstructed models failed to capture the same surface features. Similarly, the individual teeth are more distinctly separated with clearly defined gaps in the ground truth model, as opposed to the reconstructed models (indicated by yellow arrows).

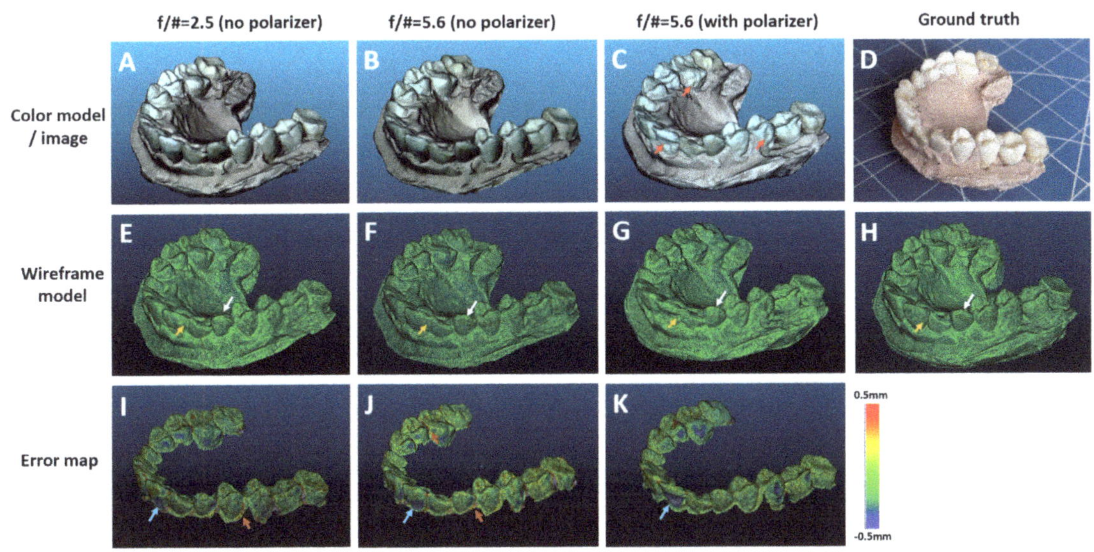

Figure 4. Parametric study of CPDI under three imaging settings: f/2.5 lens without circular polarizer (**A**,**E**,**I**), f/5.6 lens without circular polarizer (**B**,**F**,**J**), and f/5.6 lens with circular polarizer (**C**,**G**,**K**). Results were visualized in full color (**A**–**C**) and in wireframe (**E**,**F**). The reconstruction error was quantified and visualized (**I**–**K**) against the ground truth model (**H**). A color picture of the dental model is shown in (**D**).

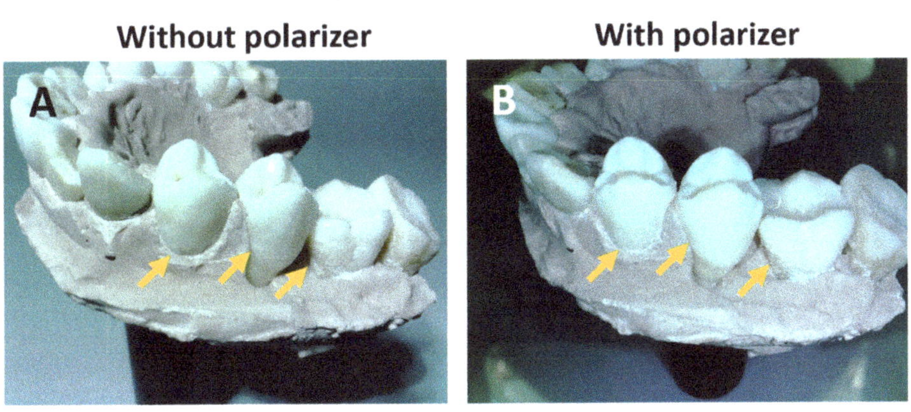

Figure 5. A circular polarizer reduces the surface contrast of a dental model (**B**) compared to that without a polarizer (**A**) as highlighted with yellow arrows.

The error maps generated from the quantitative trueness analysis aligned well with the results of the visual inspection (Figure 4I–K). Figure 4I–K reveal that areas with high reconstruction errors typically appear in the recessed regions (gaps) (indicated by orange arrows) or on the smooth surfaces (indicated by cyan arrows).

Trueness was quantified within two error ranges: ±0.1 mm and ±0.2 mm. An accuracy of 0.1 mm is generally accepted to be clinically adequate and has been used for evaluating the performance of conventional intraoral scanners [33]. A typical histogram of the reconstruction error is shown in Figure 6A. It is evident that the majority of the reconstructed points fall within the ±0.2 mm error range. Notably, some reconstructed points exhibit a larger error (>0.5 mm, shown in gray color). Without a polarizer, the f/2.5 and f/5.6 lenses achieved averages of 55.9 ± 5.4% and 57.7 ± 3.2% of the reconstructed points within the ±0.1 mm error range. The extended DoF offered by the f/5.6 lens resulted in a minor improvement in reconstruction quality. With the f/5.6 lens, the reconstruction accuracy was notably higher without a polarizer than that achieved with a polarizer, which was 44.1 ± 5.6%. When relaxing the error to ±0.2 mm, the f/2.5 and f/5.6 lenses without a polarizer achieved an averaged inlier ratio of 80.9 ± 3.1% and 82.4 ± 2.7%, which was approximately a 25% improvement over the ±0.1 mm error range. A slightly greater improvement in the inlier ratio (74.9 ± 5.2%) was realized using an f/5.6 lens and a polarizer. Overall, the reconstruction with an f/5.6 lens without a polarizer yielded the most accurate and consistent results.

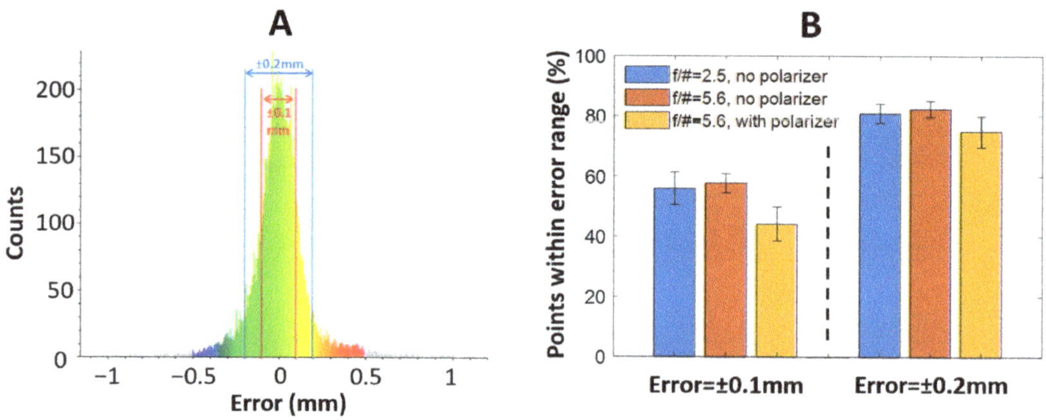

Figure 6. (**A**) Histogram of the reconstruction error with f/5.6 lens and without polarizer. (**B**) Percentage of reconstructed points within the error range of ±0.1 mm and ±0.2 mm for three imaging settings. The error bar represents the standard deviation over 3 trials.

3.3. *In Vivo Teeth Imaging*

The in vivo teeth imaging demonstrated that both the color and 3D shape of teeth can be reproduced with high fidelity with CPDI. The anterior (Figure 7A) and posterior (Figure 7B) views of the full-color 3D models effectively captured the volunteer's intraoral condition. It is worth noting that the gums were also reconstructed as part of the 3D model. Owing to the nature of the image-based reconstruction, the 3D model accurately preserved colors and intricate features, which could enable a comprehensive evaluation of dental health and the examination of gum-related conditions. For instance, the discoloration of the incisor teeth (indicated by the green arrow in Figure 7A), small blood vessels in the gums (indicated by the yellow arrow in Figure 7A), and dental calculus (indicated by the cyan arrow in Figure 7D) were all visible in high resolution. The full-color meshes (Figure 7B,E) revealed the 3D models were densely reconstructed. In the solid body view, individual teeth were reconstructed with precise morphological features and clear separation from neighboring teeth. For instance, a small chip of an incisor tooth is clearly visible (indicated by the red arrow in Figure 7C). Some imperfections, such as non-smooth surfaces, were noticeable in the solid model. However, these imperfections did not significantly impact the visualization of the teeth and gums, particularly with the full-color models. Despite the lack of a ground truth model of the volunteer for quantitative analysis, it is evident, based on the visual inspection, that CPDI is effective in in vivo 3D dental imaging.

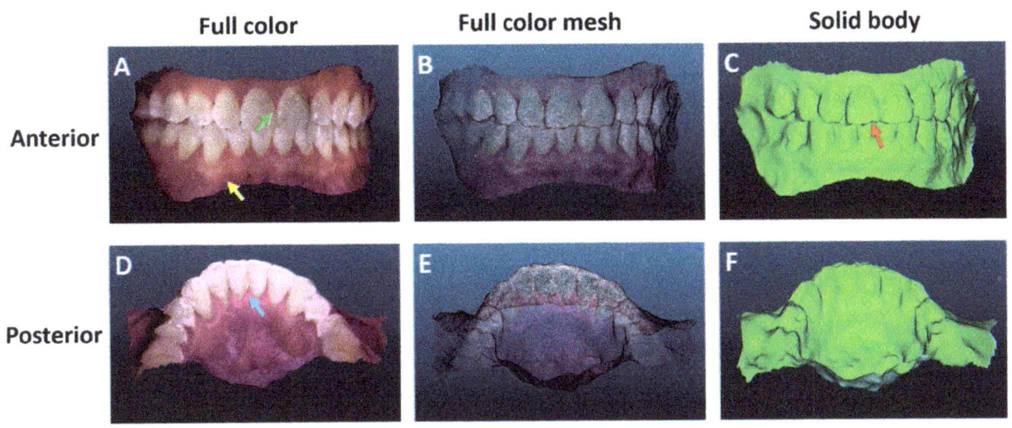

Figure 7. Full-color 3D models of (**A**) upper and lower labial arches (anterior view) and (**D**) lower labial arch (posterior view). (**B**,**E**) show the corresponding full-color meshes, and (**C**,**F**) show the solid body of the reconstructed arches. The green, yellow, and cyan arrows show tooth discoloration, gum, and dental plaque, respectively. The red arrow shows a small chip in the incisor tooth.

4. Discussion

CPDI aims to fill the gap between conventional 2D dental photography and high-precision 3D dental scanning and to improve the quality of patient care. Similar to 2D dental photography, CPDI provides rich and accurate color information in high spatial resolution, facilitating detailed close-up examinations. Furthermore, this color information is integrated into the 3D dental model, allowing it to be visualized and evaluated from arbitrary angles with the awareness of its 3D geometry. It is worth noting that while CPDI-derived models may not achieve the precision of those obtained with commercial 3D dental scanners, they faithfully reconstruct the most prominent morphological and structural dental features of the dental arch and individual teeth. This compromise is justified as CPDI is not intended for orthodontics and dental surgeries where a highly precise dental scan is critical to ensure optimal treatment outcomes. The low-cost nature of CPDI could lower the barrier for adopting this technology in clinical dental practice, potentially improving the quality of patient care.

The parametric study suggested that while the surface reflection is undesirable in an image, it appears not to significantly affect the reconstruction results. This is potentially because the same region exhibiting high reflection in one image may not do so in other images due to the change in imaging angles. With the information from multiple images, those regions can still be reconstructed. Suppressing reflection seems to deteriorate the reconstruction quality, possibly due to the loss of surface contrast, as discussed in Section 3.2. In the in vivo imaging, we had to use a polarizer to mitigate excessive reflection. The reconstruction quality, however, was better than the result presented in the parametric study with a polarizer. While the exact cause warrants further investigation, we speculate that sufficient surface contrast of the teeth was retained for this specific volunteer, even with a polarizer. This also suggests that whether or not to use a polarizer for surface reflection suppression needs to be tested and evaluated for each imaging task. A lens with a smaller aperture (offering a larger DoF) indeed improved the overall reconstruction quality, but only by a small margin, suggesting that such a lens is still preferable under adequate illumination. If illumination intensity and potential motion blur become a point of concern, opting for a lens with a larger aperture would still yield satisfactory results.

Through in vivo 3D imaging, we demonstrated accurate reconstructions of both the incisor and canine teeth. However, it proved challenging to reconstruct the pre-molar and molar teeth due to certain limitations within the current system. These limitations, in turn, provide valuable insights for guiding future improvements. Despite the overall compact size of the intraoral camera, the working distance is still too large to comfortably image the teeth positioned deep within the oral cavity. A potential improvement could be achieved by introducing a variable extension tube to the imaging system. This adaptation will allow easy adjustment of the working distance and FoV to facilitate imaging different portions of the arch, such as a larger working distance (larger FoV) for imaging the anterior arch and a shorter working distance (smaller FoV) for imaging the posterior arch. Furthermore, a warm and moist intraoral environment often leads to rapid condensation that complicates imaging. To counteract this issue, an anti-fogging coating on the lens or a heating mechanism to keep the imager warm might be necessary. Finally, the camera needs to move around the teeth in order to capture images from multiple angles. In the case of molar teeth, moving from the anterior to the posterior of the teeth will flip the imaged field vertically. Such a sudden change can disorient the operator and interrupt the imaging process. This issue could be alleviated by incorporating an inertial measurement unit (IMU) within the camera system to estimate the camera orientation and adjust the image accordingly.

The reconstruction algorithm of CPDI relies on subject contrast to triangulate and recover shape. Healthy teeth unfortunately typically lack pronounced visual contrast, which could partly explain the reconstruction errors shown in our studies, particularly on the smooth teeth surface. Surface contrast is determined by the local absorption and scattering properties, which are wavelength dependent. It would be valuable to perform a multispectral analysis of the intrinsic sources of contrast [34], and identify the most effective wavelengths in red, green, and blue spectral bands to maximize the contrast. Another inherent limitation of photogrammetry lies in its inability to recover the absolute physical scale. If a dimensional measurement is required with CPDI, a calibrator with known dimensions can be placed within the scene to serve as a reference [24].

The low-cost nature and ease of use of CPDI will potentially benefit patient care both within dental clinics and home settings. 3D full-color dental models have the potential to replace conventional dental images and augment the text-based descriptions in patient charts. CPDI could also reduce the frequency of dental visits and overall costs by enabling at-home applications [35]. For instance, dental visits after orthodontic procedures could be replaced by providing the 3D dental model generated at home to the dental care providers for progress evaluation. Similarly, a 3D dental model could aid dental examination during remote dental visits.

5. Conclusions

In this study, we demonstrated that CPDI is a viable, cost-effective technique for achieving full-color 3D dental imaging. CPDI can generate 3D scans of dental models and in vivo teeth with good representations of their morphology and dental features. Future studies will focus on a comprehensive evaluation of its clinical benefits through a large patient cohort.

Author Contributions: Conceptualization, B.Y. and J.A.V.; methodology, B.Y. and J.A.V.; software, B.Y. and A.R.; formal analysis, B.Y.; investigation, M.T., J.S. and A.R.; writing—original draft preparation, B.Y.; writing—review and editing, J.A.V. and M.T.; funding acquisition, B.Y. and J.A.V. All authors have read and agreed to the published version of the manuscript.

Funding: This research was supported by a research contract between Duquesne University and Jiesao Techno Co., Ltd. under grant number G2200064.

Institutional Review Board Statement: Not applicable.

Informed Consent Statement: Not applicable.

Data Availability Statement: The data that support the findings of this study are available from the corresponding author upon reasonable request.

Acknowledgments: The authors thank the UbiLoc research group at China University of Geosciences for providing the dental model and its 3D data.

Conflicts of Interest: B.Y. has financial interests in Jiesao Techno Ltd. The conflict of interest was managed by Duquesne University. Other authors declare no conflict of interest. The funders had no role in the design of this study; in the collection, analyses, or interpretation of data; in the writing of the manuscript; or in the decision to publish the results.

References

1. Ahmad, I. Digital dental photography. Part 1: An overview. *Br. Dent. J.* **2009**, *206*, 403–407. [CrossRef] [PubMed]
2. Ahmad, I. Digital dental photography. Part 2: Purposes and uses. *Br. Dent. J.* **2009**, *206*, 459–464. [CrossRef] [PubMed]
3. Casaglia, A.; De Dominicis, P.; Arcuri, L.; Gargari, M.; Ottria, L. Dental photography today. Part 1: Basic concepts. *Oral Implantol.* **2015**, *8*, 122.
4. Mladenović, D.; Mladenović, L.; Mladenović, S. Importance of digital dental photography in the practice of dentistry. *Sci. J. Fac. Med. Niš* **2010**, *27*, 75–79.
5. Birnbaum, N.S.; Aaronson, H.B.; Stevens, C.; Cohen, B. 3D digital scanners: A high-tech approach to more accurate dental impressions. *Inside Dent.* **2009**, *5*, 70–74.
6. Logozzo, S.; Zanetti, E.M.; Franceschini, G.; Kilpelä, A.; Mäkynen, A. Recent advances in dental optics–Part I: 3D intraoral scanners for restorative dentistry. *Opt. Lasers Eng.* **2014**, *54*, 203–221.
7. Jedliński, M.; Mazur, M.; Grocholewicz, K.; Janiszewska-Olszowska, J. 3D scanners in orthodontics—Current knowledge and future perspectives—A systematic review. *Int. J. Environ. Res. Public Health* **2021**, *18*, 1121. [CrossRef]
8. Taneva, E.; Kusnoto, B.; Evans, C.A. 3D scanning, imaging, and printing in orthodontics. *Issues Contemp. Orthod.* **2015**, *148*, 862–867.
9. Dreiseidler, T.; Neugebauer, J.; Ritter, L.; Lingohr, T.; Rothamel, D.; Mischkowski, R.A.; Zöller, J.E. Accuracy of a newly developed integrated system for dental implant planning. *Clin. Oral Implant. Res.* **2009**, *20*, 1191–1199. [CrossRef]
10. Flügge, T.V.; Nelson, K.; Schmelzeisen, R.; Metzger, M.C. Three-dimensional plotting and printing of an implant drilling guide: Simplifying guided implant surgery. *J. Oral Maxillofac. Surg.* **2013**, *71*, 1340–1346.
11. Sarment, D.P.; Sukovic, P.; Clinthorne, N. Accuracy of implant placement with a stereolithographic surgical guide. *Int. J. Oral Maxillofac. Implant.* **2003**, *18*, 571–577.
12. Mangano, F.; Gandolfi, A.; Luongo, G.; Logozzo, S. Intraoral scanners in dentistry: A review of the current literature. *BMC Oral Health* **2017**, *17*, 149. [CrossRef] [PubMed]
13. Ahn, J.S.; Park, A.; Kim, J.W.; Lee, B.H.; Eom, J.B. Development of three-dimensional dental scanning apparatus using structured illumination. *Sensors* **2017**, *17*, 1634. [CrossRef] [PubMed]
14. Geng, J. Structured-light 3D surface imaging: A tutorial. *Adv. Opt. Photonics* **2011**, *3*, 128–160. [CrossRef]
15. Wang, J.; Suenaga, H.; Hoshi, K.; Yang, L.; Kobayashi, E.; Sakuma, I.; Liao, H. Augmented reality navigation with automatic marker-free image registration using 3-D image overlay for dental surgery. *IEEE Trans. Biomed. Eng.* **2014**, *61*, 1295–1304. [CrossRef]
16. Orban, G.A.; Janssen, P.; Vogels, R. Extracting 3D structure from disparity. *Trends Neurosci.* **2006**, *29*, 466–473. [CrossRef]
17. Aguilar, J.-J.; Torres, F.; Lope, M. Stereo vision for 3D measurement: Accuracy analysis, calibration and industrial applications. *Measurement* **1996**, *18*, 193–200. [CrossRef]
18. Jiang, R.; Jáuregui, D.V.; White, K.R. Close-range photogrammetry applications in bridge measurement: Literature review. *Measurement* **2008**, *41*, 823–834. [CrossRef]
19. Luhmann, T. Close range photogrammetry for industrial applications. *ISPRS J. Photogramm. Remote Sens.* **2010**, *65*, 558–569. [CrossRef]
20. Luhmann, T.; Robson, S.; Kyle, S.; Boehm, J. *Close-Range Photogrammetry and 3D Imaging*; Walter de Gruyter: Berlin, Germany, 2013.
21. Hartley, R.I.; Sturm, P. Triangulation. *Comput. Vis. Image Underst.* **1997**, *68*, 146–157. [CrossRef]
22. Zhang, Y.-J.; Qian, S.-J.; Lai, H.-C.; Shi, J.-Y. Accuracy of photogrammetric imaging versus conventional impressions for complete-arch implant-supported fixed dental prostheses: A comparative clinical study. *J. Prosthet. Dent.* **2021**, *130*, 212–218. [CrossRef] [PubMed]
23. Fu, X.; Peng, C.; Li, Z.; Liu, S.; Tan, M.; Song, J. The application of multi-baseline digital close-range photogrammetry in three-dimensional imaging and measurement of dental casts. *PLoS ONE* **2017**, *12*, e0178858. [CrossRef]
24. Zotti, F.; Rosolin, L.; Bersani, M.; Poscolere, A.; Pappalardo, D.; Zerman, N. Digital dental models: Is photogrammetry an alternative to dental extraoral and intraoral scanners? *Dent. J.* **2022**, *10*, 24. [CrossRef]

25. Morgan, S.P.; Stockford, I.M. Surface-reflection elimination in polarization imaging of superficial tissue. *Opt. Lett.* **2003**, *28*, 114–116. [CrossRef] [PubMed]
26. Collett, E. *Field Guide to Polarization*; SPIE: Bellingham, WA, USA, 2005.
27. Colomina, I.; Molina, P. Unmanned aerial systems for photogrammetry and remote sensing: A review. *ISPRS J. Photogramm. Remote Sens.* **2014**, *92*, 79–97. [CrossRef]
28. Konecny, G. *Geoinformation: Remote Sensing, Photogrammetry and Geographic Information Systems*; CRC Press: Boca Raton, FL, USA, 2014.
29. Honkavaara, E.; Arbiol, R.; Markelin, L.; Martinez, L.; Cramer, M.; Bovet, S.; Chandelier, L.; Ilves, R.; Klonus, S.; Marshal, P. Digital airborne photogrammetry—A new tool for quantitative remote sensing?—A state-of-the-art review on radiometric aspects of digital photogrammetric images. *Remote Sens.* **2009**, *1*, 577–605. [CrossRef]
30. Remondino, F. Heritage recording and 3D modeling with photogrammetry and 3D scanning. *Remote Sens.* **2011**, *3*, 1104–1138. [CrossRef]
31. Portalés, C.; Lerma, J.L.; Navarro, S. Augmented reality and photogrammetry: A synergy to visualize physical and virtual city environments. *ISPRS J. Photogramm. Remote Sens.* **2010**, *65*, 134–142. [CrossRef]
32. Lindeberg, T. Scale invariant feature transform. *Digit. Vetensk. Ark.* **2012**, *7*, 10491. [CrossRef]
33. Tomita, Y.; Uechi, J.; Konno, M.; Sasamoto, S.; Iijima, M.; Mizoguchi, I. Accuracy of digital models generated by conventional impression/plaster-model methods and intraoral scanning. *Dent. Mater. J.* **2018**, *37*, 628–633. [CrossRef]
34. Mathys, A.; Semal, P.; Brecko, J.; Van den Spiegel, D. Improving 3D photogrammetry models through spectral imaging: Tooth enamel as a case study. *PLoS ONE* **2019**, *14*, e0220949. [CrossRef] [PubMed]
35. Zotti, F.; Rosolin, L.; Simoncelli, F.; Pappalardo, D.; Cominziolli, A.; Zerman, N. Telediagnosis of dental caries: Possible or impossible? A pilot cross-sectional study. *Clin. Exp. Dent. Res.* **2022**, *8*, 1614–1622. [CrossRef] [PubMed]

bioengineering

MDPI

Article

Using Geometric Morphometric Analysis of Magnetic Resonance Imaging to Assess the Anatomy of the Eustachian Tube in Children with and without Otitis Media

Ellen E. Fricano [1,*], Allison P. Gremba [2], Miriam S. Teixeira [3], J. Douglas Swarts [4] and Cuneyt M. Alper [4,5,*]

1 College of Osteopathic Medicine of the Pacific, Western University of Health Sciences, Pomona, CA 91766, USA
2 Doctor of Physical Therapy Program, School of Natural and Health Sciences, Seton Hill University, Greensburg, PA 15601, USA
3 Department of Graduate Medical Education, Arnot Ogden Medical Center, Elmira, NY 14905, USA
4 Department of Otolaryngology, University of Pittsburgh School of Medicine, Pittsburgh, PA 15213, USA
5 Division of Pediatric Otolaryngology, University of Pittsburgh Medical Center Children's Hospital of Pittsburgh, Pittsburgh, PA 15224, USA
* Correspondence: efricano@westernu.edu (E.E.F.); cuneyt.alper@chp.edu (C.M.A.)

Abstract: Otitis media (OM) is among the most common of childhood illnesses. It has long been hypothesized that children under age two are predisposed to OM due to differences in the anatomy of the Eustachian tube (ET), including the angle of the ET. OM in later childhood is less common but does occur, begging the question, are there shape differences in the ET that persist underlying later occurrences of OM? To answer this question, a novel method, which applied geometric and morphometric shape analysis to landmarks obtained from MRI data, was used. MRI scans were performed on 16 children (5 control, 3 cOME, and 8 rAOM) between 2011 and 2015. Sixteen landmarks representing the shape of the ET, cranial base, and palate were analyzed. The results of a Procrustes ANOVA indicate that the shape of the ET varies significantly ($p < 0.01$) between the OM and control groups. The shape differences between the OM group and the control are a medial and low attachment site of the tensor veli palatini (TVP) muscle, a posterior and high torus tubarius, and an anteriorly projected palate. These results support previous findings that a relatively horizontal ET is associated with a predisposition for OM. This study used a novel approach to examine anatomical differences in children with and without OM. First, the data set is unique in that it includes MRI scans of children with a confirmed OM diagnosis. Second, the use of MRI scans in craniofacial anatomy OM research is novel and allows for the collection of soft tissue landmarks and the visualization of soft tissue structures. Third, geometric morphometric shape analysis is a statistical method that captures shape differences, offering a more universal picture of nuanced changes within the entire set of landmarks, in contrast to more traditional linear and angular measurements used in prior OM studies examining craniofacial anatomy.

Keywords: Eustachian; pharyngotympanic; craniofacial growth; biomedical imaging; MRI scans; geometric morphometric; Procrustes analysis; canonical variate analysis; principal component analysis

Citation: Fricano, E.E.; Gremba, A.P.; Teixeira, M.S.; Swarts, J.D.; Alper, C.M. Using Geometric Morphometric Analysis of Magnetic Resonance Imaging to Assess the Anatomy of the Eustachian Tube in Children with and without Otitis Media. *Bioengineering* **2023**, *10*, 1115. https://doi.org/10.3390/bioengineering10101115

Academic Editors: Alan Wang and Zhen Cheng

Received: 26 July 2023
Revised: 6 September 2023
Accepted: 21 September 2023
Published: 23 September 2023

1. Introduction

In the wake of a global pandemic, respiratory health and predisposition for upper respiratory tract infections are at the forefront of concerns for health professionals, patients, and parents. Upper respiratory tract infections are often complicated with coinciding middle ear infections in children [1–3]. There is a range of middle ear pathologies distinguished by frequency, duration, and fluid with and without signs of infection (e.g., otitis media with effusion (OME), acute otitis media (AOM), recurrent acute otitis media (rAOM), chronic otitis media with effusion (cOME), and chronic suppurative otitis media (CSOM)). Otitis media (OM) peaks in prevalence at approximately one to two years of age; the incidence decreases after age six, although cases can be seen into adolescence and adulthood [4–8]. This correlation between age, respiratory health, and OM has been linked to an immature immune system and several anatomical and functional differences, including the relative horizontal Eustachian tube (ET) in young children [9–12].

The ET's function is several-fold: (1) pressure equalizing—when the lumen is dilated, air pressure in the middle ear can equalize with the external environment; (2) clearance—the ET provides a channel through which middle ear secretions and pathogens can drain; and (3) protection—the ET's normally closed lumen prevents particulates, pathogens, nasopharyngeal sounds, and pressure variations from easily accessing the middle ear [13,14]. The ET lumen is closed at rest but can be opened passively by either middle ear or nasopharyngeal pressures that exceed that of the paratubal tissue or action of paratubal muscles, mainly the tensor veli palatini (TVP) and levator veli palatini muscles (LVP) [15–21]. These muscles are adjacent and/or attached to the ET, and their coordinated contraction dilates the ET lumen, facilitating the clearance and pressure equalizing functions of the ET, thus contributing to the health of the middle ear cavity.

The ET, located deep within the head, is particularly challenging to assess and visualize, even with modern imaging technology, which perhaps explains the relatively few studies that directly compare its shape in groups with known diagnoses. Methodologies for evaluating the shape of this structure range from two-dimensional lateral radiographs [22] to three-dimensional quantification from dry skulls and CT imaging [23–28]. Past literature suggests that the angle of the ET changes dramatically from infancy to adulthood and that shape changes in the basicranium and nasopharynx coincide with the timing of peak OM incidence. Those changes include an increase in verticality of the cartilaginous ET, an increase in height and narrowing of the choana, and an increase in basicranial flexion from infancy [27]. These changes are a reflection of the displacements of osseous landmarks created by the attachments of ET soft tissues to bone. The inability of these imaging modalities to discriminate soft tissue differences is the primary challenge for studying the functional cartilaginous ET.

Medical radiographs and computed tomography (CT) scans often fail to capture the cartilaginous ET; even MRI is likely to miss the ET unless the scan is optimized because, under resting conditions, the ET lumen is closed [29]. Smith et al. [29] point out that high-resolution CT and MRI can both be used to assess the ET with relative accuracy, but MRI is preferred because the Ostmann's fat pad, which runs the length of the ET and has high MRI signal intensity, is particularly useful in identifying the ET [30–33].

The anatomy important for ET function and middle ear health consists of a network of structures located in the anatomical spaces of the head and neck; these spaces include the petrous temporal bone, the parapharyngeal space, the infratemporal fossa, the nasopharynx, and the oropharyngeal isthmus. The shape and orientation of additional structures (e.g., the hard palate) have been shown to play a role in OM risk, and even more structures (e.g., the maxillary sinus and nasal cavity) may play a role in the transmission of pathogens and the development of OM [34–39]. Prior studies [16,22,36] have focused on linear and angular measurements to study this interrelated network of structures, but those methods do not account for global shape differences and consequently have the potential to miss more nuanced differences that include multiple structures. Geometric morphometric shape analysis records landmarks as three-dimensional cartesian coordinates. The cartesian coordinates are then plotted and compared to look for global shape differences while minimizing the effects of size, rotation, and translation.

In the present study, we applied geometric morphometric shape analysis to T1-weighted MRI scans of children with and without a history of OM. Our goal was to study the differences in the cartilaginous ET and the muscles acting on it. We hypothesize that the shape of the basicranium and associated anatomy of the ET would be significantly different between the OM and control groups and that the OM groups would exhibit ETs that were significantly more horizontal.

2. Materials and Methods

The included children were enrolled in a larger NIH (DC007667)-funded longitudinal study investigating the relationship between age-related changes in ET anatomy and ET function. The children were recruited by advertisements in the Pittsburgh area. The enrolled subjects had an ENT exam every 6 months from 3 to 7.5 years of age. They also had regularly scheduled ET function tests and regularly scheduled measurements of craniofacial anatomy, which included anthropometry, cephalometric measurements, and dental casts. A subset of the enrolled children had a non-sedated MRI between 6 and 7 years of age. A 3 Tesla GE Signa scanner was used to capture the MRI scans. For this analysis, proton density-weighted fast-spin echo scans taken in the oblique axial plane were used (matrix 512 × 512, field of view = 12 cm, 2.0 mm slices). The location for this data scan was obtained by first imaging in the sagittal plane, then using the midsagittal slice to identify the orientation of the oblique axial scan as the plane through the incisive occlusion point and the inferior pontine sulcus (Figure 1). Exclusion criteria for enrollment included any child with (1) a syndrome that predisposed them to OM (e.g., cleft palate), (2) a significant history of orthodontic treatment, (3) a history of ear surgery, excluding ventilation tube insertion, or (4) a child unable to cooperate for testing.

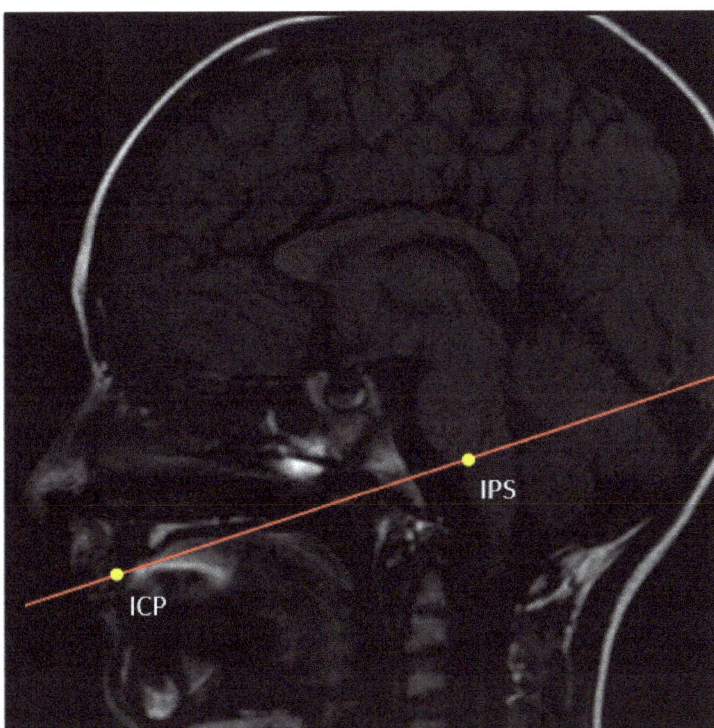

Figure 1. Midsagittal slice of the initial sagittal MRI used to specify the oblique axial scan through the plane of the inferior pontine sulcus and incisive occlusion point. IPS: inferior pontine sulcus (groove between the pons and medulla oblongata); ICP: incisive occlusion point (point of contact between the right and left upper central incisors).

Twenty-three children completed the MRI, and 16 children were included in this analysis. Seven subjects were excluded because the scan area did not include the relevant landmarks. On enrollment in the longitudinal study, the subjects were allocated into one of three groups: control, cOME, and rAOM. The control group had no history of ventilation tubes and did not meet the criteria for the cOME and rAOM groups. The cOME group met at least one of the following criteria: (1) bilateral middle ear effusion for at least 3 consecutive months; (2) unilateral middle ear effusion for at least 6 consecutive months; (3) at least 3 episodes of OM lasting at least 2 months and at least 1 episode within 12 months of enrollment in the study; or (4) ventilation tubes inserted for cOME. The rAOM group met at least one of the following criteria: (1) at least 3 episodes of AOM within one year; (2) at least 5 episodes of AOM by 3 years of age and 2 episodes of AOM within 12 months of enrollment in the study; or (3) ventilation tubes inserted for rAOM.

Of the 16 children included in the analysis, 5 were in the control group, 3 in the cOME group and 8 in the rAOM group. There were 7 males and 9 females. Race was reported as 1 black, 13 white, and 2 American Indian. The average age at the time of the MRI scan was 6.92 +/−0.33 years.

Landmarks that captured the shape of the cartilaginous ET, as well as some of the basicranial anatomy, were collected (Table 1 and Figures 2–5). Where possible, landmarks were pulled from the literature, many of which are standard craniometric points. Several novel landmarks were established for the present study that describe the soft tissue, including the anterior and posterior torus tubarius and the TVP point. These novel landmarks can only be viewed and analyzed using MRI and specific imaging. Landmarking was done in Checkpoint software on individual slices of the MRI scan and exported to MorphoJ and SPSS for analysis [24,25]. The Procrustes analysis and PCA convert the data (i.e., the three-dimensional Cartesian coordinates) into a shape for each subject. The shapes are then scaled and rotated to find the orientation that produces the least difference among subjects. PCA takes a covariance matrix of the Procrustes aligned coordinates and identifies the axes of greatest variation. Specific shape differences in the cranial base can be isolated and visualized in a scatter plot; further, PC scores can be used to test for significant correlation with an independent variable like OM diagnosis. Each PC explains a different dimension of variance, and variance is a measure of the difference from the mean (i.e., it is the spread of data and more specifically, the average of the squared differences from the mean). The first PC encompasses the greatest proportion of total shape variance, and each consecutive PC represents orthogonal vectors of descending proportions of the variance [40,41]. PCs can be plotted against one another (e.g., PC1 against PC2) or against the centroid size, the calculated size of the shape. Each point in the scatterplot represents the shape of the cranium of a single individual. The data points within the scatterplots can be assigned different colors based on the research group (e.g., control and disease). This is used as a visual representation of the distribution of the research groups. Confidence ellipses can then be used to illustrate if there are shape differences in any of the PCs between the groups.

A Procrustes ANOVA can be used to test whether shape, as described by the Procrustes distances, is significantly correlated with an independent variable, in this case, OM status. Canonical variate analysis is a complementary analysis that uses the Procrustes distances to identify the shape features that maximize the distance between groups by modifying the data so that within-group variation is isotropic; the variance is equally distributed around the mean. This was particularly useful in a study like this one, with known groupings (OM and control), to test whether the two groups can be significantly separated and identify the shape features that separate them best.

Measurement error was assessed using Procrustes ANOVA in MorphoJ. A second iteration of landmarking was completed on a subset of the sample (n = 5) with at least 1 week between trials. Results of the error study indicate that the effect of individual shape remains significant despite intraobserver error associated with landmarking ($p < 0.01$).

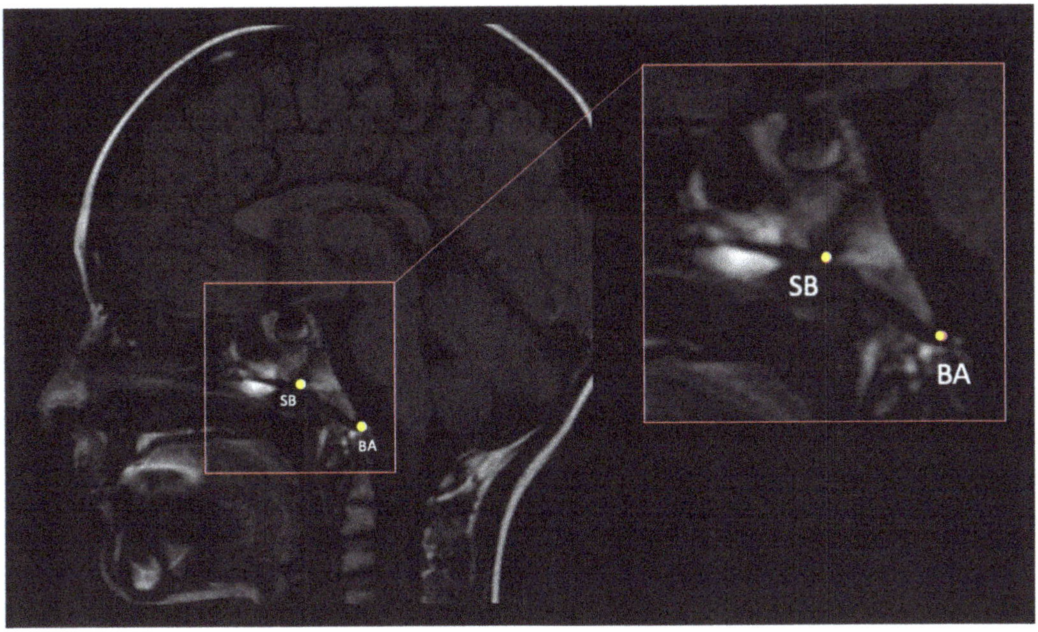

Figure 2. Midsagittal slice from an MRI scan depicting basion (BA) and sphenobasion (SB).

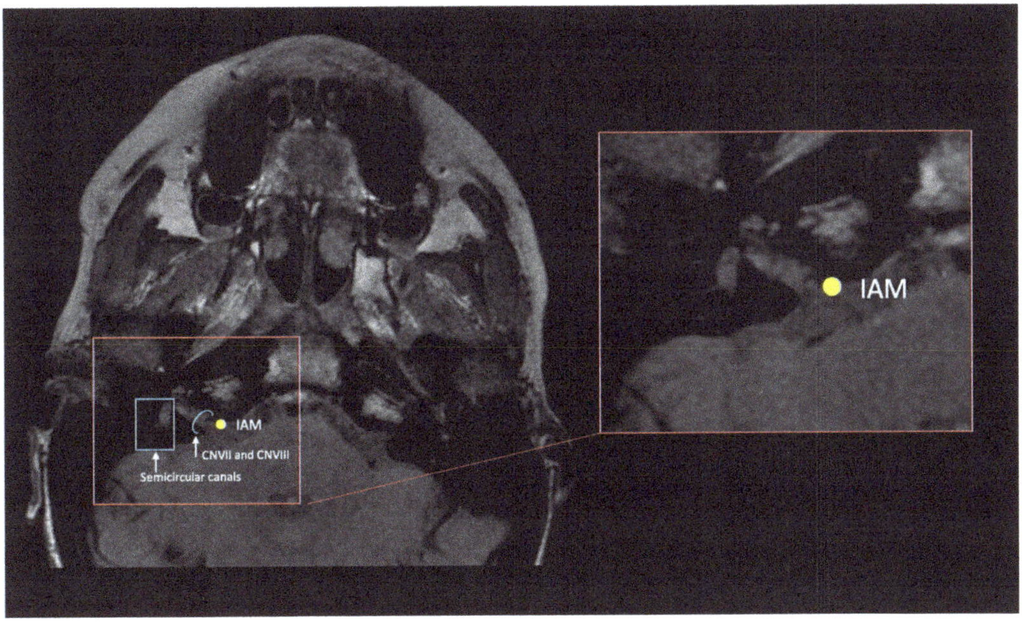

Figure 3. Oblique axial slice from an MRI scan depicting the internal acoustic meatus. CNVII: facial nerve (cranial nerve 7); CNVIII: vestibulocochlear nerve (cranial nerve 8); semicircular canals; IAM: internal acoustic meatus.

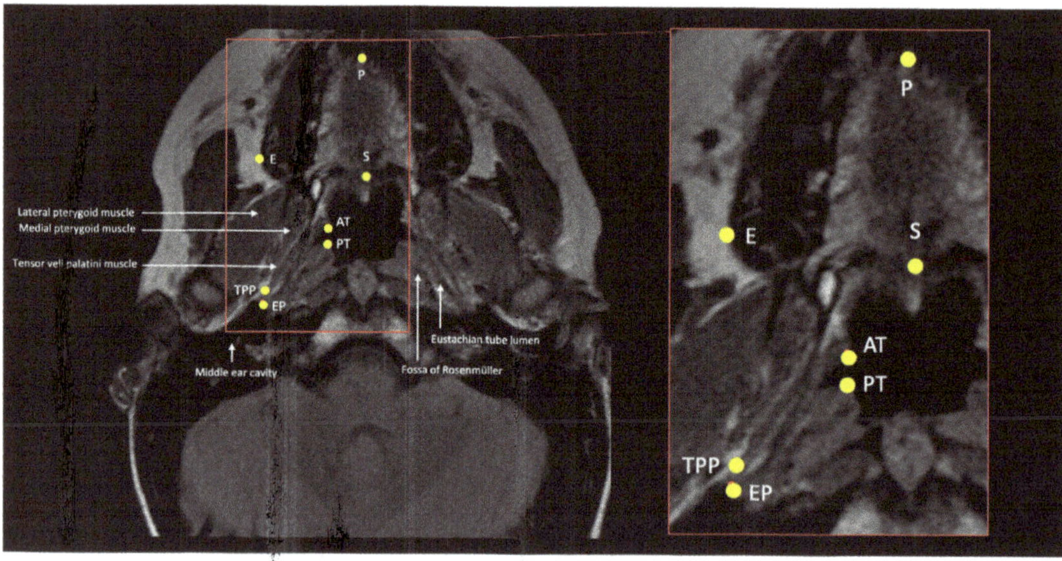

Figure 4. Oblique axial slice from an MRI scan depicting prosthion (P), staphylion(S), ectomolare (E), anterior and posterior torus tubarius (AT and PT), proximal insertion of TVP (TPP), and Eustachian point (EP). This is not the optimal slice for all these landmarks but is included because it demonstrates their relative positions.

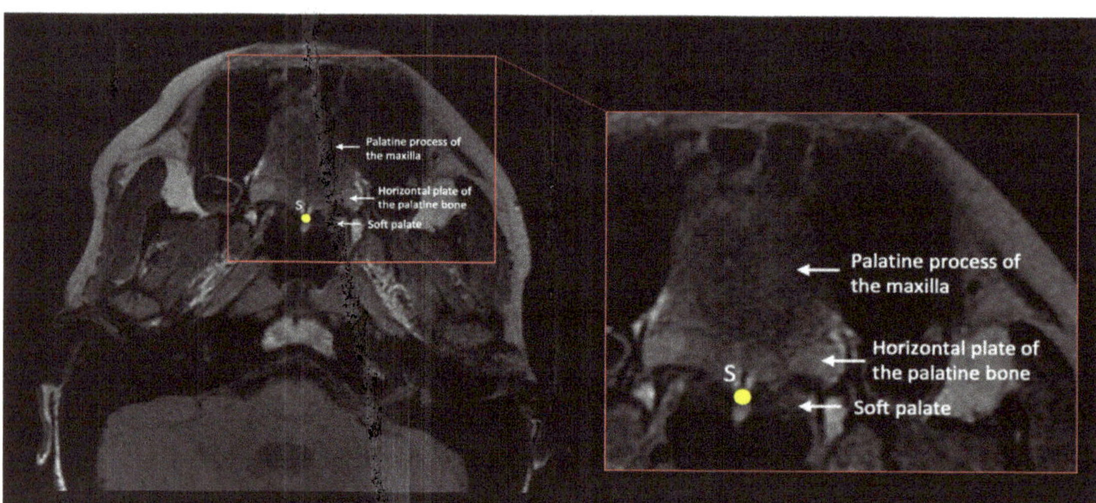

Figure 5. Oblique axial slice from an MRI scan depicting staphylion (S).

Table 1. Landmarks included in the present study.

Bone Landmark	Abbr.	Location	Definition	Description	Figure
Sphenobasion	SB	Midline	Midline of spheno-occipital synchondrosis	In a midsagittal plane, identify the clivus and place the landmark on the superior border of the clivus on the external surface in between the sphenoid and occipital bones; this is a cartilaginous plate in children.	Figure 2
Basion (exo-basion)	BA	Midline	Midline of anterior foramen magnum	In a midsagittal plane, identify the most inferior–posterior point on the clivus.	Figure 2
Staphylion	S	Midline	Posterior maxillary spine	In the oblique axial plane, identify the slice that contains the maxillary palatine process and palatine bones; these form the hard palate. Place the landmark on the most posteromedial point of the palatine palate.	Figures 4 and 5
Prosthion (endoprosthion)	P	Midline	Midline of the anterior palate	In the oblique axial plane, identify the slice that contains the cementoenamel junction of the upper central incisors (the cross-section of the incisors will be oval). Place the landmark on the lingual surface of the palate between the right and left upper central incisors at the level of the cementoenamel junction.	Figure 4
Internal acoustic meatus	IAM	Right and left	Internal opening of the petrous temporal bone	In the oblique axial plane, identify the facial and vestibulocochlear nerves emerging from the brainstem and follow the nerves medially. Locate the point of greatest curvature at the anterior entry to the IAM.	Figure 3
Eustachian point	EP	Right and left	Junction of bony and cartilaginous ET	In the oblique axial plane, identify the external acoustic meatus, and follow the space into the middle ear cavity, and place the landmark at the most anteromedial opening.	Figure 4
Proximal tensor veli palatini attachment point	TPP	Right and left	Origin of the TVP muscle at the skull base	In the oblique axial plane, identify the TVP muscle fibers and follow them proximally (toward the cranial base), and place the landmark at the most proximal insertion.	Figure 4
Anterior torus tubarius	AT	Right and left	Anterior border of the distal ET	In the oblique axial plane, identify anterior and posterior medial ends of the nasopharyngeal orifice of the ET.	Figure 4
Posterior torus tubarius	PT	Right and left	Posterior border of the distal ET		Figure 4
Ectomolare	E	Right and left	Most lateral point on the exterior surface of the alveolus	In the oblique axial plane, identify the slice with the widest alveolus and palate. Place the landmark on the external surface of the alveolus at the widest point.	Figure 4

3. Results

The results of the Procrustes ANOVA indicate that the shape and orientation of the ET significantly vary between the OM (combined rAOM, and cOME) and control groups ($p < 0.01$). Results of the CVA support these results, with significant differences found in the Procrustes distances between the OM and control groups (Wilks lambda = 0.226—a Wilk's lambda close to 0 indicates significant discrimination between groups; Figure 6). A set of ad hoc analyses were completed on a subset of landmarks, excluding those of the palate, because they showed high levels of variability. Palatal variation seems to, at least in part, be swamping out some of the finer variation in the cranial base. In the ad hoc analyses, results remained significant (Wilks lambda = 0.147).

Results of the PCA show that groups do not separate well on PCs 1 or 2 (Table 2; Figure S1). Principal component 1 (PC1) did not correlate with centroid size (Figure 7; $\beta = -0.32$); this result suggests that although PC1 captures most of the variance, it is not associated with head size. Based on the Kruskal–Wallis test, the differences between the groups (control and OM) are best summarized by PC3. PC3 describes a medial and low attachment site of the TVP muscle on the osseous ET, a posteriorly oriented and high torus tubarius, and an anteriorly projected palate in the OM group (rAOM and cOME) compared to the control group (Figures 8 and 9).

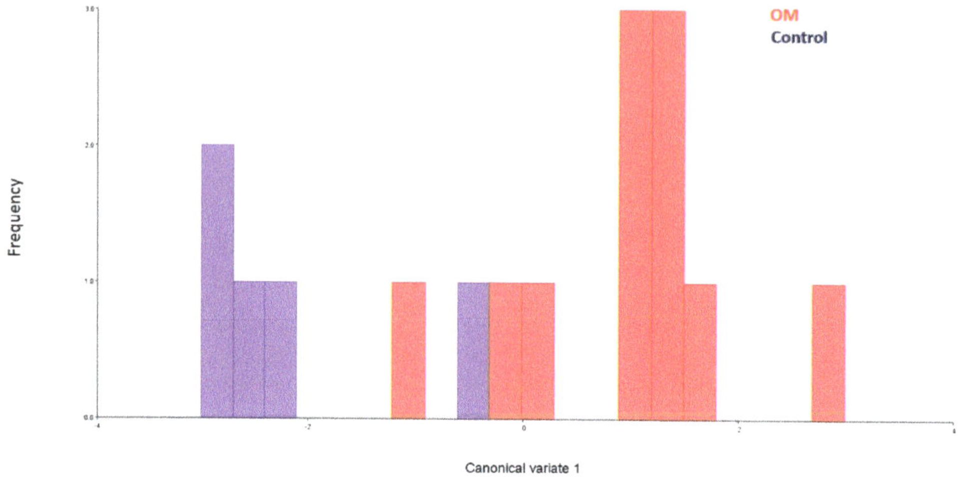

(**a**)

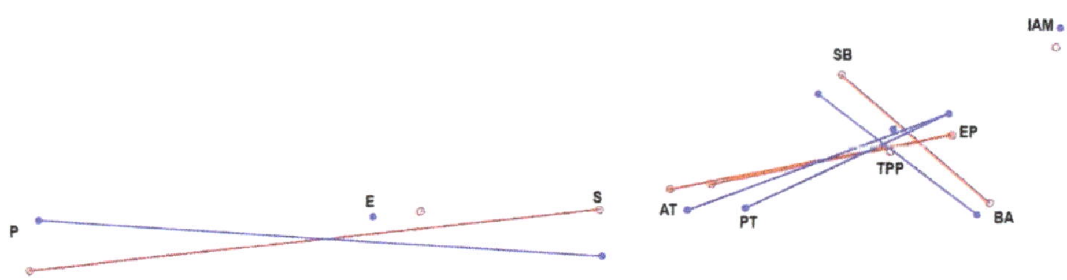

(**b**)

Figure 6. Results of CVA. The OM and control groups separated significantly on CV 1 (**a**). Canonical variate 1 captures a more horizontal ET in the OM group; a lateral view (**b**) of the shape differences illustrates the differences in ET between the OM group (red) and control (blue). Landmarks include internal acoustic meatus (IAM), basion (BA), Eustachian point (EP), tensor veli palatini proximal attachment (TPP), sphenobasion (SB), staphylion (S), ectomalare (E), endoprosthion (P), and anterior and posterior torus tubarius (AT and PT).

Table 2. Description of shape variation captured in PCs 1–3. PC 3 is the only PC that significantly correlated with OM diagnosis.

PC	% Variance	Total Variance	Cranial Base	Eustachian	Torus Tubarius
1	29.8	29.8	Wide (medial/lateral)	Lateral	Anterior
2	21.1	50.9	Tall (superior/inferior)	Inferior	High
3	11.4	62.3	Short	Inferior	Posterior, superior

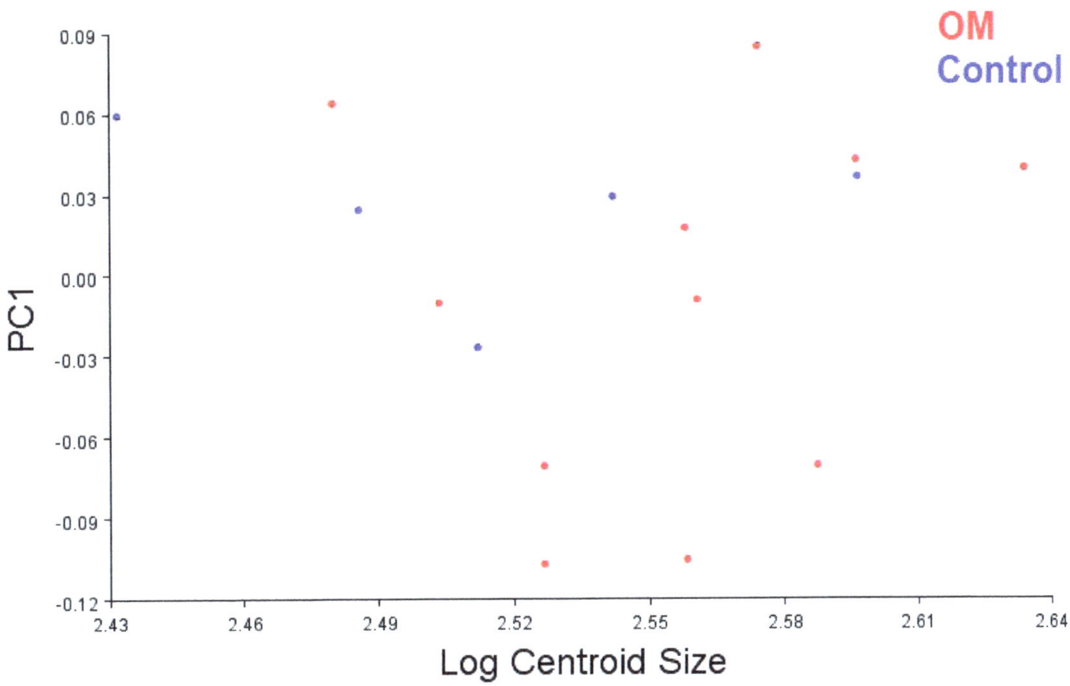

Figure 7. PC plot of PC1 and centroid size. There was no significant relationship between shape and head size.

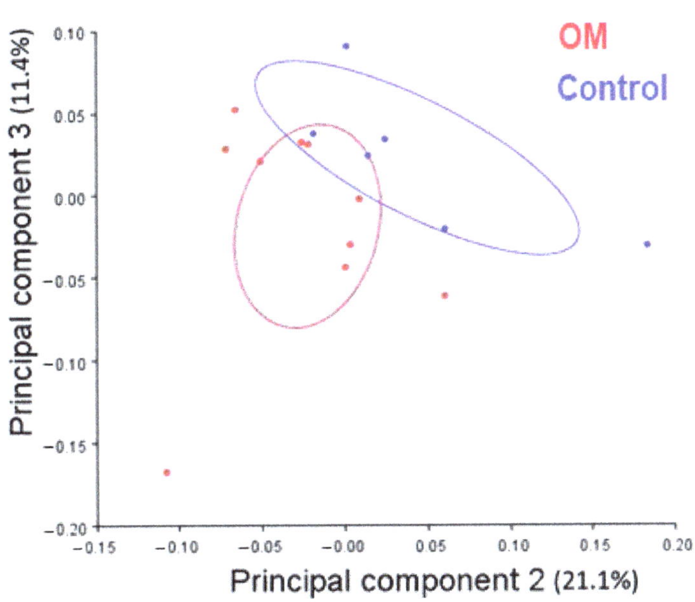

Figure 8. PC plot of PC2 and PC3. OM individuals are illustrated in red, and control in blue. Ellipses illustrate a 95% confidence interval for each group.

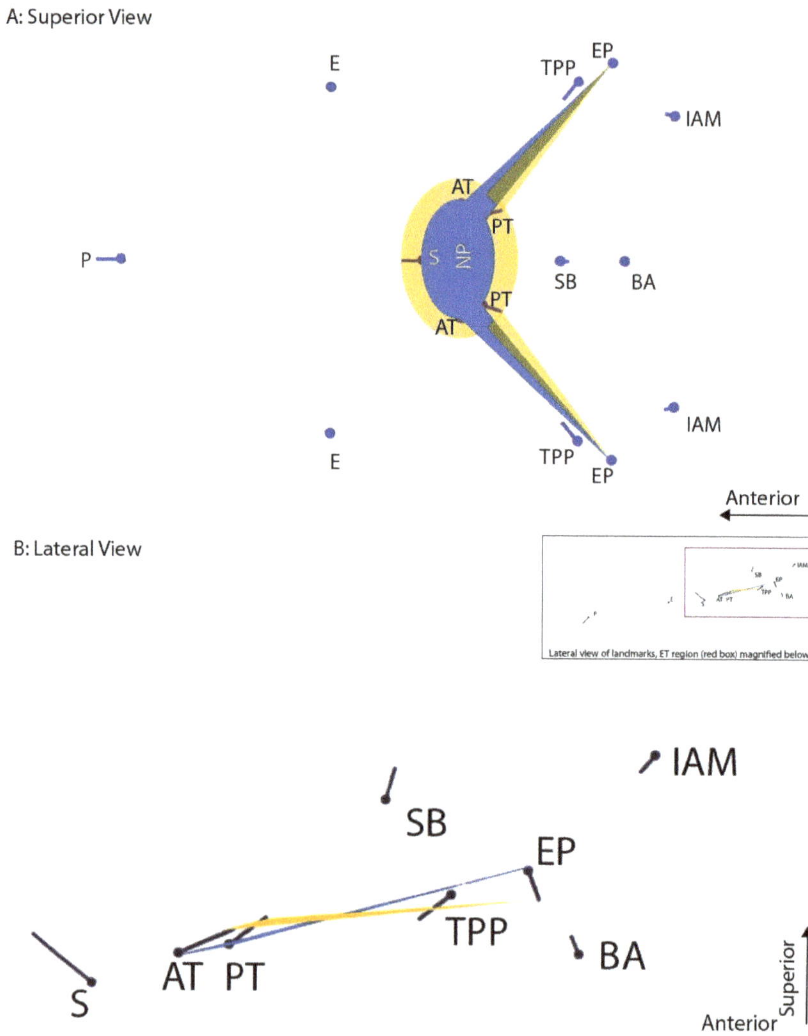

Figure 9. PC3 shape differences. PC 3 captures 11.4% of the variation. Superior (**A**) and lateral (**B**) views of basicranial shape illustrating average (nodes) and high values (lines) on PC 3. The ET angulation and nasopharynx (NP) are illustrated in blue (average) and yellow (OM). Landmarks include internal acoustic meatus (IAM), basion (BA), Eustachian point (EP), tensor veli palatini proximal attachment (TPP), sphenobasion (SB), staphylion (S), ectomalare (E), endoprosthion (P), and anterior and posterior torus tubarius (AT and PT). The node is the average of all individuals, and the vector illustrates variation in the direction of shapes associated with OM diagnosis of PC3.

4. Discussion

Otitis media is extremely common in children under age 2. Many factors contribute to OM susceptibility, including viral type and load, immune system immaturity, life history, and/or craniofacial anatomy, which includes ET shape and orientation. The anatomical and structural factors are at the foundation of OM in infancy, over which the immune system immaturity and infection are overlaid. The peak age of rAOM is around 18 months, but maturing immune defense mechanisms through acquired immunity in response to repeated viral and bacterial infections increase the causal weight of anatomical/structural factors in the etiopathogenesis [9]. In the current study, an enrollment age of 3 years and follow-up until age 7.5 years reduces the impact of infectious and immune etiologies, allowing the role of morphology (i.e., ET shape) on OM incidence to emerge. While rAOM continues to be seen in children older than 3, the prevalence of cOME becomes more dominant after this age [9]. Group assignments may be in error if they are made based on recall or retrospective history. Thus, being part of a prospectively and closely followed cohort employing validated and experienced clinicians, particularly ENT specialists, diminishes the risk of misdiagnosis. The main hypothesis relies on differentiating between children with and without the history of OM. One of the strengths of the current study is having a reliable follow-up and assessment, resulting in an accurate differentiation of study and control groups.

Many differences between adult and infant ET anatomy exist that may affect susceptibility to OM. Of greatest interest to this study is the relative horizontality of the Eustachian tube, but others are important to note as well, although they could not be tested using the current methodology [42]. Bluestone [42] classifies the pathophysiological effects of ET dysfunction into three categories: impairment of pressure regulation, loss of protective function, and impairment of clearance. The angulation of the ET falls under the category of impairment of pressure regulation due to the ET being too closed. A more horizontal angulation of the ET tube can lead to inefficient action of the TVP and, thus, ineffective opening of the ET. Conversely, ET dysfunction may be caused by the ET being too open (i.e., patulous), which can lead to a loss of the protective function. The cartilage itself is less stiff in the infant and is less supported by a smaller Ostmann's fat pad, resulting in a floppier ET. Additionally, the ET is relatively short in the infant. These factors also likely play a role in OM susceptibility.

Few studies have directly compared the shape of the craniofacial complex between children with and without a history of OM, at least partially because most children with OM do not require imaging for the diagnosis or treatment of the disease. Using lateral radiographs, Gremba et al. [22] found that the shape of the basicranium in children with rAOM had distinct craniofacial morphologies compared to the control group in their 4-year-old age group. Takasaki et al. [36] found no significant difference in the angle or length of the ET in their sample of children with and without OME. Geometric morphometric shape analysis of MRI scans has several distinct advantages over both linear measurements and X-ray imaging techniques (radiographs and CT scans). First, with MRI scans, soft tissue landmarks may be captured, which is not possible with radiographs or CT scans, and second, MRI scans allow for more accurate placement of some bony landmarks because of the visibility of relative soft tissue. One of the major advantages of this study was to analyze previously unstudied soft-tissue landmarks in relation to OM. Due to MRI scan constraints, novel soft tissue landmarks were limited to the following: the anterior and posterior torus tubarius and tensor veli palatini point. However, the presence of soft tissue structures allowed precise landmarking of the following bony landmarks as well, which would not have been possible with an X-ray or CT scan: Eustachian point and internal acoustic meatus. Therefore, the use of MRI scans (i.e., visualization of soft tissue structures) was essential in placing landmarks for 5/10 landmarks, and if we exclude the landmarks of the maxilla and cranial base, then it was essential for 5/6 ET and TVP-specific landmarks.

Geometric morphometric analysis offers its own advantages over linear measurements because it reduces differences due to size and tests the differences between groups based on the relative location of landmarks to one another. Therefore, the plane in which the data was collected does not matter. Pagano et al. [25] found that the length and shape of the cartilaginous ET, as represented by bony proxies in dry skulls, changes dynamically throughout infancy and early childhood and correlates with peak incidences of OM. OM is much less common by the age of six but can and does occur into later childhood and adulthood [29]. It is possible that some of the same variables that go into OM susceptibility in infancy persist [43]. Relatively few studies exist that directly compare children with known medical histories, and even fewer take into account soft tissue and/or three-dimensional relationships. The results of this study indicate there are significant differences in anatomical relationships between groups with and without a history of OM.

Previous studies have found that the TVP muscles are necessary for the opening of the ET lumen, and by extension, the function of these muscles may be an important protection of the middle ear that may help prevent OM [11,19,34]. Sapci et al. [44] is one of the few studies that directly compared TVP muscle electromyography in adult subjects with or without OM, and they found no difference in motor unit potential or duration of the TVP. While the present study did find differences in the position of the proximal insertion of the TVP on the cranial base, the TVP landmarks varied with the middle ear landmark Eustachian point. So, while the TVP was found to be oriented more inferiorly in the OM group, it did not vary relative to the positioning of the middle ear in the PCs, which captured a significant amount of the variation. Regardless, it is possible that the angle of the TVP relative to the hamulus of the pterygoid plates (the fulcrum around which the TVP is set) is less acute and thus may be suboptimal in the OM group. One thing that could not be tested using the method presented here is the insertion ratio or the relative length of the insertion of the TVP to the total length of the ET [45]. It may be that a lower posterior boundary of the TVP attachment would mean a lower anterior boundary as well, but it is certainly possible that the thickness of the muscle varies, not the total position.

It is interesting that the variation at the palate had to be removed in some analyses because it was "swamping out" the fine-grained differences at the ET. The relationship between OM and palate shape is frequently discussed in clinical research, as exemplified by the extensive literature concerning cleft palate and OM (e.g., [20,34,35,44,46]). A cleft palate leads to many anatomical and structural differences in the craniofacial complex that may lead to OM, including differences in TVP size as measured on MRI in adults with and without a cleft palate [47]. The shape of the palate certainly seems to vary between patients with and without OM in this study, though it was not specifically isolated and analyzed. The palate shape is integrated with the shape of the oral cavity and dentition—especially the mixed dentition, which characterizes this age group. Further, the shape of the palate reflects the general width of the face. The oral vestibule and palate are continuous with the nasopharynx posteriorly; thus, an anteriorly projected palate could be associated with a long (anteroposterior) nasopharynx. A large nasopharynx would have implications for the position of the hamulus of the pterygoid plates. The presented results provide a basis upon which future analyses of the palate shape in OM children can be based.

5. Conclusions

This study contributes to the growing literature on imaging and shape analysis of the ET. This is the first attempt that the authors could find at using MRI data to perform geometric morphometric shape analysis on the ET in any age group. In the present study, we found evidence that the shape of the Eustachian tube varies significantly between these groups of children with and without OM.

6. Limitations and Future Directions

It is important to note our limitations. The sample size is small, and a larger sample would increase the power of the statistical analyses. However, despite the small sample size, the authors found significant differences between cohorts and established novel protocols for the collection of soft tissue landmarks in the ET that can be used in future studies. The soft tissue landmarks presented here open up the types of hypotheses that can be tested using three-dimensional geometric morphometrics of the middle ear and its pathologies.

While the MRI scans capture an enormous amount of anatomy, the slice thicknesses and window sizes limit the types of landmarks collected. Certain biologically significant landmarks could not be collected in the present study due to the window size of the original scans, including landmarks around the paranasal sinus drainage points, nasal, and nasopharyngeal landmarks. Further, placing landmarks on individual slices did not allow for the use of semilandmarks that extended beyond a single plane. This can be addressed in the future by using this pilot protocol and adjusting scan settings to widen the scan windows and narrow the slice thicknesses.

In the future, this method should be tested with a larger sample and other age ranges, particularly in the older ages that have been historically under-evaluated in the literature. With a larger sample size, the authors will be able to test for potentially confounding or informative variables like sex, population affinity, and life history. Further, digital isolation of the ET and the paratubal muscles will allow for a wider range of statistical analyses.

Supplementary Materials: The following supporting information can be downloaded at: https://www.mdpi.com/article/10.3390/bioengineering10101115/s1.

Author Contributions: Conceptualization, E.E.F., A.P.G. and J.D.S.; methodology, E.E.F., A.P.G. and J.D.S.; software, E.E.F.; validation, E.E.F. and A.P.G.; formal analysis, E.E.F. and A.P.G.; investigation, E.E.F. and A.P.G.; resources, E.E.F., A.P.G., J.D.S., M.S.T. and C.M.A.; data curation, E.E.F., A.P.G., J.D.S., M.S.T. and C.M.A.; writing—original draft preparation, E.E.F. and A.P.G.; writing—review and editing, E.E.F., A.P.G., J.D.S., M.S.T. and C.M.A.; visualization, E.E.F. and A.P.G.; supervision, E.E.F. and J.D.S.; project administration, C.M.A.; funding acquisition, C.M.A. All authors have read and agreed to the published version of the manuscript.

Funding: This research was funded by the National Institute of Health, grant number NIH 1P50DC007667.

Institutional Review Board Statement: This study was reviewed and approved by the Institutional Review Board of the University of Pittsburgh—REN11040024/IRB0605009.

Informed Consent Statement: Informed consent was obtained from a parent of all subjects involved in the study.

Data Availability Statement: Data are available upon request.

Acknowledgments: Thanks to the late William J. Doyle—the original grant and project administrator—and Margaretha L. Casselbrant and Ellen M. Mandel for assigning the subject group and performing the follow-up assessments.

Conflicts of Interest: CMA is a consultant to Doya Medical Inc. The other authors declare no conflict of interest.

References

1. Chonmaitree, T.; Revai, K.; Grady, J.J.; Clos, A.; Patel, J.A.; Nair, S.; Fan, J.; Henrickson, K.J. Viral upper respiratory tract infection and otitis media complication in young children. *Clin. Infect. Dis.* **2008**, *46*, 815–823. [CrossRef] [PubMed]
2. Winther, B.; Doyle, W.J.; Alper, C.M. A high prevalence of new onset otitis media during parent diagnosed common colds. *Int. J. Pediatr. Otorhinolaryngol.* **2006**, *70*, 1725–1730. [CrossRef] [PubMed]
3. Heikkinen, T. Role of viruses in the pathogenesis of acute otitis media. *Pediatr. Infect. Dis. J.* **2000**, *19*, S17–S23. [CrossRef]
4. Alho, O.P.; Koivu, M.; Sorri, M. What is an 'otitis-prone' child? *Int. J. Pediatr. Otorhinolaryngol.* **1991**, *21*, 201–209. [CrossRef]
5. Alho, O.-P.; Koivu, M.; Sorri, M.; Rantakallio, P. The occurrence of acute otitis media in infants. A life-table analysis. *Int. J. Pediatr. Otorhinolaryngol.* **1991**, *21*, 7–14. [CrossRef]
6. Teele, D.W.; Klein, J.O.; Rosner, B.A. Epidemiology of otitis media in children. *Ann. Otol. Rhinol. Laryngol.* **1980**, *89*, 5–6. [CrossRef] [PubMed]
7. Dewey, K.G.; Heinig, M.J.; Nommsen-Rivers, L.A. Differences in morbidity between breast-fed and formula-fed infants. *J. Pediatr.* **1995**, *126*, 696–702. [CrossRef]
8. Bentdal, Y.E.; Nafstad, P.; Karevold, G.; Kværner, K.J. Acute otitis media in schoolchildren: Allergic diseases and skin prick test positivity. *Acta Otolaryngol.* **2007**, *127*, 480–485. [CrossRef]
9. Bluestone, C.D. Impact of evolution on the eustachian tube. *Laryngoscope* **2008**, *118*, 522–527. [CrossRef]
10. Proctor, B. Anatomy of the Eustachian Tube. *Arch. Otolaryngol.* **1973**, *97*, 2–8. [CrossRef]
11. Proctor, B. Embryology and Anatomy of the Eustachian Tube. *J. Craniofacial Surg.* **1967**, *86*, 503–514. [CrossRef]
12. Shim, H.J.; Choi, A.Y.; Yoon, S.W.; Kwon, K.H.; Yeo, S.G. The value of measuring eustachian tube aeration on temporal bone ct in patients with chronic otitis media. *Clin. Exp. Otorhinolaryngol.* **2010**, *3*, 59–64. [CrossRef]
13. Elner, Å.; Ingelstedt, S.; Ivarsson, A. The normal function of the eustachian tube: A study of 102 cases. *Acta Otolaryngol.* **1971**, *72*, 320–328. [CrossRef] [PubMed]
14. Seibert, J.W.; Danner, C.J. Eustachian tube function and the middle ear. *Otolaryngol. Clin. N. Am.* **2006**, *39*, 1221–1235. [CrossRef] [PubMed]

15. Alper, C.M.; Teixeira, M.S.; Swarts, J.D. Correlations between videoendoscopy and sonotubometry of eustachian tube opening during a swallow. *Laryngoscope* **2016**, *126*, 2778–2784. [CrossRef] [PubMed]
16. Alper, C.M.; Teixeira, M.S.; Swarts, J.D.; Doyle, W.J. Quantitative description of eustachian tube movements during swallowing as visualized by transnasal videoendoscopy. *JAMA Otolaryngol. Head Neck Surg.* **2015**, *141*, 160–168. [CrossRef]
17. Alper, C.M.; Teixeira, M.S.; Swarts, J.D. Eustachian tube function in adults with ventilation tubes inserted for otitis media with effusion. *J. Int. Adv. Otol.* **2018**, *14*, 255–262. [CrossRef]
18. Honjo, I.; Okazaki, N.; Kumazawa, T. Opening mechanism of the eustachian tube. A clinical and experimental study. *Ann. Otol. Rhinol. Laryngol. Suppl.* **1980**, *89*, 25–27. [CrossRef]
19. Honjo, I.; Okazaki, N.; Kumazawa, T. Experimental study of the eustachian tube function with regard to its related muscles. *Acta Otolaryngol.* **1979**, *87*, 84–89. [CrossRef]
20. Huang, M.H.S.; Lee, S.T.; Rajendran, K. A Fresh Cadaveric Study of the Paratubal Muscles: Implications for Eustachian Tube Function in Cleft Palate. *Plast. Reconstr. Surg.* **1997**, *100*, 833–842. [CrossRef]
21. Rood, S.R.; Doyle, W.J. Morphology of tensor veli palatini, tensor tympani, and dilatator tubae muscles. *Ann. Otol. Rhinol. Laryngol.* **1978**, *87*, 202–210. [CrossRef] [PubMed]
22. Gremba, A.P.; Weinberg, S.M.; Swarts, J.D.; Casselbrant, M.L. Craniofacial shape in children with and without a positive otitis media history. *Int. J. Pediatr. Otorhinolaryngol.* **2016**, *84*, 110–115. [CrossRef] [PubMed]
23. Tarabichi, M.; Najmi, M. Visualization of the eustachian tube lumen with Valsalva computed tomography. *Laryngoscope* **2015**, *125*, 724–729. [CrossRef] [PubMed]
24. Yoshida, H.; Kobayashi, T.; Morikawa, M.; Hayashi, K.; Tsujii, H.; Sasaki, Y. CT imaging of the patulous eustachian tube—Comparison between sitting and recumbent positions. *Auris Nasus Larynx* **2003**, *30*, 135–140. [CrossRef] [PubMed]
25. Pagano, A.S.; Laitman, J.T. Three-Dimensional Geometric Morphometric Analysis of the Nasopharyngeal Boundaries and Its Functional Integration with the Face and External Basicranium Among Extant Hominoids. *Anat. Rec.* **2015**, *298*, 85–106. [CrossRef] [PubMed]
26. Pagano, A.S.; Márquez, S.; Laitman, J.T. Reconstructing the Neanderthal Eustachian Tube: New Insights on Disease Susceptibility, Fitness Cost, and Extinction. *Anat. Rec.* **2019**, *302*, 2109–2125. [CrossRef]
27. Pagano, A.S.; Wang, E.; Yuan, D.; Fischer, D.; Bluestone, C.; Marquez, S.; Laitman, J. Cranial Indicators Identified for Peak Incidence of Otitis Media. *Anat. Rec.* **2017**, *300*, 1721–1740. [CrossRef]
28. Alper, C.M.; Rath, T.J.; Teixeira, M.S.; Swarts, J.D. A Novel Imaging Method for the Cartilaginous Eustachian Tube Lumen: Computerized Tomography During the Forced Response Test. *Ann. Otol. Rhinol. Laryngol.* **2018**, *127*, 13–20. [CrossRef]
29. Smith, M.E.; Scoffings, D.J.; Tysome, J.R. Imaging of the Eustachian tube and its function: A systematic review. *Neuroradiology* **2016**, *58*, 543–556. [CrossRef]
30. Terzi, S.; Beyazal Çeliker, F.; Özgür, A.; Çeliker, M.; Beyazal, M.; Demirci, M.; Dursun, E. The evaluation of eustachian tube paratubal structures using magnetic resonance imaging in patients with chronic suppurative otitis media. *Acta Otolaryngol.* **2016**, *136*, 673–676. [CrossRef]
31. Oshima, T.; Kikuchi, T.; Hori, Y.; Kawase, T.; Kobayashi, T. Magnetic resonance imaging of the eustachian tube cartilage. *Acta Otolaryngol.* **2008**, *128*, 510–514. [CrossRef] [PubMed]
32. Alper, C.M.; Sabo, D.L.; Doyle, W.J. Validation by magnetic resonance imaging of tympanometry for diagnosing middle ear effusion. *Otolaryngol. Head Neck Surg.* **1999**, *121*, 523–527. [CrossRef] [PubMed]
33. Alper, C.M.; Doyle, W.J. MRI validation of the accuracy of tympanometric gradient for the diagnosis of OME. *Br. J. Audiol.* **1999**, *33*, 233–239. [CrossRef]
34. Shprintzen, R.J.; Croft, C.B. Abnormalities of the eustachian tube orifice in individuals with cleft palate. *Int. J. Pediatr. Otorhinolaryngol.* **1981**, *3*, 15–23. [CrossRef] [PubMed]
35. Maue-Dickson, W.; Dickson, D.R.; Rood, S.R. Anatomy of the Eustachian Tube and Related Structures in Age-Matched Human Fetuses with and without Cleft Palate. *Trans. Am. Acad. Ophthalmol. Otolarygngol.* **1976**, *82*, 159–164.
36. Takasaki, K.; Takahashi, H.; Miyamoto, I.; Yoshida, H.; Yamamoto-Fukuda, T.; Enatsu, K.; Kumagami, H. Measurement of Angle and Length of the Eustachian Tube on the CT Using Multiplanar Reconstruction Technique. *Laryngoscope* **2007**, *117*, 1251–1254. [CrossRef]
37. Wald, E.R. Acute otitis media and acute bacterial sinusitis. *Clin. Infect. Dis.* **2011**, *52*, S277–S283. [CrossRef]
38. Mills, R.P.; Irani, B.S.; Vaughan-Jones, R.J.; Padgham, N.D. Maxillary sinusitis in children with otitis media with effusion. *J. Laryngol. Otol.* **1994**, *108*, 842–844. [CrossRef]
39. Güçlü, O.; Şahin, E.M.; Tekin, K.; Dereköy, F.S. Evaluation of nasal airways by objective methods in chronic otitis media. *Eur. Arch. Oto-Rhino-Laryngol.* **2013**, *270*, 1263–1266. [CrossRef]
40. Zelditch, M.L.; Swiderski, D.L.; Sheets, H.D. *Geometric Morphometrics for Biologists: A Primer*; Academic Press: New York, NY, USA, 2012.
41. Bookstein, F.L. *Morphometric Tools for Landmark Data: Geometry and Biology*; Cambridge University Press: Cambridge, UK, 1997.
42. Bluestone, C. *Eustachian Tube: Structure, Function, and Role in Otitis Media*; PMPH-USA: Shelton, CT, USA, 2005; pp. 67–90.
43. Teixeira, M.S.; Banks, J.; Swarts, J.D.; Alper, C.M.; Doyle, W.J. Eustachian tube opening measured by sonotubometry is poorer in adults with a history of past middle ear disease. *Int. J. Pediatr. Otorhinolaryngol.* **2014**, *78*, 593–598. [CrossRef]

44. Sapci, T.; Mercangoz, E.; Evcimik, M.F.; Karavus, A.; Gozke, E. The evaluation of the tensor veli palatini muscle function with electromyography in chronic middle ear diseases. *Eur. Arch. Oto-Rhino-Laryngol.* **2008**, *265*, 271–278. [CrossRef] [PubMed]
45. Matsune, S.; Sando, I.; Takahashi, H. Insertion of the tensor veli palatini muscle into the eustachian tube cartilage in cleft palate cases. *Ann. Otol. Rhinol. Laryngol.* **1991**, *100*, 439–446. [CrossRef] [PubMed]
46. Dayan, J.H.; Smith, D.; Oliker, A.; Haring, J.; Cutting, C.B. A virtual reality model of eustachian tube dilation and clinical implications for cleft palate repair. *Plast. Reconstr. Surg.* **2005**, *116*, 236–241. [CrossRef] [PubMed]
47. George, T.N.; Kotlarek, K.J.; Kuehn, D.P.; Sutton, B.P.; Perry, J.L. Differences in the tensor veli palatini between adults with and without cleft palate using high-resolution 3-dimensional magnetic resonance imaging. *Cleft Palate-Craniofacial J.* **2018**, *55*, 697–705. [CrossRef] [PubMed]

bioengineering

Article

Super-Resolution Imaging of Neuronal Structures with Structured Illumination Microscopy

Tristan C. Paul, Karl A. Johnson and Guy M. Hagen *

UCCS BioFrontiers Center, University of Colorado Colorado Springs, 1420 Austin Bluffs Parkway, Colorado Springs, CO 80918, USA; tpaul@uccs.edu (T.C.P.); k9johnso@ucsd.edu (K.A.J.)
* Correspondence: ghagen@uccs.edu

Abstract: Super-resolution structured illumination microscopy (SR-SIM) is an optical fluorescence microscopy method which is suitable for imaging a wide variety of cells and tissues in biological and biomedical research. Typically, SIM methods use high spatial frequency illumination patterns generated by laser interference. This approach provides high resolution but is limited to thin samples such as cultured cells. Using a different strategy for processing raw data and coarser illumination patterns, we imaged through a 150-micrometer-thick coronal section of a mouse brain expressing GFP in a subset of neurons. The resolution reached 144 nm, an improvement of 1.7-fold beyond conventional widefield imaging.

Keywords: fluorescence microscopy; structured illumination; brain; Bayesian methods; super-resolution

Citation: Paul, T.C.; Johnson, K.A.; Hagen, G.M. Super-Resolution Imaging of Neuronal Structures with Structured Illumination Microscopy. *Bioengineering* **2023**, *10*, 1081. https://doi.org/10.3390/bioengineering10091081

Academic Editor: Cuneyt M. Alper

Received: 2 September 2023
Accepted: 9 September 2023
Published: 13 September 2023

1. Introduction

Recently developed methods for surpassing the diffraction limit in optical fluorescence microscopy include stimulated emission depletion microscopy (STED) [1], stochastic optical reconstruction microscopy (STORM) [2], photoactivated localization microscopy (PALM) [3], super-resolution optical fluctuation imaging (SOFI) [4], and structured illumination microscopy (SIM) [5,6]. These methods have had large impacts in many fields, with super-resolution microscopy previously being used in many applications, including imaging the a mouse brain using STED [7], STORM [8], and SIM approaches [9]. SIM methods have been used in many situations, including the rapid imaging of clinical samples [10].

SIM is a method in which sets of images are acquired with shifting illumination patterns. The subsequent processing of these image sets results in images with optical sectioning, resolutions beyond the diffraction limit (super-resolution), or both [5,6,11–14]. Since its emergence over two decades ago [15], SIM has matured as an imaging technique, with multiple proposed methods for generating the structured illumination patterns [12–25] and processing the image data [6,13,26–30]. Compared to other super-resolution techniques, the speed, high signal-to-noise ratio, and low excitation light intensities characteristic to SIM make it a good choice for imaging a variety of samples in three dimensions. As shown here, SIM is accomplished with a standard fluorescence microscope with some additional required elements, whereas other approaches such as light-sheet microscopy require more specialized setups. There is an increasing interest in the methods and applications of SIM, with the field seeing many recent (2022–2023) improvements [31–37], including new methods involving deep learning approaches [38,39].

The imaging method we used, maximum a posteriori probability SIM (MAP-SIM), uses a Bayesian framework to reconstruct super-resolution SIM images [26,29,40]. This method has advantages including flexibility in the range of SIM illumination patterns which can be used and, in our case, the ability to use patterns with lower spatial frequencies. This, in turn, allows imaging deeper into samples in which scattering degrades the high spatial frequency patterns which are more commonly used in SIM. When the SIM pattern is out of focus, it blurs rapidly with increasing depth, producing a high intensity of out of focus light. This results in reduced pattern contrast in the acquired images. Because of this, traditional super-resolution SIM methods are typically limited to an imaging depth of 10–20 μm [41,42]. Here, we used MAP-SIM to image a fixed, optically cleared, ~150-μm-thick mouse brain coronal slice expressing a neuronal GFP marker while achieving a lateral resolution of 144 nm.

To overcome the challenges of imaging deeper into brain tissues, SIM has previously been combined with two-photon excitation [16] or with adaptive optics for in vivo studies [43,44]. These methods have offered impressive results, but they do involve additional costs and require additional optical devices and expertise, reducing the number of labs that can use these approaches. Here, we used a simpler and more economical approach with a non-laser light source and open-source software for SIM [45].

2. Methods

The sample used for this work was an optically cleared, green fluorescent protein (GFP) -labeled coronal mouse brain slice. The slice was approximately 150 μm thick and was obtained from SunJin Lab (Hsinchu City, Taiwan). The supplier used a Thy1-GFP mouse strain, and they stated that the sample was prepared as follows:

1. cardiac perfusion with cold, freshly prepared 4% paraformaldehyde (PFA)
2. fixation of the dissected brain with a 4% PFA solution on an orbital shaker overnight at 4 °C followed by washing three times with phosphate-buffered saline (PBS) at room temperature
3. sectioning the brain manually using a vibratome followed by clearing of the slice with RapiClear 1.52 (SunJin Lab) overnight at room temperature
4. mounting of the cleared sample with fresh RapiClear 1.52 reagent in a 0.25-mm-deep iSpacer microchamber (SunJin Lab)

For the SIM imaging, we used a home-built set-up based on the same design as described previously [20,26,40,46]. The current SIM system was based on an IX83 microscope equipped with several objectives (Olympus, Tokyo, Japan). Illumination was provided by a liquid light guide-coupled Spectra-X light source (Lumencor, Beaverton, OR, USA) using the cyan channel, which had an emission maximum of 470 nm. The illumination was collimated by an achromatic 50 mm focal length lens (Thor labs, Newton, NJ, USA) and vertically polarized with a linear polarizer (Edmund Optics, Barrington, NJ, USA) before entering a polarized beam splitter (PBS) cube (Thor Labs) and reflecting onto a liquid-crystal on silicon (LOCS) microdisplay (Forth Dimension Displays, Dalgety Bay, Scotland, UK). This device is a ferroelectric reflective-type spatial light modulator. The pixels, which were turned on, rotated the polarization of the light by ~90 degrees, converting vertical polarization to horizontal polarization. The horizontally polarized output of the microdisplay then passed through the PBS and was imaged into the microscope using a 180 mm focal-length lens (SWTLU-C, Olympus). The emitted fluorescent light was filtered (using a GFP filter set with dichroic T495lpxr and ET525/50 emission filters; Chroma, Bellows Falls, VT, USA) and then imaged with an sCMOS camera (Zyla 4.2+, Andor). The illumination power density on the sample used a 100× objective that was measured at 2.542 W/cm^2 without SIM patterning (widefield illumination) and at 0.214 W/cm^2 with the SIM pattern active. Sample movements and focusing were controlled by an XY piezo Z stage (Applied Scientific Instrumentation, Eugene, OR, USA).

The microdisplay was used to produce the SIM patterns, and it was controlled by the software supplied with the device (MetroCon, Forth Dimension Displays). Various SIM patterns were used as shown in the supplementary material in Table S6. The pattern position was shifted by one pixel after each image was acquired such that the sum of all illumination masks resulted in homogenous illumination. Figure 1 shows a simplified diagram of the SIM optical system and a connection diagram illustrating how the microdisplay system was synchronized with the camera using IQ software (Andor) and a digital input/output computer card (DDA06/16, Measurement Computing, Concord, NH, USA). More details about the SIM system are given in the supplementary material. Table S1 shows a list of the components we used along with the manufacturer, part number, and vendor website. Tables S2–S5 show some of the relevant optical and performance characteristics of the camera, microdisplay, and light source. These details should be useful for those wishing to build their own SIM systems of this type. The supplementary text explains, and Figure S7 shows, a schematic of the timing scheme used by the SIM system, and they illustrate the function of the AND gates shown in Figure 1. The supplementary text also explains, and Figure S8 shows, additional details about the operation of the microdisplay.

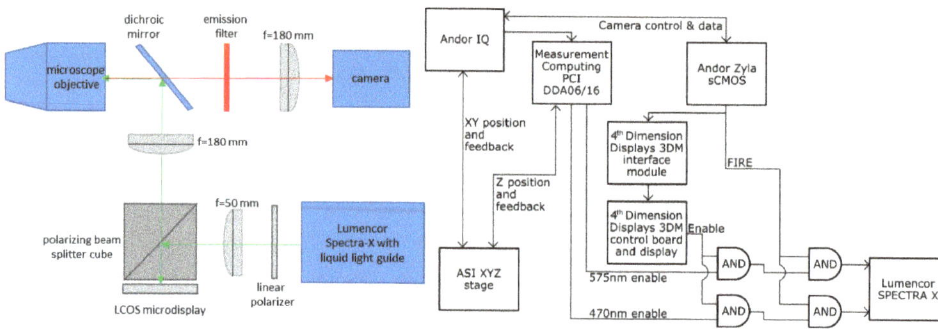

Figure 1. Simplified optical diagram (**left**) and connection diagram (**right**). The connection setup for the two-wavelength acquisition is shown, and in this study, only 470 nm illumination was used.

3. Data Analysis

3.1. Optical Sectioning SIM (OS-SIM)

Several data processing methods are possible for generating optically sectioned images from SIM data (OS-SIM) [20,47]. The most commonly used implementation of this technique was introduced in 1997 by Neil et al. [15]. Their method worked by projecting a line illumination pattern onto a sample, followed by the acquisition of a set of three images with the pattern shifted by the relative spatial phases 0, 2π/3, and 4π/3, respectively. Using this method, an optically sectioned image can be recovered computationally as follows:

$$I_{OS-SIM} = \left[(I_1 - I_2)^2 + (I_1 - I_3)^2 + (I_2 - I_3)^2\right]^{1/2}, \quad (1)$$

where $I_{OS\text{-}SIM}$ is an optically sectioned image and I_1, I_2, and I_3 are the three images acquired with the different pattern positions. This type of optically sectioned image is expected to be similar to that obtained with a laser scanning confocal microscope. If the sum of the individual SIM patterns results in homogeneous illumination, as was the case in our setup, a widefield (WF) image can also be recovered from the SIM data by taking the average of all images I_n, as follows:

$$I_{WF} = \frac{1}{N}\sum_{n=1}^{N} I_n. \quad (2)$$

This was the approach we used throughout this study to generate conventional widefield images.

Instead of using Equation (1), in this study, we used a method originally shown by Neil et al. [15] and later elaborated upon [20,47], as follows:

$$I_{OS-SIM} = \left|\sum_{n=1}^{N} I_n \exp\left(2\pi i \frac{n}{N}\right)\right|. \quad (3)$$

We found that this method provided consistent results and could be applied when using any number of patterns instead of the three patterns used in the original work. The actual positions of the illumination patterns in the camera images were determined using a calibrated camera according to our previous work [15], and this was based on a well-known method for the spatial calibration of a camera [48].

3.2. SIM with Maximum a Posteriori Probability Estimation

MAP-SIM has been described previously [26]. In our study, the imaging process could be denoted as follows:

$$y_k = HM_k x + n_k, \tag{4}$$

where M_k is a matrix in which the elements represent the k-th illumination pattern; y_k denotes a low-resolution image acquired using the k-th illumination pattern; x is an unknown, high-resolution image; and n_k is (Gaussian) additive noise. H is a matrix that models the convolution between the high-resolution image and the point-spread function (PSF) of the system. Each SIM image acquired generates an Equation (4) with a different illumination pattern (k). The linear system of Equation (4) produced in a SIM experiment must be solved in order to reconstruct a high-resolution image. This reconstruction can be defined as the inversion of the system of equations. In the presence of noise (n_k), the inversion becomes unstable and is considered an ill-posed problem. This means we need to add a constraint which stabilizes the inversion of the system and ensures the uniqueness of the solution. In this imaging model, the low-resolution images (y_k), high-resolution image (x), and nose (n_k) are measurement-dependent.

We modeled the PSF as an Airy disk which, in Fourier space, would lead to an optical transfer function (OTF) of the form [49] as follows:

$$OTF(f) = \frac{1}{\pi}\left[2\cos^{-1}\left(\frac{f}{f_c}\right) - \sin\left(2\cos^{-1}\left(\frac{f}{f_c}\right)\right)\right], \tag{5}$$

where f is the spatial frequency. We estimated the cut-off frequency (f_c) by calculating the radial average of the power spectral density (PSD) of a widefield image of 100 nm fluorescent beads [50]. This could also be calculated by taking the Rayleigh limit of the resolution $d = 0.61\lambda/NA$ and expressing this value in terms of spatial frequency ($1/d$).

Using a Bayesian approach [26,27,29,51–54], high-resolution image estimation can be expressed as a minimized cost function according to the following:

$$x_{\mathrm{HR-MAP}} = \underset{x}{\operatorname{argmin}}\left[\sum_{k=1}^{K} \|y_k - HM_k x\|^2 + \lambda\Gamma(x)\right]. \tag{6}$$

The cost function in Equation (6) consists of two terms. The first term describes the mean square error between the estimated HR image and the observed LR images. The second term ($\lambda\Gamma(x)$) is a regularization term. To ensure positivity and promote a smoothness condition, we relied on quadratic regularization [54]. The contribution of $\Gamma(x)$ was controlled by the parameter λ, a small positive constant defining the strength of the regularization (typically, $\lambda = 0.01$). We solved Equation (6) using gradient descent methods [54].

3.3. Spectral Merging

MAP estimation of high-resolution images obtained with structured illumination enables the reconstruction of high-resolution images (HR-MAP) with details that are unresolvable in a widefield microscope. However, MAP estimation, as described above, does not suppress out-of-focus light. On the other hand, the processing method according to Equation (3) used in optical sectioning SIM [15,20] provides images (LR-HOM) with optical sectioning. Noting that the unwanted, out-of-focus light was dominant at low spatial frequencies, we merged the LR-HOM and HR-MAP images in the frequency domain to obtain the final HR image (MAP-SIM). For 3D data, this is completed in a slice-by-slice fashion, resulting in a Z-stack of SIM images. Frequency-domain Gaussian low-pass filtering was applied to the LR-HOM image, and a complementary high-pass filter was applied to the HR-MAP image. We used a weighting scheme that could be described by the following equation:

$$x_{\mathrm{MAP-SIM}} = \mathcal{F}^{-1}\left\{(1-\beta)\mathcal{F}\{x_{\mathrm{LR-HOM}}\}\exp\left(-\frac{f^2}{2\sigma^2}\right) + \beta\mathcal{F}\{x_{\mathrm{HR-MAP}}\}\left(1-\exp\left(-\frac{f^2}{2\sigma^2}\right)\right)\right\}, \quad (7)$$

where $\mathcal{F}, \mathcal{F}^{-1}$ denotes the Fourier transform operator and its inverse, respectively, and f is the spatial frequency, σ is the standard deviation of the Gaussian filter, and β is a weighting coefficient. Usually, we would set β to 0.85. We would typically use a standard incoherent apodizing function to shape the MAP-SIM spectrum before the final inverse FFT.

4. Results

To acquire an overview of the slice with SIM methods, we first imaged using a 10×/0.4 NA water immersion objective. We acquired 60 image positions with a 20-percent overlap between each position and with 12 z-planes. In this image, the Z-plane spacing was 20 µm. Image stitching was accomplished using our lab's methods and an ImageJ plugin [55], as shown in [46]. A composite image of the slice is shown in Figure 2, and it is color-coded based on depth using the isolum color table [56]. This image was acquired in 5 min and 30 s, with an additional 15 min and 35 s required for the OS-SIM processing, according to Equation (3). The final image was 8.4 GB in size, and it had 16,859 × 10,378 × 12 pixels.

This slice was matched to Paxinos and Franklin's mouse brain atlas [57] to identify which section of the brain was being imaged. Our slice was visually matched with slice 64. We further matched our sample to slice 92 of 132 in the Allen brain atlas [58,59]. Second order polynomial fits were made for both the horizontal and vertical directions using the edges and the central aqueduct as reference points. This allowed any point on this brain slice, recorded from the microscope stage coordinates, to be translated into the coordinates of the atlas. This method placed the neuron shown in Figure 3 in the temporal association area (TeA) of the mouse brain isocortex, as indicated by the yellow box in Figure 2a.

Imaging Deep Neurons

To demonstrate MAP-SIM's ability to image deeper into the sample than traditional SR-SIM, a TeA neuron 41–66 µm deep was imaged. The depth was measured using the closed-loop piezo stage. A 100×/1.4 NA oil immersion objective was used with an exposure time of 300 ms per SIM phase. This image is shown in Figure 3. The profile of a dendric spine neck was also measured (Figure 3d,e). The profile was fit in MatLab using a Gaussian function weighted by the square root of the counts, with nonlinear least squares methods. The full width at half-max (FWHM) was determined to be 164.0 ± 4.9 nm. To determine the image resolution, we calculated the power spectral density (PSD) as previously described [50]. We found that the WF image had a resolution of 247.6 nm while the MAP-SIM image had a resolution of 143.6 nm, an improvement of ~1.7-fold. These results are summarized in Table 1. Figures S1–S3 show additional images of cortical neurons imaged with MAP-SIM and a resolution analysis by a Fourier ring correlation (FRC) [60,61] and the PSD methods. The FRC measurements indicated a MAP-SIM resolution of approximately 150–160 nm, in good agreement with the 144 nm measured by the PSD methods.

A comparison of widefield, basic OS-SIM (Equation (3)), and MAP-SIM (Equations (6) and (7)) for this same cortical neuron is shown in Figure 4 and further analyzed in Table 1. As is evident in the figure, widefield had the largest background due to out-of-focus light, with basic OS-SIM providing optical sectioning and MAP-SIM providing both optical sectioning and super-resolution. The imaging depth of 41–66 µm exceeded the depth limit of traditional SIM by approximately three-fold.

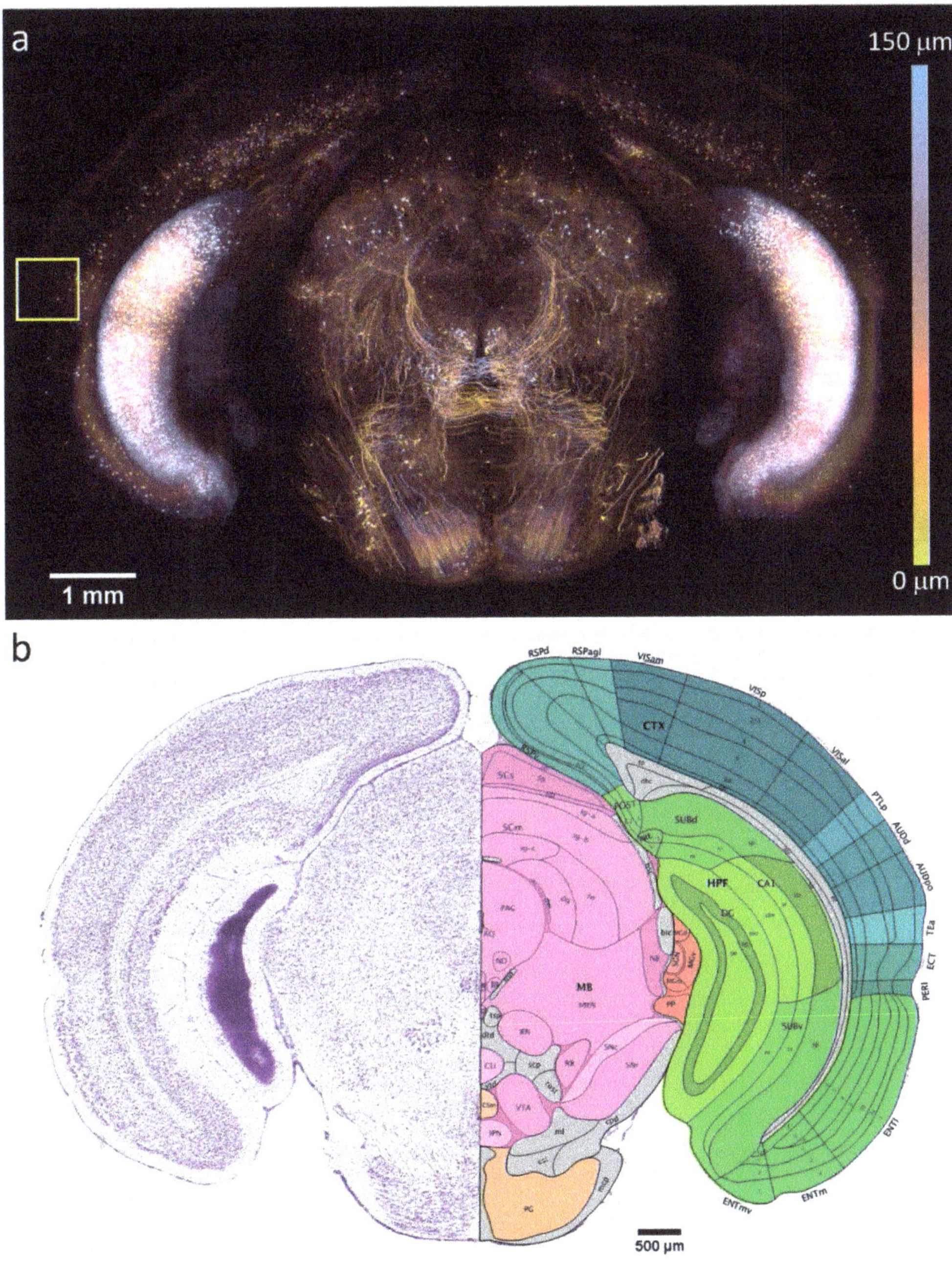

Figure 2. (**a**) Overview of the OS-SIM image. The yellow box indicates the temporal association area where the neurons were imaged with super-resolution MAP-SIM. (**b**) Nissl (**left**) and anatomical annotations (**right**) from the Allen mouse brain atlas and the Allen Reference Atlas—Mouse Brain, at the same slice position as (**a**) (slice 92 of 132, Allen Mouse Brain Atlas, mouse.brain-map.org and atlas.brain-map.org).

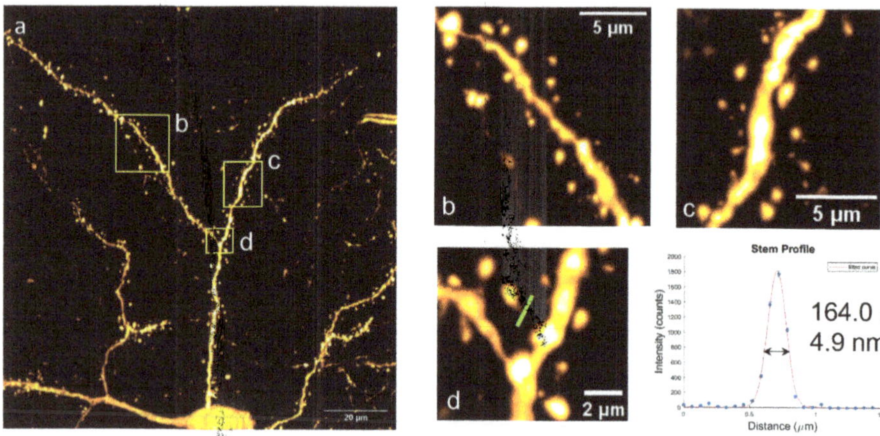

Figure 3. (**a**) TeA neuron imaged at a depth of 41 μm to 66 μm using a 100×/1.4 NA oil immersion objective. (**b**,**c**) Zoomed in views of the selected areas indicated in (**a**) by yellow boxes. The width of the spine neck, selected in (**d**), was fit to a Gaussian function (FWHM 164.0 ± 4.9 nm).

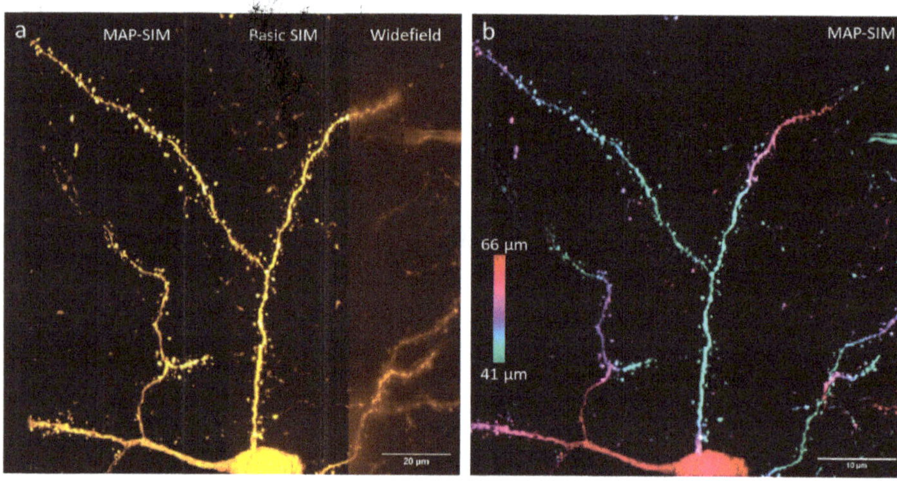

Figure 4. (**a**) TeA neuron shown in widefield, basic OS-SIM, and MAP-SIM. (**b**) MAP-SIM image color-coded by depth.

Table 1. SNR and resolution measurements.

	SNR (dB)	Resolution (nm)
Widefield	43.87	247.6
Basic SIM	29.33	251.6
MAP-SIM	39.27	143.6

We further imaged a neuron at a depth of 71–83 μm. This is shown in Figure 5. In this particular image, the resolution, measured by calculating the PSD, was 161 nm. While this was a decrease in resolution from the shallower neuron shown in Figure 3, it still surpassed the diffraction limit. Imaging at approximately 100 μm using these methods often resulted in images with large amounts of noise, and so the maximum imaging depth with a 100× objective in this sample appeared to be approximately 85 μm. Using a 60× objective, we were able to image up to 113 μm, as shown in Figure 6.

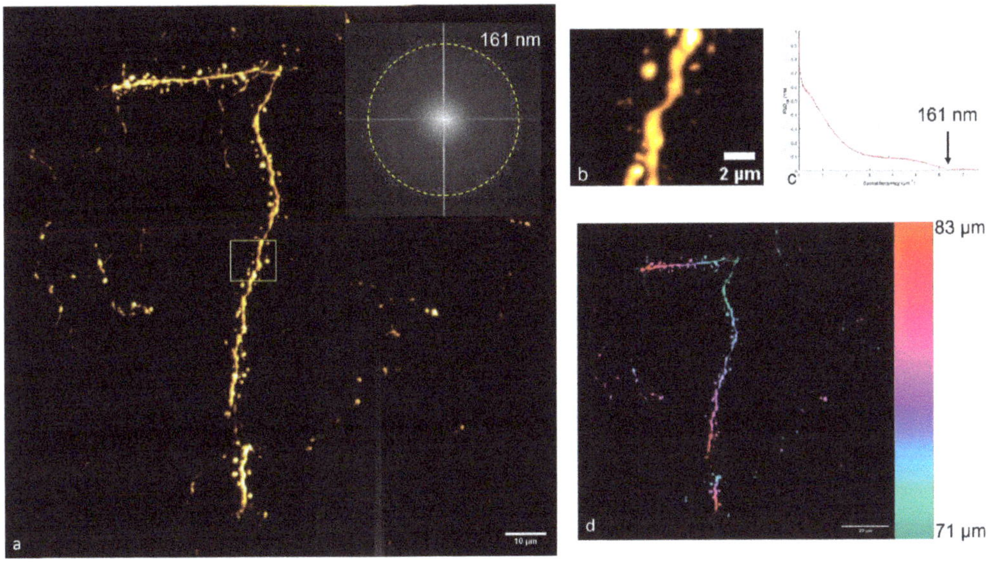

Figure 5. (**a**) TeA neuron imaged at a depth of 71 µm to 83 µm using a 100×/1.4 NA oil immersion objective. The inset shows the fast Fourier transform (FFT) of the image in (**a**), the boundary of which indicates the resolution. (**b**) Zoomed-in view of the selected area indicated in (**a**) by a yellow box. (**c**) A measurement of the resolution determined by measuring the power spectral density. (**d**) The MAP-SIM image color-coded by depth.

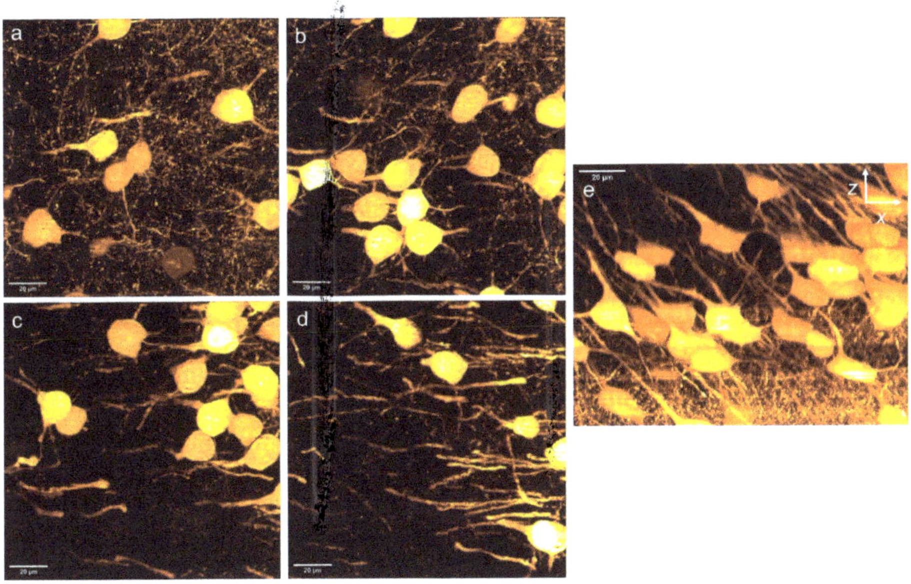

Figure 6. Neurons of the subiculum, ventral part, pyramidal layer (SUBv-sp), with an imaging depth of 0 to 113 µm (60×/1.42 NA oil immersion objective). The maximum-intensity projections of the imaged area have depths of (**a**) 0.2–28.4 µm, (**b**) 28.6–56.6 µm, (**c**) 56.8–84.8 µm, and (**d**) 85.0–113.0 µm. (**e**) X-Z projection of the imaged area.

In addition to the cortical neurons imaged in Figures 3–5 (and in Figures S1, S2 and S4), we also imaged an area of the brain in which a higher proportion of the neurons expressed the GFP marker (SUBv-sp subiculum, ventral part, pyramidal layer, also see Figure S5). This is shown in Figure 6. The maximum-intensity projection images, shown at different depths, showed good imaging at all depths. In addition, the imaging quality remained high even at depths past 100 µm using this objective (60× oil immersion).

Typically, SIM uses high-frequency patterns to maximize the obtainable resolution, but it has limited imaging depth due to scattering and the generation of large amounts of background fluorescence. The pattern used here used a lower spatial frequency to penetrate deeper into the mouse brain while maintaining the pattern integrity. A comparison of (cropped) images acquired using a high-frequency pattern (i.e., one out of three microdisplay pixels was activated) and our lower-frequency pattern (i.e., two out of ten microdisplay pixels were activated) is shown in Figure 7. Also shown is a plot of the measured modulation of the SIM pattern vs. the depth for various SIM patterns using a thick fluorescent plastic slide (obtained from Chroma). The modulation, measured as the average of the (max − min/max + min) in a region of interest, fell as the pattern spatial frequency increased and as the depth increased. This was expected in the case of incoherent illumination, as we used here, because the incoherent optical transfer function applied [49]. The higher-frequency pattern resulted in a weaker signal and poorer image reconstruction when imaging deep into the sample, as shown in Figure 7c.

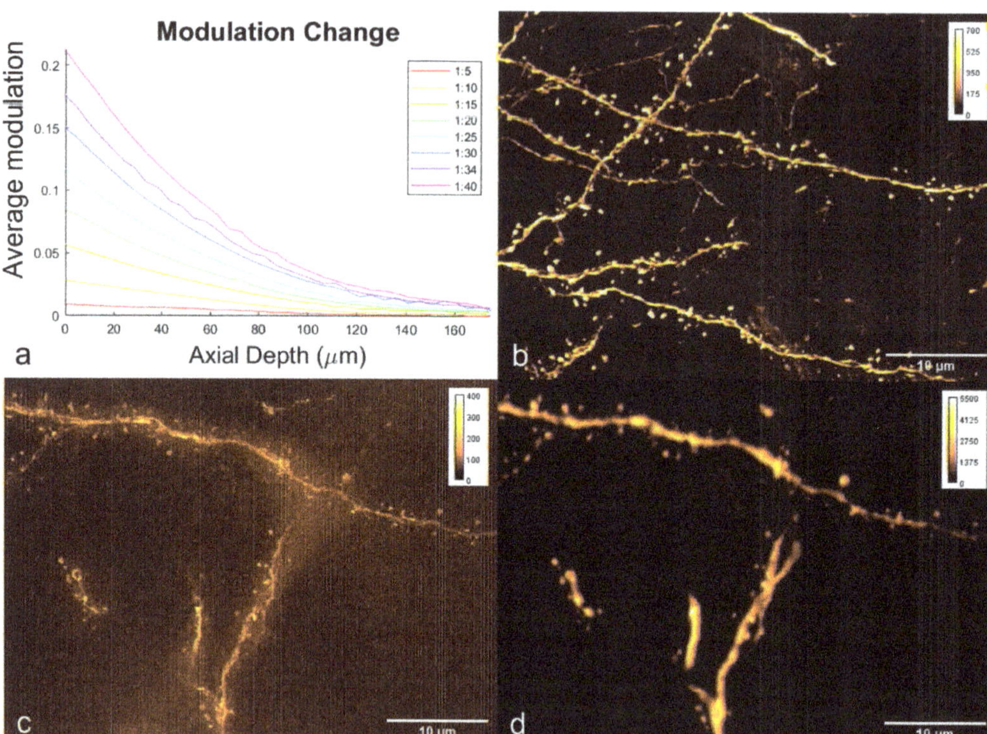

Figure 7. (**a**) Plot of modulation vs. axial depth for the different SIM patterns. (**b**) High-frequency pattern imaging of the TeA cortical neurons at the surface of the slice (0–10 µm). (**c**) High-frequency pattern imaging of the TeA cortical neurons at a depth of 41–45 µm. (**d**) Low-frequency pattern imaging of the same field of view shown in (**c**).

5. Discussion

By combining a structured illumination microscope with a large field-of-view and an image reconstruction method based on Bayesian statistics, we demonstrated synapse-resolving meso- and micro-scale volumetric imaging in an optically cleared coronal slice of adult mouse brain. The use of MAP-SIM and sample-optimized illumination patterns allowed us to collect super-resolution images well beyond the typical depth limit for SIM.

Compared to other super-resolution methods, SIM has poorer resolution. For example, the 144 nm lateral resolution achieved here is worse in comparison to the approximately 20 nm resolution that is typically achieved with STORM. However, SIM requires ~15 (or less, depending on the method used) images to reconstruct a super-resolution image. This is far lower than the 20,000 (or more) images usually required for STORM, making SIM imaging much faster and, therefore, a possibility for use when imaging live cells. The excitation power needed for SIM is much lower than that used in STORM. Here, we used 0.214 W/cm^2 for SIM compared to the 2 kW/cm^2 we previously used for STORM [62]. We found that the photobleaching in our experiments was minimal (Figure S6).

Most of the progress in super-resolution SIM has been in the acquisition and processing of images, but SIM was used in a detailed study on dendritic spines [9] where the authors developed a method for reconstructing and measuring the surface geometries of dendritic spines from 3D-SIM images. By adopting more flexible strategies for image acquisition and processing, such as the methods shown here, SIM is expected to be used more frequently in biological studies, including those on dense tissues such as brain tissue.

Supplementary Materials: The following supporting information can be downloaded at: https://www.mdpi.com/article/10.3390/bioengineering10091081/s1, Table S1. Main components of the SIM system; Table S2. Camera parameters; Table S3. Light source parameters; Table S4. Microdisplay parameters; Table S5. Camera parameters for the machine vision camera used (only) in Figures S4 and S5; Table S6. Parameters of the imaging data; Figure S1: Cortical neuron; Figure S2: Enlarged views; Figure S3: Resolution analysis; Figure S4: Cortical neuron; Figure S5: Neurons of the midbrain; Figure S6: Photobleaching analysis; Figure S7: SIM system details; Figure S8: SIM system diagrams.

Author Contributions: T.C.P.: acquired the data, analyzed the data, and wrote the paper; K.A.J.: acquired the data and analyzed the data; G.M.H.: conceived the project, acquired the data, analyzed the data, supervised the research, and wrote the paper. All authors have read and agreed to the published version of the manuscript.

Funding: The research reported in this publication was supported by the National Institute of General Medical Sciences of the National Institutes of Health under award number 2R15GM128166-02. This work was also supported by the UCCS BioFrontiers center.

Institutional Review Board Statement: Not applicable.

Informed Consent Statement: Not applicable.

Data Availability Statement: The data are available upon request.

Conflicts of Interest: The authors declare no conflict of interest.

References

1. Hell, S.W.; Wichmann, J. Breaking the diffraction resolution limit by stimulated emission: Stimulated-emission-depletion fluorescence microscopy. *Opt. Lett.* **1994**, *19*, 780. [CrossRef] [PubMed]
2. Rust, M.J.; Bates, M.; Zhuang, X. Sub-diffraction-limit imaging by stochastic optical reconstruction microscopy (STORM). *Nat. Methods* **2006**, *3*, 793–795. [CrossRef] [PubMed]
3. Betzig, E.; Patterson, G.H.; Sougrat, R.; Lindwasser, O.W.; Olenych, S.; Bonifacino, J.S.; Davidson, M.W.; Lippincott-Schwartz, J.; Hess, H.F. Imaging intracellular fluorescent proteins at nanometer resolution. *Science* **2006**, *313*, 1642–1645. [CrossRef] [PubMed]
4. Dertinger, T.; Xu, J.; Naini, O.; Vogel, R.; Weiss, S. SOFI-based 3D superresolution sectioning with a widefield microscope. *Opt. Nanoscopy* **2012**, *1*, 2. [CrossRef] [PubMed]
5. Gustafsson, M.G.L. Surpassing the lateral resolution limit by a factor of two using structured illumination microscopy. *J. Microsc.* **2000**, *198*, 82–87. [CrossRef]

6. Heintzmann, R.; Cremer, C. Laterally modulated excitation microscopy: Improvement of resolution by using a diffraction grating. *Proc. SPIE* **1998**, *3568*, 185–196.
7. Berning, S.; Willig, K.I.; Steffens, H.; Dibaj, P.; Hell, S.W. Nanoscopy in a Living Mouse Brain. *Science* **2012**, *335*, 551. [CrossRef]
8. Mlodzianoski, M.J.; Cheng-Hathaway, P.J.; Bemiller, S.M.; McCray, T.J.; Liu, S.; Miller, D.A.; Lamb, B.T.; Landreth, G.E.; Huang, F. Active PSF shaping and adaptive optics enable volumetric localization microscopy through brain sections. *Nat. Methods* **2018**, *15*, 583–586. [CrossRef]
9. Kashiwagi, Y.; Higashi, T.; Obashi, K.; Sato, Y.; Komiyama, N.H.; Grant, S.G.N.; Okabe, S. Computational geometry analysis of dendritic spines by structured illumination microscopy. *Nat. Commun.* **2019**, *10*, 1285. [CrossRef]
10. Wang, M.; Tulman, D.B.; Sholl, A.B.; Kimbrell, H.Z.; Mandava, S.H.; Elfer, K.N.; Luethy, S.; Maddox, M.M.; Lai, W.; Lee, B.R.; et al. Gigapixel surface imaging of radical prostatectomy specimens for comprehensive detection of cancer-positive surgical margins using structured illumination microscopy. *Sci. Rep.* **2016**, *6*, 27419. [CrossRef]
11. Gao, L.; Shao, L.; Higgins, C.D.; Poulton, J.S.; Peifer, M.; Davidson, M.W.; Wu, X.; Goldstein, B.; Betzig, E. Noninvasive imaging beyond the diffraction limit of 3D dynamics in thickly fluorescent specimens. *Cell* **2012**, *151*, 1370–1385. [CrossRef]
12. Kner, P.; Chhun, B.B.; Griffis, E.R.; Winoto, L.; Gustafsson, M.G.L. Super-resolution video microscopy of live cells by structured illumination. *Nat. Methods* **2009**, *6*, 339–342. [CrossRef]
13. Gustafsson, M.G.L.; Shao, L.; Carlton, P.M.; Wang, C.J.R.; Golubovskaya, I.N.; Cande, W.Z.; Agard, D.A.; Sedat, J.W. Three-dimensional resolution doubling in wide-field fluorescence microscopy by structured illumination. *Biophys. J.* **2008**, *94*, 4957–4970. [CrossRef]
14. Schermelleh, L.; Carlton, P.M.; Haase, S.; Shao, L.; Winoto, L.; Kner, P.; Burke, B.; Cardoso, M.C.; Agard, D.A.; Gustafsson, M.G.L.; et al. Subdiffraction multicolor imaging of the nuclear periphery with 3D structured illumination microscopy. *Science* **2008**, *320*, 1332–1336. [CrossRef]
15. Neil, M.A.A.; Juškaitis, R.; Wilson, T. Method of obtaining optical sectioning by using structured light in a conventional microscope. *Opt. Lett.* **1997**, *22*, 1905. [CrossRef]
16. Pilger, C.; Pospíšil, J.; Müller, M.; Ruoff, M.; Schütte, M.; Spiecker, H.; Huser, T. Super-resolution fluorescence microscopy by line-scanning with an unmodified two-photon microscope. *Philos. Trans. R. Soc. A Math. Phys. Eng. Sci.* **2021**, *379*, 20200300. [CrossRef] [PubMed]
17. Liu, W.; Liu, Q.; Zhang, Z.; Han, Y.; Kuang, C.; Xu, L.; Yang, H.; Liu, X. Three-dimensional super-resolution imaging of live whole cells using galvanometer-based structured illumination microscopy. *Opt. Express* **2019**, *27*, 7237. [CrossRef] [PubMed]
18. Brown, P.T.; Kruithoff, R.; Seedorf, G.J.; Shepherd, D.P. Multicolor structured illumination microscopy and quantitative control of polychromatic light with a digital micromirror device. *Biomed. Opt. Express* **2021**, *12*, 3700. [CrossRef] [PubMed]
19. Chang, B.J.; Meza, V.D.P.; Stelzer, E.H.K. csiLSFM combines light-sheet fluorescence microscopy and coherent Structured illumination for a lateral resolution below 100 nm. *Proc. Natl. Acad. Sci. USA* **2017**, *114*, 4869–4874. [CrossRef] [PubMed]
20. Křížek, P.; Raška, I.; Hagen, G.M. Flexible structured illumination microscope with a programmable illumination array. *Opt. Express* **2012**, *20*, 24585. [CrossRef] [PubMed]
21. Rossberger, S.; Best, G.; Baddeley, D.; Heintzmann, R.; Birk, U.; Dithmar, S.; Cremer, C. Combination of structured illumination and single molecule localization microscopy in one setup. *J. Opt.* **2013**, *15*, 094003. [CrossRef]
22. Young, L.J.; Ströhl, F.; Kaminski, C.F. A Guide to Structured Illumination TIRF Microscopy at High Speed with Multiple Colors. *J. Vis. Exp.* **2016**, *111*, e53988.
23. Poher, V.; Zhang, H.X.; Kennedy, G.T.; Griffin, C.; Oddos, S.; Gu, E.; Elson, D.S.; Girkin, M.; French, P.M.W.; Dawson, M.D.; et al. Optical sectioning microscope with no moving parts using a micro-stripe array light emitting diode. *Opt. Express* **2007**, *15*, 11196–11206. [CrossRef] [PubMed]
24. Pospíšil, J.; Wiebusch, G.; Fliegel, K.; Klíma, M.; Huser, T. Highly compact and cost-effective 2-beam super-resolution structured illumination microscope based on all-fiber optic components. *Opt. Express* **2021**, *29*, 11833. [CrossRef]
25. Hinsdale, T.A.; Stallinga, S.; Rieger, B. High-speed multicolor structured illumination microscopy using a hexagonal single mode fiber array. *Biomed. Opt. Express* **2021**, *12*, 1181. [CrossRef]
26. Lukeš, T.; Křížek, P.; Švindrych, Z.; Benda, J.; Ovesný, M.; Fliegel, K.; Klíma, M.; Hagen, G.M. Three-dimensional super-resolution structured illumination microscopy with maximum a posteriori probability image estimation. *Opt. Express* **2014**, *22*, 29805–29817. [CrossRef]
27. Orieux, F.; Sepulveda, E.; Loriette, V.; Dubertret, B.; Olivo-Marin, J.C. Bayesian estimation for optimized structured illumination microscopy. *IEEE Trans. Image Process.* **2012**, *21*, 601–614. [CrossRef]
28. Huang, X.; Fan, J.; Li, L.; Liu, H.; Wu, R.; Wu, Y.; Wei, L.; Mao, H.; Lal, A.; Xi, P.; et al. Fast, long-term, super-resolution imaging with Hessian structured illumination microscopy. *Nat. Biotechnol.* **2018**, *36*, 451–459. [CrossRef]
29. Lukeš, T.; Hagen, G.M.; Křížek, P.; Švindrych, Z.; Fliegel, K.; Klíma, M. Comparison of image reconstruction methods for structured illumination microscopy. *Proc. SPIE* **2014**, *9129*, 91293J.
30. Chakrova, N.; Rieger, B.; Stallinga, S. Deconvolution methods for structured illumination microscopy. *J. Opt. Soc. Am. A* **2016**, *33*, B12. [CrossRef]
31. Chen, X.; Zhong, S.; Hou, Y.; Cao, R.; Wang, W.; Li, D.; Dai, Q.; Kim, D.; Xi, P. Superresolution structured illumination microscopy reconstruction algorithms: A review. *Light Sci. Appl.* **2023**, *12*, 172. [CrossRef] [PubMed]

32. Mo, Y.; Wang, K.; Li, L.; Xing, S.; Ye, S.; Wen, J.; Duan, X.; Luo, Z.; Gou, W.; Chen, T.; et al. Quantitative structured illumination microscopy via a physical model-based background filtering algorithm reveals actin dynamics. *Nat. Commun.* **2023**, *14*, 3089. [CrossRef] [PubMed]
33. Cao, R.; Li, Y.; Chen, X.; Ge, X.; Li, M.; Guan, M.; Hou, Y.; Fu, Y.; Xu, X.; Jiang, S.; et al. Open-3DSIM: An open-source three-dimensional structured illumination microscopy reconstruction platform. *Nat. Methods* **2023**, *20*, 1183–1186. [CrossRef] [PubMed]
34. Hannebelle, M.T.; Raeth, E.; Leitao, S.M.; Lukeš, T.; Pospíšil, J.; Toniolo, C.; Venzin, O.F.; Chrisnandy, A.; Swain, P.P.; Ronceray, N.; et al. OpenSIM: Open source microscope add-on for structured illumination microscopy. *bioRxiv* **2023**, arXiv:2023.06.16.545316.
35. Wen, G.; Li, S.; Liang, Y.; Wang, L.; Zhang, J.; Chen, X.; Jin, X.; Chen, C.; Tang, Y.; Li, H. Spectrum-optimized direct image reconstruction of super-resolution structured illumination microscopy. *PhotoniX* **2023**, *4*, 19. [CrossRef]
36. Li, X.; Wu, Y.; Su, Y.; Rey-Suarez, I.; Matthaeus, C.; Updegrove, T.B.; Wei, Z.; Zhang, L.; Sasaki, H.; Li, Y.; et al. Three-dimensional structured illumination microscopy with enhanced axial resolution. *Nat. Biotechnol.* **2023**, *2023*, 1–13. [CrossRef]
37. Johnson, K.A.; Noble, D.; Machado, R.; Paul, T.C.; Hagen, G.M. Flexible Multiplane Structured Illumination Microscope with a Four-Camera Detector. *Photonics* **2022**, *9*, 501. [CrossRef]
38. Luo, F.; Zeng, J.; Shao, Z.; Zhang, C. Fast structured illumination microscopy via transfer learning with correcting. *Opt. Lasers Eng.* **2023**, *162*, 107432. [CrossRef]
39. Burns, Z.; Liu, Z.; Liu, Z.; Liu, Z.; Liu, Z. Untrained, physics-informed neural networks for structured illumination microscopy. *Opt. Express* **2023**, *31*, 8714–8724. [CrossRef]
40. Pospíšil, J.; Lukeš, T.; Bendesky, J.; Fliegel, K.; Spendier, K.; Hagen, G.M. Imaging tissues and cells beyond the diffraction limit with structured illumination microscopy and Bayesian image reconstruction. *Gigascience* **2018**, *8*, giy126. [CrossRef]
41. Wu, Y.; Shroff, H. Faster, sharper, and deeper: Structured illumination microscopy for biological imaging. *Nat. Methods* **2018**, *15*, 1011–1019. [CrossRef] [PubMed]
42. Mandula, O.; Kielhorn, M.; Wicker, K.; Krampert, G.; Kleppe, I.; Heintzmann, R. Line scan—Structured illumination microscopy super-resolution imaging in thick fluorescent samples. *Opt. Express* **2012**, *20*, 24167–24174. [CrossRef] [PubMed]
43. Li, Z.; Zhang, Q.; Chou, S.W.; Newman, Z.; Turcotte, R.; Natan, R.; Dai, Q.; Isacoff, E.Y.; Ji, N. Fast widefield imaging of neuronal structure and function with optical sectioning in vivo. *Sci. Adv.* **2020**, *6*, eaaz3870. [CrossRef]
44. Lu, R.; Liang, Y.; Meng, G.; Zhou, P.; Svoboda, K.; Paninski, L.; Ji, N. Rapid mesoscale volumetric imaging of neural activity with synaptic resolution. *Nat. Methods* **2020**, *17*, 291–294. [CrossRef] [PubMed]
45. Křížek, P.; Lukeš, T.; Ovesný, M.; Fliegel, K.; Hagen, G.M. SIMToolbox: A MATLAB toolbox for structured illumination fluorescence microscopy. *Bioinformatics* **2015**, *32*, 318–320. [CrossRef]
46. Johnson, K.A.; Hagen, G.M. Artifact-free whole-slide imaging with structured illumination microscopy and Bayesian image reconstruction. *Gigascience* **2021**, *9*, giaa035. [CrossRef]
47. Heintzmann, R. Structured illumination methods. In *Handbook of Biological Confocal Microscopy*; Pawley, J.B., Ed.; Springer: New York, NY, USA, 2006; pp. 265–279.
48. Šonka, M.; Hlaváč, V.; Boyle, R. *Image Processing Analysis and Machine Vision*, 2nd ed.; PWS Publishing: Boston, MA, USA, 1998.
49. Goodman, J.W. Frequency Analysis of Optical Imaging Systems. In *Introduction to Fourier Optics*; McGraw-HIll Int.: New York, NY, USA, 1968; pp. 126–171. ISBN 0-07-024254-2.
50. Pospíšil, J.; Fliegel, K.; Klíma, M. Assessing resolution in live cell structured illumination microscopy. In *Proceedings of SPIE—The International Society for Optical Engineering*; Páta, P., Fliegel, K., Eds.; SPIE: Prague, Czech Republic, 2017; Volume 10603, p. 39.
51. Verveer, P.J.; Jovin, T.M. Efficient superresolution restoration algorithms using maximum a posteriori estimations with application to fluorescence microscopy. *JOSA A* **1997**, *14*, 1696. [CrossRef]
52. Verveer, P.J.; Gemkow, M.J.; Jovin, T.M. A comparison of image restoration approaches applied to three-dimensional confocal and wide-field fluorescence microscopy. *J. Microsc.* **1999**, *193*, 50–61. [CrossRef]
53. Vermolen, B.J.; Garini, Y.; Young, I.T. 3D restoration with multiple images acquired by a modified conventional microscope. *Microsc. Res. Tech.* **2004**, *64*, 113–125. [CrossRef]
54. Chaudhuri, S. *Super-Resolution Imaging*; Milanfar, P., Ed.; CRC Press: Boca Raton, FL, USA, 2011; ISBN 978-1-4398-1931-9.
55. Preibisch, S.; Saalfeld, S.; Tomancak, P. Globally optimal stitching of tiled 3D microscopic image acquisitions. *Bioinformatics* **2009**, *25*, 1463–1465. [CrossRef]
56. Geissbuehler, M.; Lasser, T. How to display data by color schemes compatible with red-green color perception deficiencies. *Opt. Express* **2013**, *21*, 9862. [CrossRef] [PubMed]
57. Paxinos, G.; Franklin, K.B.J. *The Mouse Brain in Stereotaxic Coordinates*, 2nd ed.; Academic Press: New York, NY, USA, 2001; ISBN 9780128161579.
58. Allen Institute for Brain Science (2004). Allen Institute for Brain Science (2011). Allen Mouse Brain Atlas [Dataset]. Available online: https://mouse.brain-map.org/ (accessed on 24 May 2023).
59. Lein, E.S.; Hawrylycz, M.J.; Ao, N.; Ayres, M.; Bensinger, A.; Bernard, A.; Boe, A.F.; Boguski, M.S.; Brockway, K.S.; Byrnes, E.J.; et al. Genome-wide atlas of gene expression in the adult mouse brain. *Nature* **2007**, *445*, 168–176. [CrossRef]
60. Nieuwenhuizen, R.P.J.; Lidke, K.A.; Bates, M.; Puig, D.L.; Grünwald, D.; Stallinga, S.; Rieger, B. Measuring image resolution in optical nanoscopy. *Nat. Methods* **2013**, *10*, 557–562. [CrossRef] [PubMed]

61. Culley, S.; Albrecht, D.; Jacobs, C.; Pereira, P.M.; Leterrier, C.; Mercer, J.; Henriques, R. Quantitative mapping and minimization of super-resolution optical imaging artifacts. *Nat. Methods* **2018**, *15*, 263–266. [CrossRef] [PubMed]
62. Smirnov, E.; Borkovec, J.; Kováčik, L.; Svidenská, S.; Schröfel, A.; Skalníková, M.; Švindrych, Z.; Křížek, P.; Ovesný, M.; Hagen, G.M.; et al. Separation of replication and transcription domains in nucleoli. *J. Struct. Biol.* **2014**, *188*, 259–266. [CrossRef] [PubMed]

bioengineering

MDPI

Article

Self-Supervised Learning Application on COVID-19 Chest X-ray Image Classification Using Masked AutoEncoder

Xin Xing [1,2], Gongbo Liang [3], Chris Wang [4], Nathan Jacobs [5] and Ai-Ling Lin [2,6,7,*]

1 Department of Computer Science, University of Kentucky, Lexington, KY 40506, USA
2 Department of Radiology, University of Missouri, Columbia, MO 65212, USA
3 Department of Computing and Cyber Security, Texas A&M University-San Antonio, San Antonio, TX 78224, USA
4 Department of Computer Science, University of Missouri, Columbia, MO 65211, USA
5 Department of Computer Science & Engineering, Washington University in St. Louis, St. Louis, MO 63130, USA
6 Department of Biological Sciences, University of Missouri, Columbia, MO 65211, USA
7 Institute for Data Science and Informatics, University of Missouri, Columbia, MO 65211, USA
* Correspondence: ai-ling.lin@health.missouri.edu

Abstract: The COVID-19 pandemic has underscored the urgent need for rapid and accurate diagnosis facilitated by artificial intelligence (AI), particularly in computer-aided diagnosis using medical imaging. However, this context presents two notable challenges: high diagnostic accuracy demand and limited availability of medical data for training AI models. To address these issues, we proposed the implementation of a Masked AutoEncoder (MAE), an innovative self-supervised learning approach, for classifying 2D Chest X-ray images. Our approach involved performing imaging reconstruction using a Vision Transformer (ViT) model as the feature encoder, paired with a custom-defined decoder. Additionally, we fine-tuned the pretrained ViT encoder using a labeled medical dataset, serving as the backbone. To evaluate our approach, we conducted a comparative analysis of three distinct training methods: training from scratch, transfer learning, and MAE-based training, all employing COVID-19 chest X-ray images. The results demonstrate that MAE-based training produces superior performance, achieving an accuracy of 0.985 and an AUC of 0.9957. We explored the mask ratio influence on MAE and found ratio = 0.4 shows the best performance. Furthermore, we illustrate that MAE exhibits remarkable efficiency when applied to labeled data, delivering comparable performance to utilizing only 30% of the original training dataset. Overall, our findings highlight the significant performance enhancement achieved by using MAE, particularly when working with limited datasets. This approach holds profound implications for future disease diagnosis, especially in scenarios where imaging information is scarce.

Keywords: vision transformer (ViT); self-supervised learning; chest X-ray image; image classification

Citation: Xing, X.; Liang, G.; Wang, C.; Jacobs, N.; Lin, A.-L. Self-Supervised Learning Application on COVID-19 Chest X-ray Image Classification Using Masked AutoEncoder. *Bioengineering* **2023**, *10*, 901. https://doi.org/10.3390/bioengineering10080901

Academic Editor: Cuneyt M. Alper

Received: 28 May 2023
Revised: 19 July 2023
Accepted: 26 July 2023
Published: 29 July 2023

1. Introduction

The COVID-19 pandemic has brought attention to the vital role of artificial intelligence (AI) in combating infectious diseases, specifically through the analysis of lung images, such as X-ray chest images. Computer-aided diagnosis (CAD) has emerged as a promising tool for accurate and rapid diagnosis in this context. deep learning (DL) models, known for their exceptional performance in computer vision tasks such as image recognition [1–4], semantic segmentation [5,6], and object detection [7–9], have increasingly been adopted for CAD and other healthcare applications [10–13].

Despite the significant potential of DL models in medical data analysis, there are several practical challenges impeding their widespread adoption. First, medical datasets are often smaller compared to those used for natural image analysis, such as the widely used ImageNet dataset [14]. DL models have numerous parameters and require substantial data for effective training. Consequently, training such models with limited datasets can be challenging and may lead to overfitting [15,16]. Second, labeling medical data is a resource-intensive and time-consuming process. CAD relies on labeled data for training, but labeling medical data necessitates specialized medical knowledge and expertise, making it more demanding than labeling natural images. Lastly, current DL models for medical image analysis primarily rely on convolutional neural networks (CNNs) [17,18]. While CNNs excel at capturing local features, they may not be optimally suited for capturing global information across an entire image. Therefore, further research is necessary to develop more effective DL network architectures capable of capturing both local and global features for medical image analysis, particularly when dealing with limited datasets.

To address these challenges, various methods have been proposed. For example, transfer learning [19] has gained widespread usage, wherein DL models are pretrained on large-scale natural image datasets such as ImageNet and subsequently fine-tuned on smaller medical datasets. This approach helps mitigate the overfitting issue caused by limited medical datasets, although it may not fully bridge the gap between natural and medical images. Regarding labeling issues, weakly supervised learning (WSL) [20,21] has become popular, where models are trained using only image-level labels for tasks such as object detection [22]. However, WSL may not be suitable for classification tasks that still require image-level labels. Recently, novel DL models such as the Vision Transformer (ViT) [23] and its variants [24,25] have demonstrated promising results in capturing global information from medical images. Nevertheless, these models often necessitate extensive amounts of data for effective training.

In our study, we explored a novel method for medical image analysis that addresses the challenges associated with training strategy and limited medical datasets. We propose the utilization of self-supervised learning (SSL) [26], a method that leverages the intrinsic attributes of the data as pretraining tasks, eliminating the reliance on labeled data. SSL implementation involves utilizing attributes such as image rotation prediction [27], patch localization [28], and image reconstruction [29], which can be accessed without manual labeling. Additionally, by substituting the conventional CNN backbone with a Vision Transformer (ViT) model, our method effectively captures both local and global features of medical images. To accomplish this, we employ a Masked Autoencoder (MAE) model [30], a self-supervised learning approach that utilizes the ViT model as its backbone. By combining SSL with ViT through the MAE model, we anticipate that this method can contribute to more accurate and efficient medical image analysis.

To our best knowledge, we are the first to apply MAE on COVID-19 X-ray imaging. During the exploration, we found an innovative application of MAE on the limited dataset, which is not studied by the previous work [30]. We demonstrated the superior performance of the MAE model compared to baseline models, explored the influence of mask ratios on the MAE model's performance, and evaluated the MAE model's performance using different proportions of limited training data. The contributions of our study are as follows:

- We conducted a comparative analysis of various training strategies using the same public COVID-19 dataset and observed that the MAE model outperformed other approaches, demonstrating superior performance.
- To further investigate the impact of different mask ratios on the MAE model's performance, we examined how varying mask ratios affected the effectiveness of the model. Our experiments revealed that the model achieved its best performance with a mask ratio of 0.4.
- Through extensive evaluations, we examined the applicability of the MAE model across different proportions of available training data. Remarkably, the MAE model achieved comparable performance even when trained with only 30% of the available data.

2. Materials and Methods

2.1. Data

Our study adopted the chest X-ray classification dataset: COVIDxCXR-3 [31], which is a public dataset with more than 29,000 chest X-ray images, for positive/negative detection. COVIDxCXR-3 collects the data from different public data sources: covid-chest x-ray-dataset [32], Figure 1 COVID-19 Chest X-ray Dataset Initiative [33], Actualmed COVID-19 Chest X-ray Dataset Initiative [34], COVID-19 Radiography Database—Version 3 [35], RSNA Pneumonia Detection Challenge [36], RSNA International COVID-19 Open Radiology Database (RICORD) [37], BIMCV-COVID19+: a large annotated dataset of RX and CT images of COVID-19 patients [38], and Stony Brook University COVID-19 Positive Cases (COVID-19-NY-SBU) [39]. Figure 1 visualizes the positive/negative image samples of the COVID-19 subject. Table 1 shows the details of the COVIDxCXR-3 dataset distribution. The dataset has a multinational cohort of over 16,600 patients. The whole dataset is split into training and testing sets by the dataset authors. The training dataset has 13,992 negative and 16,490 positive images. The testing dataset has 200 negative and positive images, respectively.

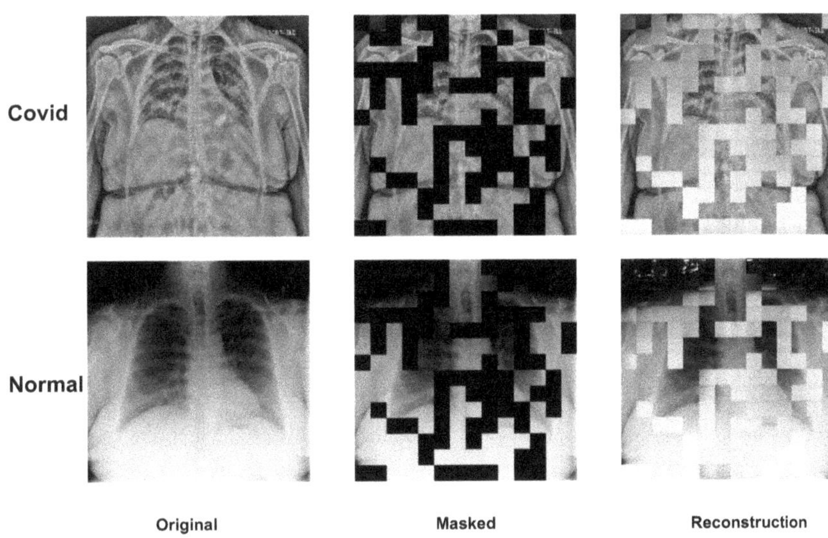

Figure 1. The visualization of the chest X-ray image of the COVIDxCXR-3. The first row shows the negative subjects and the second row shows the positive subjects. The input image size is 224 × 224. We normalize the image pixel from 0 to 255.

Table 1. The images and patients distribution of the dataset COVIDxCXR-3.

Type	Negative	Positive	Total
	Images Distribution		
Train	13,992	15,994	29,986
Test	200	200	400
	Patients Distribution		
Train	13,850	2808	16,648
Test	200	178	378

During the implementation of our experiment, we configured the image size to 224 × 224 pixels, while normalizing the image pixel values within the range of 0 to 255. Under normal circumstances, in order to provide an unbiased evaluation of a model, cross-validation is typically conducted during the training process. However, as demonstrated in Table 1, there exists an inequality in both image and patient distributions. There are 15,994 positive images for COVID-19, derived from 2808 COVID-19-positive patients, indicating multiple X-ray images per patient in the dataset (approximately 6 images per patient). Meanwhile, the dataset authors have not provided the specific subject information, precluding the possibility of conducting subject-level cross-validation. Concurrently, cross-validation at the image level would result in data leakage. Therefore, we chose to adopt the train/test split as defined by the dataset authors in these particular circumstances.

2.2. Vision Transformer

Since ViT is built upon the self-attention mechanism and many works adopt multi-head attention in the implementation, we first introduce the basics of the attention mechanism, then describe the ViT architecture.

The attention mechanism includes three inputs: a query (Q), a key (K), and a value (V). The attention operation is defined as Equation (1):

$$\text{Attention}(Q, K, V) = \text{Softmax}(\frac{QK^T}{\sqrt{d_k}})V \tag{1}$$

Afterward, the multi-head attention is defined as Equation (2):

$$\begin{aligned} \text{MultiHeadAttn}(Q, K, V) &= \text{Concat}(head_1, ..., head_n)W^o \\ head_i &= \text{Attention}(QW_i^Q, KW_i^K, VW_i^V) \end{aligned} \tag{2}$$

where W_i^Q, W_i^K, and W_i^V are learnable projection matrices.

Figure 2 illustrates the ViT structure, which has a patch embedding module and $N\times$ stacked transformer encoder blocks. Each transformer encoder block contains a multi-head attention (MSA), two layer normalization (LN) [40], and a multi-layer perceptron (MLP).

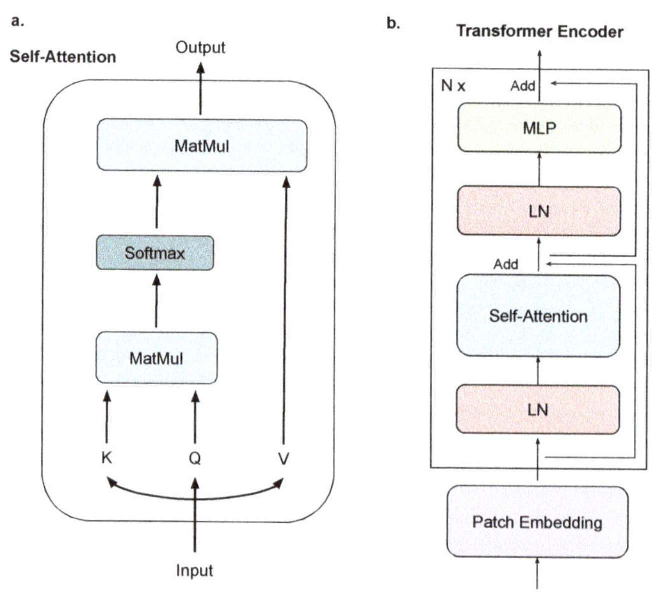

Figure 2. The structure of Vision Transformer. Sub-figure (**a**) illustrates the structure of the self-attention module. Sub-figure (**b**) shows the architecture of the Vision Transformer encoder.

In the implementation, the input image I is first transformed into a series of patch embeddings:

$$z_0 = \text{PatchEmbedding}(I) \tag{3}$$

The patch embeddings are forwarded through the transformer encoder under the following operations:

$$z^l = MLP(Norm(MSA(Norm(z^{l-1})))) \tag{4}$$

where z^{l-1} and z^l are the l-th transformer encoder block input and output, respectively.

2.3. MAE

Originally, MAE is a proficient self-supervised learning method employed in Natural Language Processing (NLP) tasks. Building upon its initial foundations, the MAE has significantly broadened its application scope beyond the NLP field, marking its presence in the field of computer vision. This widening of its application is facilitated by the universality of its core principle—the strategic "masking" operation. This operation forms the core of the self-supervised learning methodology, by selectively omitting sections of input data to create a challenge for the model to reconstruct these masked elements. This process allows the model to develop a robust understanding of the intrinsic structure and properties of the target dataset, optimizing its ability to generate representative features during the pretraining phase. This initial phase is followed by a fine-tuning stage, which utilizes labeled input data to further refine the model's comprehension of the dataset, thereby improving its overall performance. This two-tiered approach equips the model with the essential tools to tackle novel data and perform reliably on the target dataset. The successful adaptation of the MAE methodology to the realm of computer vision was achieved by employing techniques parallel to its NLP counterpart. Images are decomposed into a multitude of patches, a subset of which are randomly masked. Subsequently, the model is trained to perform the reconstruction pretraining task, effectively learning to predict the obscured sections of the image. Following this, a fine-tuning phase is undertaken with labeled data, ensuring the model's efficient performance on the target dataset.

When considering the choice of backbone for the MAE model, ViT emerges as an optimal option when compared to the convolutional neural network (CNN). As previously mentioned, the ViT architecture offers distinct advantages. One notable feature of the ViT model is its initial operation, where the input image is segmented into various patches. This patch-based approach can easily utilize the masking of random regions. Given these advantages, it is judicious to adopt a ViT architecture as the backbone for the MAE model.

Figure 3 illustrates the comprehensive workflow of a self-supervised learning system based on the MAE, encompassing two essential stages: the pretext task of image reconstruction and the subsequent fine-tuning stage. When it comes to the architecture of the model during the pretraining stage, it incorporates a Vision Transformer (ViT) as both the encoder and decoder. Serving as an encoder, the ViT is applied to mask certain segments of the input image patches. In its role as a decoder, it is tasked with the restoration of the masked patches. Upon transitioning to the fine-tuning phase, the pretrained ViT encoder is trained further with samples and labels from the target dataset. In the context of our implementation, we elected to employ a ViT-small structure with a hidden size of 768 as the encoder and the standard decoder within the framework of the MAE. The overall pipeline, founded on block representation, is presented in Figure 4.

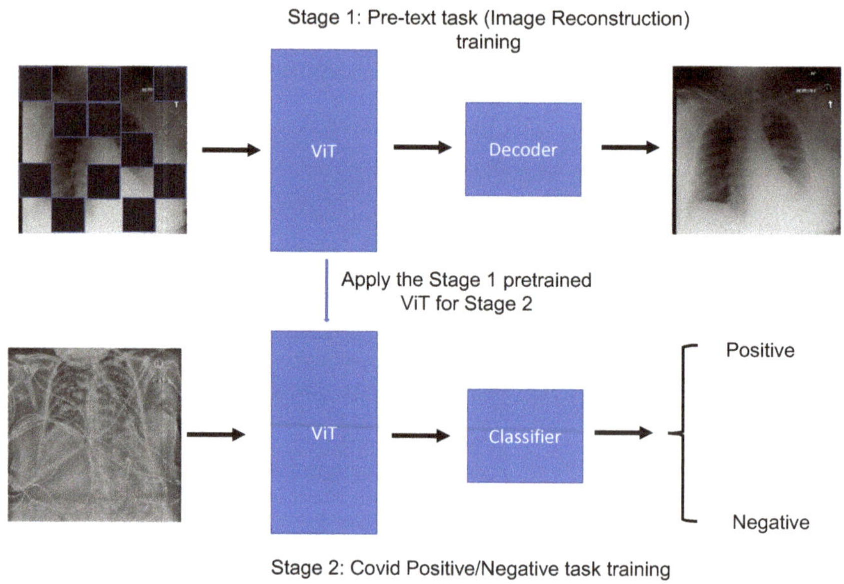

Figure 3. The workflow of the MAE method on the COVID-19 classification task. There are two stages for MAE training. The first stage is the image reconstruction pretraining stage, with the ViT backbone as the image encoder. The second stage is a fine-tuning stage, with the ViT backbone as the feature extractor for the labeled images.

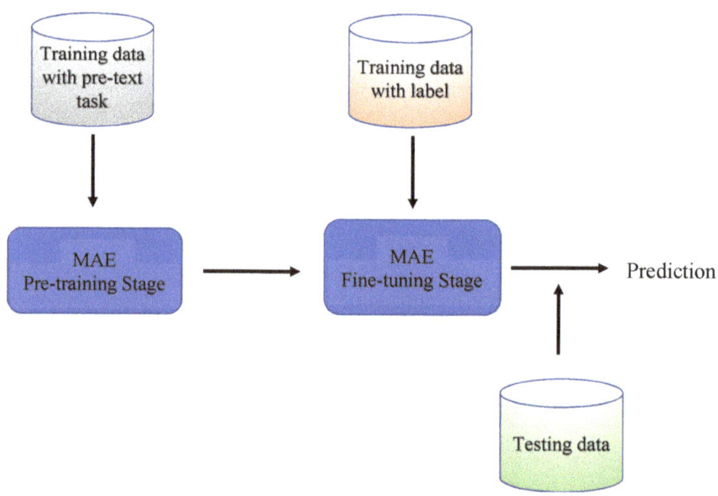

Figure 4. The block presentation of the MAE pipeline.

2.4. Loss Function

Since our work concentrates on binary classification, the overall loss function is a binary cross-entropy. For a chest X-ray image V with label l and probability prediction $p(l|V)$, the loss function is

$$loss(l, V) = l\log(p(l|V)) + (1-l)\log(1-p(l|V)) \quad (5)$$

where the label $l = 0$ indicates a negative sample and $l = 1$ indicates a positive sample, respectively.

2.5. Implementation and Metrics

We implemented the experiment models using PyTorch. We trained and tested the models based on the default setting of the dataset. For the pretrained baseline, the model is pretrained on ImageNet [14]. For the model training, we set the batch size to 16. Adam optimizer [41] with $beta1 = 0.9$, $beta2 = 0.999$, and a learning rate of 1×10^{-4} was used during the training. For the SSL model, we pretrained the model with 100 epochs. In the fine-tuning stage, we trained all the models for 40 epochs.

To evaluate the performance of our model, we used accuracy (Acc), area under the curve of receiver operating characteristics (AUC), F1 score (F1), Precision, Recall, and Average Precision (AP) as our evaluation metrics. We evaluated the training computation cost by the average epoch training time (e-Time). The accuracy is calculated with the following Equation (6):

$$\text{Accuracy} = \frac{\text{TP} + \text{TN}}{\text{TP} + \text{TN} + \text{FP} + \text{FN}} \tag{6}$$

where TP is the True Positive, TN is the True Negative, FP is the False Positive, and FN is the False Negative.

The precision is calculated by the following Equation (7):

$$\text{precision} = \frac{\text{TP}}{\text{TP} + \text{FP}} \tag{7}$$

The recall is calculated by the following Equation (8):

$$\text{recall} = \frac{\text{TP}}{\text{TP} + \text{FN}} \tag{8}$$

The F1-score is calculated by the following Equation (9):

$$\text{F1-score} = 2 \times \frac{\text{precision} \cdot \text{recall}}{\text{precision} + \text{recall}} \tag{9}$$

AUC curves compare the true positive rate and the false positive rate at different decision thresholds. AP summarizes a precision–recall curve as the weighted mean of precision achieved at each threshold.

3. Results

3.1. Model Performance Increasing by MAE

We conducted training experiments on the ViT model architecture using three different approaches: ViT-scratch, ViT-pretrain, and ViT-MAE. In the ViT-scratch approach, the ViT model was trained directly on the medical image data. The ViT-pretrain approach involved fine-tuning a pretrained ViT model on ImageNet using the medical image data. ViT-MAE refers to training the ViT model using the MAE pipeline. Accuracy was chosen as the performance metric. As depicted in Table 2, ViT-MAE achieved a remarkable accuracy of 0.985 in COVID-19 positive/negative detection, surpassing the other approaches (ViT-scratch accuracy = 0.7075 and ViT-pretrain accuracy = 0.9350) on the same dataset. To further compare ViT-MAE with CNN models, namely, ResNet50 and DenseNet121, we conducted additional experiments. It was observed that ViT-MAE outperformed both ResNet50 and DenseNet121 in terms of all metrics, except for AUC, where the difference was minimal. We think this minimal difference is due to (1) the model experiment's randomness, a common characteristic of machine learning models, and (2) the size of the test dataset. The size of the test dataset would overestimate/underestimate the model. In our experiments, compared with the training dataset, the test dataset is relatively small with only 400 images, which may overestimate the models. However, in terms of the same ViT backbone, we think ViT-MAE models exhibit comparable performance in our experiments.

Figure 5 illustrates the AUC curves for the three training approaches, clearly demonstrating that ViT-MAE outperforms the other strategies in terms of AUC performance.

We studied statistical tests that compare the ViT-MAE performance with ViT-scratch and ViT-pretrain. The metric chosen to evaluate their performance was accuracy. To ensure robustness, we conducted four independent experiments for each ViT model, employing different random seeds. The statistical summary of the three pretraining methods yielded the following mean and standard deviation values: For ViT-scratch, the mean was 0.7135 with a standard deviation of 0.0142. For ViT-pretrain, the mean was 0.9293 with a standard deviation of 0.0207. Lastly, for ViT-MAE, the mean was 0.9775 with a standard deviation of 0.006. In order to assess the significance of the differences in performance, we conducted two t-tests: ViT-scratch vs. ViT-MAE and ViT-pretrain vs. ViT-MAE. The resulting p-values for the two group t-tests were found to be less than 0.001 and 0.02, respectively. Our analysis revealed that ViT-MAE significantly outperformed ViT-scratch, indicating the critical influence of the training strategy on model performance. Additionally, we observed a relatively narrow performance gap between ViT-MAE and ViT-pretrain. These findings suggest that while ViT-MAE exhibits superior performance compared to ViT-scratch, the disparity in performance between ViT-MAE and ViT-pretrain is comparatively smaller.

Table 2. The performance of different training strategies over the ViT model. ViT-MAE model outperforms the other two training strategies.

Type	Acc	AUC	F1	Precision	Recall	AP
DenseNet121	0.9775	**0.9970**	0.9771	0.9948	0.96	0.9750
ResNet50	0.9650	0.9969	0.9641	0.9894	0.94	0.9601
ViT-scratch	0.7075	0.7808	0.7082	0.7065	0.7100	0.6466
ViT-pretrain	0.9350	0.9783	0.9340	0.9484	0.9200	0.9125
ViT-MAE	**0.9850**	0.9957	**0.9850**	**0.9950**	**0.9850**	**0.9859**

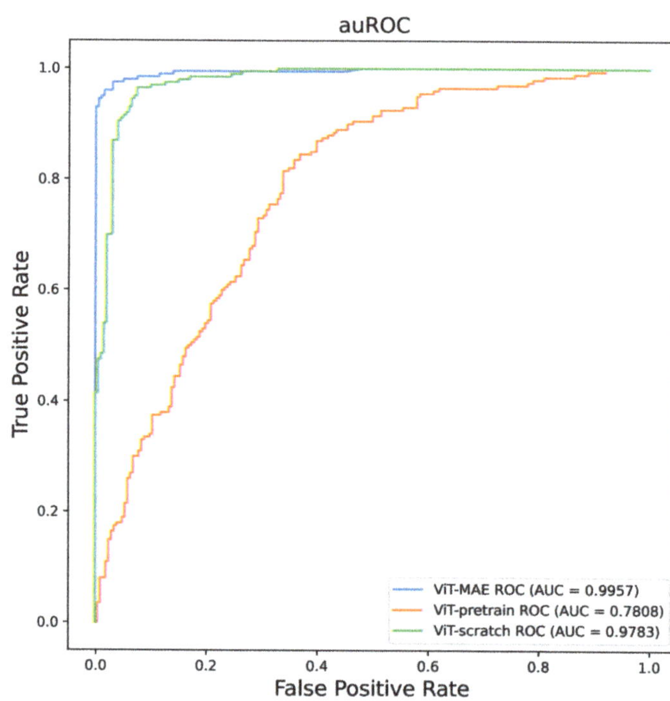

Figure 5. The AUC plot of the different training strategies for ViT model.

3.2. Mask Ratio Influence on MAE Performance

Since the pretraining of ViT-MAE is a reconstruction task, the mask ratio of the input image is a parameter that may affect the final performance. In this section, we study the mask ratio influence on ViT-MAE training. Table 3 shows the performance of different mask ratios over the MAE pretraining stage. Figure 6 illustrates the AUC curves of different mask ratios. The 40% percentage mask ratio outperforms the other mask ratio situations with Acc = 0.9850 and AUC = 0.9957. The mask ratio result indicates that a large mask ratio may decrease the final performance for the medical image dataset, while the large mask ratio (mask ratio = 0.75) shows good performance in the natural dataset. We think this may due to the difference between the medical and natural datasets, and the reconstruction results on the medical image may not show better performance than the natural image. We prove our thoughts in Section 3.4.

Table 3. The performance of different mask ratios over the MAE pretraining stage. The pre-training mask ratio = 0.4 of MAE outperforms the other pretraining strategies.

Ratio	Acc	AUC	F1	Precision	Recall	AP
0.4	0.9850	0.9957	0.9850	0.9850	0.9850	0.9559
0.5	0.9100	0.9653	0.9086	0.9277	0.8950	0.8783
0.6	0.8875	0.9579	0.8819	0.9282	0.8400	0.8597
0.7	0.8900	0.9502	0.8894	0.8939	0.8850	0.8486
0.8	0.8925	0.9516	0.8900	0.9110	0.8700	0.8576

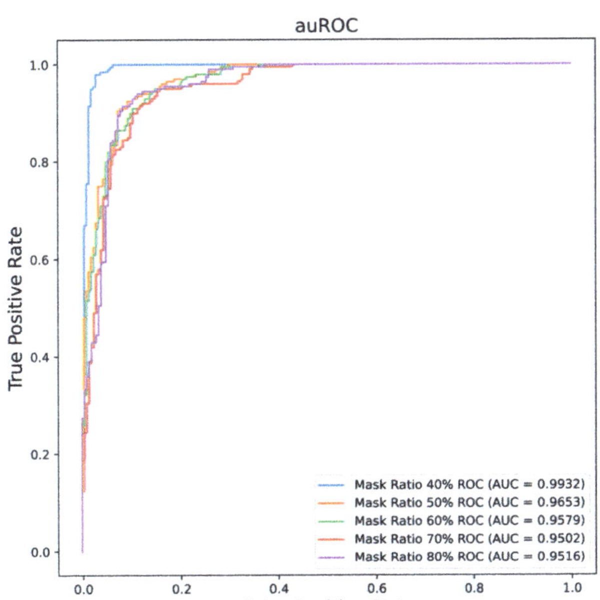

Figure 6. The AUC plot of the different mask ratios for ViT-MAE pretraining.

3.3. MAE Performance on the Limited Training Dataset

One advantage of self-supervised learning is that people can use a small labeled dataset to train a large DL model well. To explore the potential of SSL, we conduct experiments on the limited training dataset. We randomly split the partial training dataset to train our model from 10% to 90% . Table 4 shows the performance under different percentage splitting, and Figure 7 shows the AUC curves of different percentage situations. It appears that using only 30% of the training dataset is sufficient to achieve better performance than that of the ViT-pretrain scenario with the whole training dataset (94.25 vs. 93.5). Meanwhile, it is clear that increasing the labeled training percentage will contribute to better performance of the model. The promising results provide the potential training procedure for small medical dataset training on DL models.

Table 4. The performance of different percentages of training datasets at the MAE pretraining stage. The pretraining mask ratio = 0.4 of MAE outperforms the other pretraining strategies.

Percentage (%)	Acc	AUC	F1	Precision	Recall	AP
10	0.8800	0.9394	0.8776	0.8958	0.8600	0.8404
20	0.8925	0.9588	0.8877	0.9290	0.8500	0.8646
30	0.9425	0.9772	0.9415	0.9585	0.9250	0.9242
40	0.9600	0.9866	0.9602	0.9554	0.9650	0.9395
50	0.9675	0.9879	0.9673	0.9746	0.9600	0.9556
60	0.9650	0.9876	0.9645	0.9794	0.9500	0.9554
70	0.9675	0.9896	0.9669	0.9845	0.9500	0.9602
80	0.9775	0.9978	0.9771	0.9948	0.9600	0.9750
90	0.9825	0.9944	0.9823	0.9949	0.9700	0.9800

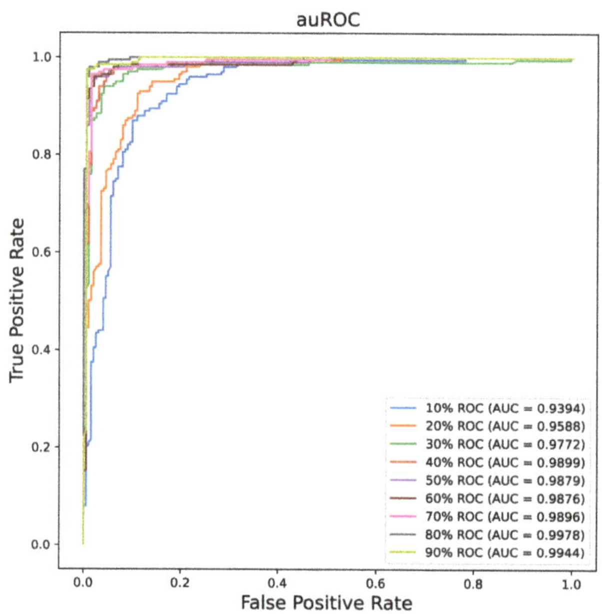

Figure 7. The AUC plot of the different percentages of training dataset for ViT-MAE model.

3.4. *Visualization of MAE on Image Reconstruction*

We present a visualization of the X-ray image reconstruction phase. As demonstrated in Figure 8, the depiction includes both Covid and Normal subjects, showcasing the original, masked, and reconstructed images. Upon visual comparison, it can be observed that the reconstructed images are relatively coarse. However, it should be emphasized that our primary objective is not to achieve pixel-perfect image reconstruction but to ensure that the deep learning model's parameters are properly initialized for the fine-tuning process on the specific dataset. Concurrently, the rough outcome of the reconstruction task implies that increasing the mask ratio will not contribute to enhanced model performance during the fine-tuning stage. This is due to the model's inability to extract additional learning during the reconstruction stage.

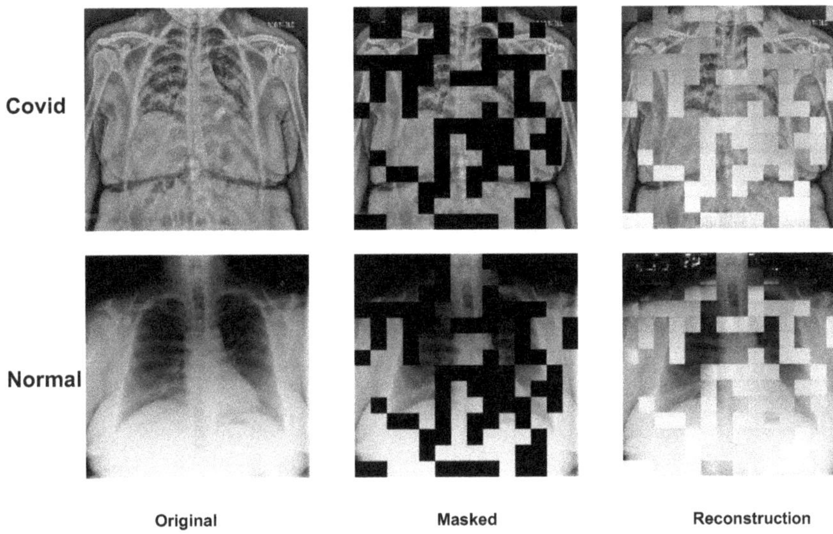

Figure 8. The visualization of the MAE for the image reconstruction pretraining. From the left column to the right are the original input image, the random masked image, and the reconstruction image. Even though the final reconstruction is not well-defined, the target of the pretraining stage is boosting the initial parameters of the ViT model.

4. Discussion

Our study involved an in-depth exploration of the MAE through a comprehensive series of experiments utilizing a publicly available COVID-19 Chest X-ray image dataset. Our study is innovative as we applied MAE to an X-ray imaging dataset for COVID-19 diagnosis, which has not been reported before. Further, we demonstrated that MAE exhibits remarkable efficiency when applied to labeled data, delivering comparable performance to utilizing only 30% of the original training dataset. The findings may have profound implications for the diagnosis of various diseases with the limited imaging dataset in the future, given that we showed that the accuracy can be maintained even with a reduced, smaller dataset. Our findings also yield several significant insights on the application of MAE in medical imaging as follows: First, by leveraging self-supervised learning with MAE, we observed notable improvements in model performance compared to alternative training methods. This underscores the efficacy of MAE in the context of medical image analysis. Second, the performance of the MAE model on medical images was found to be influenced by the masked ratio employed during training. Notably, we achieved optimal results with a masked ratio of 0.4 in our implementation. This indicates the importance of carefully selecting the appropriate ratio to achieve the best performance. Finally, our study demonstrates that MAE operates as a labeled data-efficient model, showcasing comparable performance even when trained on a partial dataset. This finding highlights the potential of MAE in situations where acquiring large quantities of labeled data may be challenging or resource-intensive.

In our implementation, we utilized a Vision Transformer (ViT) as the DL model. Compared to traditional CNN models, the ViT has shown promising performance across a range of tasks, but it is prone to being data-hungry during the training phase. We conducted experiments on the same DL model and training setting but with different training strategies: ViT-scratch, ViT-pretrain, and ViT-MAE. The results demonstrate the efficacy of self-supervised learning, yielding an accuracy of 0.985 and an AUC of 0.9957. Meanwhile, we compared the performance of the MAE model with the CNN-based models [42], which shows MAE outperforms the CNN models.

In our experiments, we explored the association between the mask ratio, a hyperparameter in the masked token reconstruction task, and the model's performance. We set the mask ratio from 0.4 to 0.8 and found that increasing the mask ratio led to a decrease in performance, which is different from the mask ratio result (0.75) of MAE on natural images [30]. To explain this trend, we visualized the original images, masked images, and reconstructed images. Comparing the original and reconstructed images, we observed that the reconstructed images were blurrier. The goal of reconstruction pretraining is to initialize the model parameters and enhance the model's understanding of the medical dataset. However, a high mask ratio may hinder the reconstruction process and weaken the model's understanding ability. Therefore, the mask ratio is a crucial factor in practical implementation.

To further highlight the advantages of self-supervised learning, we conducted limited dataset experiments. We randomly sampled the training dataset from 10% to 90% and applied the sample to conduct the reconstruction pretraining and fine-tuning of the model. The results strongly indicate the advantage of self-supervised learning on limited data. For example, using only a 30% sample of the training dataset, the ViT-MAE model still achieved 0.9425 accuracy, comparable to the performance of the ViT-pretrain model using the entire training dataset. This is particularly important in clinical applications, where datasets are often limited. Training large DL models on limited data can be challenging and can easily lead to overfitting due to the large number of parameters in the model. By this pretraining stage, the model can learn the good representation of the target dataset. Compared to the pretained model by natural image datasets, such as the ImageNet dataset, the pretraining stage MAE model has a narrow gap for the target dataset (i.e., small medical dataset) and is suitable for the later fine-tuning stage. Therefore, the MAE is suitable for the limited dataset. Additionally, using a smaller training dataset to train a large DL model reduces the cost of labeling, as traditional supervised DL training requires a large labeled training dataset to ensure model convergence. However, data labeling can be another issue when the dataset size is large, as in the case of the ImageNet dataset with one million images. Furthermore, medical image labeling often requires professional domain knowledge, such as an X-ray radiologist, to ensure accurate labeling.

The limitations of our study include the focus on COVID-19 alone. For future work, we plan to extend our work in two directions: first, we will extend the MAE model to handle 3D medical images, such as 3D brain imaging for Alzheimer's Disease [43–45]; second, we will explore the potential of the MAE for other tasks, such as image segmentation and localization [6,46], beyond image classification.

In conclusion, we applied MAE to the X-ray imaging dataset for COVID-19 diagnosis and illustrated that MAE exhibits remarkable efficiency when applied to labeled data, delivering comparable performance to utilizing only 30% of the original training dataset. Overall, our findings highlight the significant performance enhancement achieved by using MAE, particularly when working with limited datasets. This approach holds profound implications for future disease diagnosis, especially in scenarios where imaging information is scarce.

Author Contributions: X.X., N.J. and A.-L.L.; Conceptualization. X.X. and N.J.; Methodology. X.X.; Software. X.X. and G.L.; Validation. X.X., N.J. and A.-L.L.; Formal analysis. A.-L.L.; Investigation. N.J. and A.-L.L.; Resources. X.X.; Data curation. X.X. and A.-L.L.; Writing—original draft. G.L., C.W., N.J. and A.-L.L.; Review and Editing. X.X.; Visualization; N.J. and A.-L.L.; Supervision. A.-L.L.; Project administration and Funding acquisition. All authors listed have made a substantial, direct, and intellectual contribution to the work and approved it for publication. All authors have read and agreed to the published version of the manuscript.

Funding: This research was supported by NIH grants R01AG054459 and RF1AG062480 to A.-L.L.

Institutional Review Board Statement: Ethical review and approval were waived for this study due to the public dataset.

Informed Consent Statement: Patient consent was waived due to the public dataset.

Data Availability Statement: Data used in this article are from the dataset: COVIDxCXR-3. The collected data are available here: https://www.kaggle.com/datasets/andyczhao/covidx-cxr2 (2 June 2022).

Conflicts of Interest: The authors declare no conflicts of interest.

References

1. Tan, M.; Le, Q. Efficientnet: Rethinking model scaling for convolutional neural networks. In Proceedings of the International Conference on Machine Learning, Long Beach, CA, USA, 9–15 June 2019; pp. 6105–6114.
2. Huang, G.; Liu, Z.; Van Der Maaten, L.; Weinberger, K.Q. Densely connected convolutional networks. In Proceedings of the IEEE Conference on Computer Vision and Pattern Recognition, Honolulu, HI, USA, 21–26 July 2017; pp. 4700–4708.
3. He, K.; Zhang, X.; Ren, S.; Sun, J. Deep residual learning for image recognition. In Proceedings of the IEEE Conference on Computer Vision and Pattern Recognition, Las Vegas, NV, USA, 26 June–1 July 2016; pp. 770–778.
4. Xing, X.; Peng, C.; Zhang, Y.; Lin, A.L.; Jacobs, N. AssocFormer: Association Transformer for Multi-label Classification. In Proceedings of the 33rd British Machine Vision Conference, London, UK, 21–24 November 2022.
5. Fu, J.; Liu, J.; Tian, H.; Li, Y.; Bao, Y.; Fang, Z.; Lu, H. Dual attention network for scene segmentation. In Proceedings of the IEEE Conference on Computer Vision and Pattern Recognition, Long Beach, CA, USA, 15–20 June 2019; pp. 3146–3154.
6. Ronneberger, O.; Fischer, P.; Brox, T. U-net: Convolutional networks for biomedical image segmentation. In Proceedings of the Medical Image Computing and Computer-Assisted Intervention—MICCAI 2015: 18th International Conference, Munich, Germany, 5–9 October 2015; Springer: Berlin/Heidelberg, Germany, 2015; pp. 234–241.
7. Redmon, J.; Divvala, S.; Girshick, R.; Farhadi, A. You only look once: Unified, real-time object detection. In Proceedings of the IEEE Conference on Computer Vision and Pattern Recognition, Las Vegas, NV, USA, 26 June–1 July 2016; pp. 779–788.
8. Ren, S.; He, K.; Girshick, R.; Sun, J. Faster r-cnn: Towards real-time object detection with region proposal networks. In Proceedings of the Advances in Neural Information Processing Systems, Montreal, QC, Canada, 7–12 December 2015.
9. Ranjbarzadeh, R.; Jafarzadeh Ghoushchi, S.; Anari, S.; Safavi, S.; Tataei Sarshar, N.; Babaee Tirkolaee, E.; Bendechache, M. A deep learning approach for robust, multi-oriented, and curved text detection. *Cogn. Comput.* **2022**, 1–13. [CrossRef]
10. Anari, S.; Tataei Sarshar, N.; Mahjoori, N.; Dorosti, S.; Rezaie, A. Review of deep learning approaches for thyroid cancer diagnosis. *Math. Probl. Eng.* **2022**, *2022*, 5052435. [CrossRef]
11. Xing, X.; Liang, G.; Zhang, Y.; Khanal, S.; Lin, A.L.; Jacobs, N. Advit: Vision transformer on multi-modality pet images for alzheimer disease diagnosis. In Proceedings of the 2022 IEEE 19th International Symposium on Biomedical Imaging (ISBI), Kolkata, India, 28–31 March 2022; pp. 1–4.
12. Xing, X.; Rafique, M.U.; Liang, G.; Blanton, H.; Zhang, Y.; Wang, C.; Jacobs, N.; Lin, A.L. Efficient Training on Alzheimer's Disease Diagnosis with Learnable Weighted Pooling for 3D PET Brain Image Classification. *Electronics* **2023**, *12*, 467. [CrossRef] [PubMed]
13. Liang, G.; Xing, X.; Liu, L.; Zhang, Y.; Ying, Q.; Lin, A.L.; Jacobs, N. Alzheimer's disease classification using 2d convolutional neural networks. In Proceedings of the 2021 43rd Annual International Conference of the IEEE Engineering in Medicine & Biology Society (EMBC), Virtual, 1–5 November 2021; pp. 3008–3012.
14. Deng, J.; Dong, W.; Socher, R.; Li, L.J.; Li, K.; Fei-Fei, L. ImageNet: A Large-Scale Hierarchical Image Database. In Proceedings of the CVPR09, Miami, FL, USA, 20–25 June 2009.
15. Ying, X. An overview of overfitting and its solutions. *J. Phys. Conf. Ser.* **2019**, *1168*, 022022. [CrossRef]
16. Wang, X.; Liang, G.; Zhang, Y.; Blanton, H.; Bessinger, Z.; Jacobs, N. Inconsistent performance of deep learning models on mammogram classification. *J. Am. Coll. Radiol.* **2020**, *17*, 796–803. [CrossRef] [PubMed]
17. Krizhevsky, A.; Sutskever, I.; Hinton, G.E. Imagenet classification with deep convolutional neural networks. *Commun. ACM* **2017**, *60*, 84–90. [CrossRef]
18. Xing, X.; Liang, G.; Blanton, H.; Rafique, M.U.; Wang, C.; Lin, A.L.; Jacobs, N. Dynamic image for 3d mri image alzheimer's disease classification. In Proceedings of the Computer Vision—ECCV 2020 Workshops, Glasgow, UK, 23–28 August 2020; Springer: Berlin/Heidelberg, Germany, 2021; pp. 355–364.
19. Zhuang, F.; Qi, Z.; Duan, K.; Xi, D.; Zhu, Y.; Zhu, H.; Xiong, H.; He, Q. A comprehensive survey on transfer learning. *Proc. IEEE* **2020**, *109*, 43–76. [CrossRef]
20. Durand, T.; Mordan, T.; Thome, N.; Cord, M. Wildcat: Weakly supervised learning of deep convnets for image classification, pointwise localization and segmentation. In Proceedings of the IEEE Conference on Computer Vision and Pattern Recognition, Honolulu, HI, USA, 21–26 July 2017; pp. 642–651.
21. Zhou, Z.H. A brief introduction to weakly supervised learning. *Natl. Sci. Rev.* **2018**, *5*, 44–53. [CrossRef]
22. Liang, G.; Wang, X.; Zhang, Y.; Jacobs, N. Weakly-supervised self-training for breast cancer localization. In Proceedings of the 2020 42nd Annual International Conference of the IEEE Engineering in Medicine & Biology Society (EMBC), Montreal, QC, Canada, 20–24 July 2020; pp. 1124–1127.
23. Dosovitskiy, A.; Beyer, L.; Kolesnikov, A.; Weissenborn, D.; Zhai, X.; Unterthiner, T.; Dehghani, M.; Minderer, M.; Heigold, G.; Gelly, S.; et al. An Image is Worth 16 × 16 Words: Transformers for Image Recognition at Scale. *arXiv* **2020**, arXiv:2010.11929.

24. Liu, Z.; Lin, Y.; Cao, Y.; Hu, H.; Wei, Y.; Zhang, Z.; Lin, S.; Guo, B. Swin transformer: Hierarchical vision transformer using shifted windows. In Proceedings of the IEEE/CVF International Conference on Computer Vision, Montreal, QC, Canada, 11–17 October 2021; pp. 10012–10022.
25. Wu, H.; Xiao, B.; Codella, N.; Liu, M.; Dai, X.; Yuan, L.; Zhang, L. Cvt: Introducing convolutions to vision transformers. In Proceedings of the IEEE/CVF International Conference on Computer Vision, Nashville, TN, USA, 20–25 June 2021; pp. 22–31.
26. Goyal, P.; Caron, M.; Lefaudeux, B.; Xu, M.; Wang, P.; Pai, V.; Singh, M.; Liptchinsky, V.; Misra, I.; Joulin, A.; et al. Self-supervised pretraining of visual features in the wild. *arXiv* **2021**, arXiv:2103.01988.
27. Gidaris, S.; Singh, P.; Komodakis, N. Unsupervised representation learning by predicting image rotations. *arXiv* **2018**, arXiv:1803.07728.
28. Doersch, C.; Gupta, A.; Efros, A.A. Unsupervised visual representation learning by context prediction. In Proceedings of the IEEE International Conference on Computer Vision, Santiago, Chile, 7–13 December 2015; pp. 1422–1430.
29. Grill, J.B.; Strub, F.; Altché, F.; Tallec, C.; Richemond, P.; Buchatskaya, E.; Doersch, C.; Avila Pires, B.; Guo, Z.; Gheshlaghi Azar, M.; et al. Bootstrap your own latent-a new approach to self-supervised learning. *Adv. Neural Inf. Process. Syst.* **2020**, *33*, 21271–21284.
30. He, K.; Chen, X.; Xie, S.; Li, Y.; Dollár, P.; Girshick, R. Masked autoencoders are scalable vision learners. In Proceedings of the IEEE/CVF Conference on Computer Vision and Pattern Recognition, New Orleans, LA, USA, 18–24 June 2022; pp. 16000–16009.
31. Wang, L.; Lin, Z.Q.; Wong, A. COVID-Net: A tailored deep convolutional neural network design for detection of COVID-19 cases from chest X-ray images. *Sci. Rep.* **2020**, *10*, 19549. [CrossRef] [PubMed]
32. Cohen, J.P.; Morrison, P.; Dao, L. COVID-19 image data collection. *arXiv* **2020**, arXiv:2003.11597.
33. Figure 1-COVID-19 Chest X-ray Dataset Initiative. Available online: https://github.com/agchung/Figure1-COVID-chestxray-dataset (accessed on 8 May 2020).
34. Actualmed COVID-19 Chest X-ray 71 Dataset Initiative. Available online: https://github.com/agchung/Actualmed-COVID-chestxray-dataset (11 November 2020).
35. COVID-19 Radiography Database. Available online: https://www.kaggle.com/datasets/tawsifurrahman/covid19-radiography-database (accessed on 31 March 2021).
36. RSNA Pneumonia Detection Challenge. Available online: https://www.kaggle.com/c/rsna-pneumonia-detection-challenge (accessed on 17 October 2018).
37. RSNA International COVID-19 Open Radiology Database. Available online: https://wiki.cancerimagingarchive.net/pages/viewpage.action?pageId=70230281 (accessed on 15 January 2021).
38. BIMCV-COVID19+. Available online: https://bimcv.cipf.es/bimcv-projects/bimcv-covid19/ (accessed on 20 October 2020).
39. COVID-19-NY-SBU. Available online: https://wiki.cancerimagingarchive.net/pages/viewpage.action?pageId=89096912 (accessed on 11 August 2021).
40. Ba, J.L.; Kiros, J.R.; Hinton, G.E. Layer normalization. *arXiv* **2016**, arXiv:1607.06450.
41. Kingma, D.P.; Ba, J. Adam: A method for stochastic optimization. *arXiv* **2014**, arXiv:1412.6980.
42. Hasan, N.; Bao, Y.; Shawon, A.; Huang, Y. DenseNet convolutional neural networks application for predicting COVID-19 using CT image. *SN Comput. Sci.* **2021**, *2*, 389. [CrossRef] [PubMed]
43. Hammond, T.C.; Xing, X.; Wang, C.; Ma, D.; Nho, K.; Crane, P.K.; Elahi, F.; Ziegler, D.A.; Liang, G.; Cheng, Q.; et al. β-amyloid and tau drive early Alzheimer's disease decline while glucose hypometabolism drives late decline. *Commun. Biol.* **2020**, *3*, 352. [CrossRef] [PubMed]
44. Hammond, T.C.; Xing, X.; Yanckello, L.M.; Stromberg, A.; Chang, Y.H.; Nelson, P.T.; Lin, A.L. Human Gray and White Matter Metabolomics to Differentiate APOE and Stage Dependent Changes in Alzheimer's Disease. *J. Cell. Immunol.* **2021**, *3*, 397. [PubMed]
45. Ying, Q.; Xing, X.; Liu, L.; Lin, A.L.; Jacobs, N.; Liang, G. Multi-modal data analysis for alzheimer's disease diagnosis: An ensemble model using imagery and genetic features. In Proceedings of the 2021 43rd Annual International Conference of the IEEE Engineering in Medicine & Biology Society (EMBC), Virtual, 1–5 November 2021; pp. 3586–3591.
46. Zhao, Y.; Zeng, K.; Zhao, Y.; Bhatia, P.; Ranganath, M.; Kozhikkavil, M.L.; Li, C.; Hermosillo, G. Deep learning solution for medical image localization and orientation detection. *Med. Image Anal.* **2022**, *81*, 102529. [CrossRef] [PubMed]

bioengineering

Article

Effects of Alterations in Resting-State Neural Networks on the Severity of Neuropathic Pain after Spinal Cord Injury

Eunhee Park [1,2,†], Jang Woo Park [3,†], Eunji Kim [3], Yu-Sun Min [1,2], Hui Joong Lee [4,5], Tae-Du Jung [1,2,*] and Yongmin Chang [5,6,7,*]

1 Department of Rehabilitation Medicine, School of Medicine, Kyungpook National University, Daegu 41944, Republic of Korea; ehmdpark@knu.ac.kr (E.P.); ssuni119@naver.com (Y.-S.M.)
2 Department of Rehabilitation Medicine, Kyungpook National University Chilgok Hospital, Daegu 41404, Republic of Korea
3 Korea Radioisotope Center for Pharmaceuticals, Korea Institute of Radiological & Medical Sciences, Seoul 01812, Republic of Korea; giantstar.jw@gmail.com (J.W.P.); geekimeun@gmail.com (E.K.)
4 Department of Radiology, School of Medicine, Kyungpook National University, Daegu 41944, Republic of Korea; leehuijoong@knu.ac.kr
5 Department of Radiology, Kyungpook National University Hospital, Daegu 41944, Republic of Korea
6 Department of Molecular Medicine, School of Medicine, Kyungpook National University, Daegu 41944, Republic of Korea
7 Department of Medical & Biological Engineering, Kyungpook National University, Daegu 41944, Republic of Korea
* Correspondence: teeed0522@knu.ac.kr (T.-D.J.); ychang@knu.ac.kr (Y.C.); Tel.: +82-53-200-2167 (T.-D.J.); +82-53-420-5471 (Y.C.)
† These authors contributed equally to this work.

Abstract: Neuropathic pain (NP) following spinal cord injury (SCI) is refractory to pain control strategies, and the underlying neuronal mechanisms remain poorly understood. This study aimed to determine the brain regions engaged in maintaining a spontaneous resting state and the link between those regions and the severity of NP in patients with incomplete SCI. Seventy-three subjects (41 patients and 32 age- and sex-matched healthy controls) participated in this retrospective study. Regarding the neurological level of injury, patients with incomplete SCI experienced at-level or below-level NP. The severity of NP was evaluated using a visual analog scale (VAS), and patients were divided into mild and moderate–severe NP groups based on VAS scores. Graph theory and fractional amplitude of low-frequency fluctuation (fALFF) analyses were performed to compare resting-state functional magnetic resonance imaging (fMRI) analysis results among the three groups. Graph theory analysis was performed through a region of interest (ROI)-to-ROI analysis and then fALFF analysis was performed in the brain regions demonstrating significant differences among the three groups analyzed using the graph theory. We evaluated whether the brain regions showing significant differences using graph theory and fALFF correlated with the VAS scores. Patients with moderate–severe NP showed reduced node degree and fALFF in the left middle frontal gyrus compared with those with mild NP and healthy controls. Furthermore, patients with severe NP demonstrated increased average path lengths and reduced fALFF values in the posterior cingulate gyrus. This study found that changes in intrinsic oscillations of fMRI signals in the middle frontal gyrus and posterior cingulate gyrus were significant considering the severity of NP.

Keywords: neuropathic pain; resting-state fMRI; spinal cord injury

Citation: Park, E.; Park, J.W.; Kim, E.; Min, Y.-S.; Lee, H.J.; Jung, T.-D.; Chang, Y. Effects of Alterations in Resting-State Neural Networks on the Severity of Neuropathic Pain after Spinal Cord Injury. *Bioengineering* **2023**, *10*, 860. https://doi.org/10.3390/bioengineering10070860

Academic Editor: Cuneyt M. Alper

Received: 8 June 2023
Revised: 17 July 2023
Accepted: 17 July 2023
Published: 20 July 2023

1. Introduction

Approximately 50% of patients with spinal cord injury (SCI) suffer from neuropathic pain (NP) [1,2]. The presence and severity of NP are associated with substantial physical and emotional functioning impairment, thereby affecting quality of life [3]. Reducing NP is essential to improve the quality of life of patients with SCI. However, provoking or relieving factors associated with NP following SCI remain poorly understood. Furthermore, NP is usually refractory to pain control strategies, such as pharmaceutical, behavioral, and neurological approaches [4,5].

Recent neuroimaging studies have indicated the possibility of anatomical and functional changes in brain regions associated with NP following SCI. In a diffusion tensor imaging study, SCI with NP showed significant differences in mean diffusivity (MD) values in pain-related areas compared to SCI without NP [6]. MD values increased in regions, such as the posterior parietal cortex, dorsolateral prefrontal cortex, anterior insula, and premotor cortex. Conversely, MD values decreased in the ventroposterior thalamus and amygdala [6]. A functional magnetic resonance imaging (fMRI) study in patients with SCI and NP demonstrated functional reorganization in the primary somatosensory cortex (S1) [7]. Furthermore, a recent study using resting-state fMRI (rs-fMRI) on an SCI animal model found that the development of mechanical hypersensitivity was strongly associated with increased functional connectivity between the thalamus and the S1 cortical regions [8]. Another rs-fMRI study investigated the mutual effects of motor- and pain-related networks in different brain regions on motor disability and NP intensity [9]. However, functional changes in brain regions associated with NP following SCI remain poorly understood. Moreover, brain regions that are related to the severity of NP following SCI are rarely investigated.

Low-frequency blood oxygenation level-dependent (BOLD) signal oscillations, which are intrinsic components of brain activity, are represented by rs-fMRI [10]. Several analyses are used to process rs-fMRI data to explore oscillatory BOLD signal dynamics associated with changes in neural activity. A graph theory analysis helps elucidate the organization of functional connections and their efficient integration of neural information in a whole brain. In a graph theory analysis, a whole-brain network is described as a graph comprising a collection of nodes and edges between nodes, revealing that the brain network is organized according to an efficient small-world organization [11,12]. A recent systematic review found that differences between patients with chronic pain and healthy controls were mainly observed in terms of the global graph-based connectivity; however, there are no particularly affected brain regions [13]. Fractional amplitude of low-frequency fluctuation (fALFF) analysis characterizes the frequency distribution of signal variance in a time series in particular brain regions [14,15]. The fALFF is an index that quantifies spontaneous neuronal activity related to the regional metabolic level of glucose [16].

In this study, we used a novel method by combining graph theory and fALFF analyses to investigate the possible alterations in brain networks associated with the severity of NP after SCI. We hypothesized that increased NP severity following SCI is related to less efficient connections in a whole-brain region assessed using graph theory analysis. Moreover, these inefficient brain regions would be associated with less spontaneous neuronal activity assessed using fALFF analysis to modulate more severe pain conditions. Therefore, this study is the first to simultaneously investigate less efficient connections and spontaneous neuronal activities at these connections. Our findings can provide a useful reference on brain responses to the severity of NP following SCI.

2. Materials and Methods

2.1. Subjects

The following subjects were included in this retrospective cross-sectional study: (1) diagnosed with SCI more than 3 months after a traumatic injury; (2) diagnosed with incomplete SCI, defined as some degree of retained motor or sensory function below the site of injury, including sacral root segments (S4–5), according to the American Spinal Injury Association Impairment Scale (AIS) [17,18]; and (3) diagnosed with persistent at- or below-level NP according to the International Association for the Study of Pain (IASP) definition [19,20]. The exclusion criteria were as follows: (1) history of nociceptive musculoskeletal pain after SCI, (2) history of traumatic brain hemorrhage or contusion on brain computed tomography, (3) history of any peripheral nervous system disorder, (4) history of any neurologic or neuropsychiatric condition, or (5) history of alcohol or drug misuse.

Seventy-three subjects (41 patients and 32 healthy controls) participated in this study. The ages of the subjects in the control (n = 32), mild NP (n = 24), and moderate–severe NP (n = 17) groups were 50.12 $\pm$ 13.33, 49.87 $\pm$ 14.74, and 53.82 $\pm$ 12.18 years old, respectively, with no significant age difference among the groups (p = 0.60). The sex of the subjects also did not differ significantly between the control (male 20, female 12), mild NP (male 15, female 9), and moderate–severe NP (male 13, female 4) groups (p = 0.57).

This study was approved by the Institutional Review Board of Kyungpook National University Chilgok Hospital (No. 2018-12-004).

2.2. Clinical Assessments

Based on the use of the International Standards for Neurological Classification of SCI worksheet, the AIS comprises an anorectal examination, a dermatome-based sensory examination, and a myotome-based motor test [17,18]. The most caudal functioning root level with intact motor and sensory functions is identified as the neurologic level of injury (NLI).

The IASP describes NP as a burning, stabbing, or shooting sensation occurring spontaneously or in response to a central nervous system injury or disease [19]. NP is characterized as either at-level or below-level NP depending on the type of NLI. At-level NP is felt in a segmental pattern anywhere along the dermatome of the NLI, whereas below-level NP is felt in more than three dermatomes beneath the dermatome of the NLI [19,20]. The average NP intensity over the previous 7 days was measured using a visual analog scale (VAS) ranging 0–100 at the time of rs-fMRI acquisition [21]. The higher the VAS score is, the more severe the NP is. The patients were then divided into two groups based on the severity of NP. A VAS score of 34 was used as the cutoff for dividing the patients into the mild and moderate–severe NP groups [22].

The Beck Depression Inventory (BDI)-II was used to measure the severity of depressive mood, and it comprised 21 items, each answer being graded from 0 to 3, with higher BDI scores indicating greater severity [23].

2.3. rs-fMRI Data Acquisition

The whole-brain function was acquired using a Discovery MR750w 3.0 T (GE Healthcare, Milwaukee, WI, USA) with a 24-channel head coil. Participants were given no instructions other than to close their eyes during rs-fMRI scanning to avoid falling asleep. The 240 image volumes were acquired using a T2-weighted echoplanar imaging pulse sequence with the parameters of repetition time (TR, 2000 ms), echo time (TE, 30 ms), field of view (FOV, 23 cm), matrix (64 $\times$ 64), and slice thickness:gap (4:0 mm) and no gap for resting-state imaging. T1-weighted fast spoiled gradient echo sequence (FSPGR) with a TR of 8.5 ms, TE of 3.2 ms, FA of 13°, FOV of 25.6 cm, acquisition matrix of 256 $\times$ 256, and iso-voxel resolution of 1 mm was used to obtain structural brain images.

2.4. rs-fMRI Data Preprocessing

Image data were preprocessed and statistically analyzed using the Statistical Parametric Mapping program (SPM12; Wellcome Centre for Human Neuroimaging, London, UK) and MATLAB (The MathWorks, Inc., Natick, MA, USA). Preprocessing included slice timing, realignment, coregistration on individual T1 structural image volume and normalization into the standard stereotaxic coordinate space (Montreal Neurological Institute). Normalized images were spatially smoothed with an 8-mm Gaussian kernel. The preprocessed rs-fMRI data were then temporally band-pass filtered (0.008–0.09 Hz) to eliminate low-frequency drift and high-frequency noise. A component-based noise correction method (CompCor) was used in the denoising part of the CONN toolbox to identify and eliminate components, such as physiological influences on fMRI data (https://www.nitrc.org/projects/conn/) (accessed on 12 March 2020) [24].

2.5. Graph Theory and fALFF Analyses

Graph theory and fALFF analyses were performed using the CONN toolbox. A total of 140 regions of interest (ROIs) provided by the CONN toolbox were used for the analyses. First, graph theory analysis was performed via ROI-to-ROI analysis, and the brain region exhibiting a significant difference among the three groups was confirmed using analysis of variance (ANOVA). Subsequently, fALFF analysis was performed in the brain regions demonstrating significant differences (identified using ANOVA) among the three groups analyzed using graph theory. We then investigated whether the graph theory and fALFF analyses values in the region with significant differences correlated with VAS scores (an index of pain). Finally, in regions that correlated with VAS scores, we observed whether graph theory and fALFF analyses values were correlated.

2.5.1. Graph Theory Analysis

Graph theory analysis was performed within 140 ROIs provided using the CONN toolbox [25]. A network can be represented as a graph by G (n, k), indicating the number of nodes (n) and the number of edges (k). The threshold for the ROI-to-ROI connectivity matrix for each subject is at a set level (cost = 0.15, two-sided). The normalized Z-scores of raw connectivity values can be used to calculate this threshold, resulting in graphs with fixed network-level costs. Supra-threshold connectivity values are used to create an adjacency matrix that represents a graph with nodes representing ROIs and edges representing the intensity of their functional connectivity. The degree of a node is defined as the number of its connected links. The clustering coefficient of a node is defined as the ratio of the actual number of links between neighbors and the maximum possible number of links between these neighbors. The average path length is defined by the integration of a network and the easy flow of information within this network. For each node n in a graph G, the cost is defined as the proportion of connected neighbors, global efficiency is defined as the average inverse shortest path distance from node n to all other nodes in the graph G, and local efficiency is defined as the average global efficiency across all nodes in the local subgraph of node n [26].

The graph theory analysis of each node revealed a region with significant differences among the three groups and correlation with VAS, and this region was used for fALFF analysis. The significance level was set at $p < 0.005$ (uncorrected).

2.5.2. fALFF Analysis

The rs-fMRI data with preprocessing were subjected to fALFF analysis using the CONN toolbox. The value of fALFF is equal to the sum of amplitudes within a low-frequency band divided by the sum of amplitudes across the entire frequency band [15]. To standardize raw power measures, Z-transformation was performed for the fALFF analysis, which can improve the subsequent statistical analyses on the group level. The fALFF value was obtained for the regions identified via the graph theory analysis and was used to evaluate the fALFF difference among the three groups and the correlation with VAS.

2.6. Statistical Analysis

All statistical analyses were performed using the SPSS software version 23 (SPSS, Inc., Armonk, NY, USA); a p value of <0.05 was considered statistically significant. After verifying normality by performing the Shapiro–Wilk test, one-way ANOVA was performed if normality was satisfied. For post-hoc analysis, Scheffe's multiple comparisons were performed if equality of variance was satisfied after verification of equal variance through the Levene's test. If the equality of variance is not satisfied, Bunnett T3 verification was performed. If the normality was not satisfied, analysis was performed using the Kruskal–Wallis t-test.

3. Results

3.1. General Characteristics

The VAS and BDI scores were evaluated in patients with NP. The VAS score differed significantly between the mild and moderate–severe NP groups ($p < 0.001$), whereas the BDI score did not ($p = 0.67$). The demographic and clinical characteristics of patients with incomplete SCI are summarized in Table 1 and Table S1.

Table 1. Demographic of the incomplete spinal cord injury patients.

Group	Age	VAS	BDI
Mild NP	49.87 ± 14.74	20.58 ± 8.59	13.37 ± 8.85
Moderate–Severe NP	53.82 ± 12.18	54.35 ± 14.92	14.82 ± 11.99

Neuropathic pain, NP; visual analog scale, VAS; Beck depression inventory, BDI.

3.2. Graph Theory Analysis According to the Severity of NP Following SCI

Figure 1, Tables 2 and 3 shows the brain regions analyzed by one-way ANOVA in the three groups based on graph theory analysis (uncorrected $p < 0.005$). The left middle frontal gyrus (MidFG) and the left superior division of lateral occipital cortex (sLOC) showed differences among the three groups in global efficiency, cost, average path length, and degree. The moderate–severe NP group showed lower values than the control and mild NP groups in global efficiency, cost, and degree of the left MidFG and higher values in the average path length. No difference was observed between the control and mild NP groups. The posterior division of the cingulate gyrus (PCC) showed differences among the three groups in average path length and betweenness centrality. The moderate–severe NP group demonstrated the highest value of the average path length in the PCC area, which differed significantly from that of the control group. However, no significant difference was found between the mild NP and the other two groups.

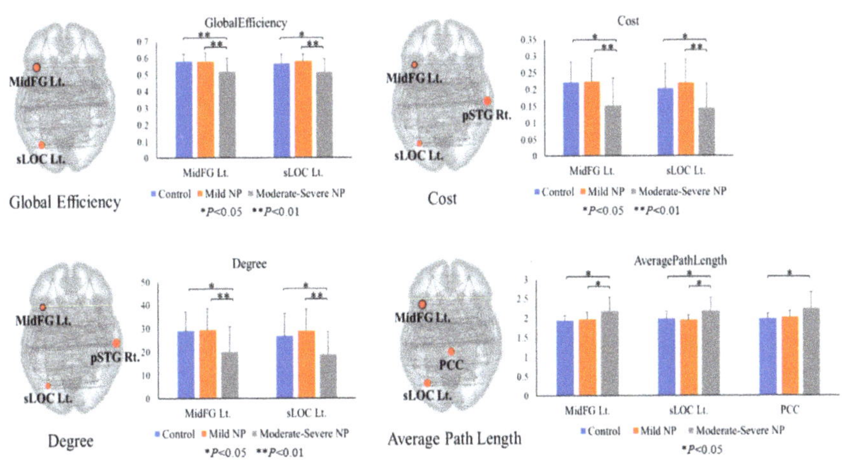

Figure 1. Graph theory analysis at 165 nodes. The nodes exhibited a significant difference (uncorrected $p < 0.005$) among the three groups and correlated with the VAS score ($p < 0.05$). Neuropathic pain, NP; left, Lt.; middle frontal gyrus, MidFG; superior division of lateral occipital cortex, sLOC; posterior division of cingulate gyrus, PCC; right, Rt.; posterior division of superior temporal gyrus, pSTG.

Table 2. The ANOVA result of graph theory analysis among three groups and the result of correlation with VAS.

Measure	ROI	ANOVA		Correlation with VAS	
		F	*p*	r	*p*
Global Efficiency	MidFG Lt.	7.35	0.001	−0.3629	0.001
	sLOC Lt.	6.45	0.002	−0.2524	0.031
Cost	MidFG Lt.	6.48	0.002	−0.3371	0.003
	sLOC Lt.	5.89	0.004	−0.2549	0.029
Degree	MidFG Lt.	6.48	0.000	−0.3371	0.003
	sLOC Lt.	5.84	0.004	−0.2549	0.029
Average Path Length	MidFG Lt.	8.01	0.000	0.3948	0.000
	sLOC Lt.	7.11	0.001	0.2751	0.018
	PCC	7.08	0.001	0.3379	0.003

Visual analog scale, VAS; left, Lt.; middle frontal gyrus, MidFG; superior division of lateral occipital cortex, sLOC; posterior division of cingulate gyrus, PCC.

Table 3. Post hoc analysis of graph theory results in the area where there is a statistical difference among the three groups and is correlated with VAS.

Measure	ROI	Control vs. Mild NP	Control vs. Moderate-Severe NP	Mild NP vs. Moderate-Severe NP
Global Efficiency	MidFG Lt.	0.9990	0.0030	0.0060
	sLOC Lt.	0.6970	0.0190	0.0040
Cost	MidFG Lt.	1.0000	0.0140	0.0080
	sLOC Lt.	0.7700	0.0280	0.0060
Degree	MidFG Lt.	1.0000	0.0140	0.0080
	sLOC Lt.	0.7170	0.0280	0.0060
Average Path Length	MidFG Lt.	1.0000	0.0110	0.0130
	sLOC Lt.	1.0000	0.0360	0.0140
	PCC	1.0000	0.0310	0.3130

Neuropathic pain, NP; visual analog scale, VAS; left, Lt.; middle frontal gyrus, MidFG; superior division of lateral occipital cortex, sLOC; posterior division of cingulate gyrus, PCC.

In the left MidFG, the global efficiency, cost, average path length, and degree showed significantly high correlations with the VAS score ($p < 0.01$), and in the left sLOC, these parameters showed negative correlations with the VAS score (Figure 2). However, the left sLOC showed a significance level of $p < 0.05$, which was lower than that of the left MidFG. In the left MidFG and left sLOC, the average path lengths demonstrated positive correlations with VAS, but other results of graph theory analysis showed negative correlations. In the PCC, the average path length was highly significantly positively correlated with the VAS score ($p < 0.01$).

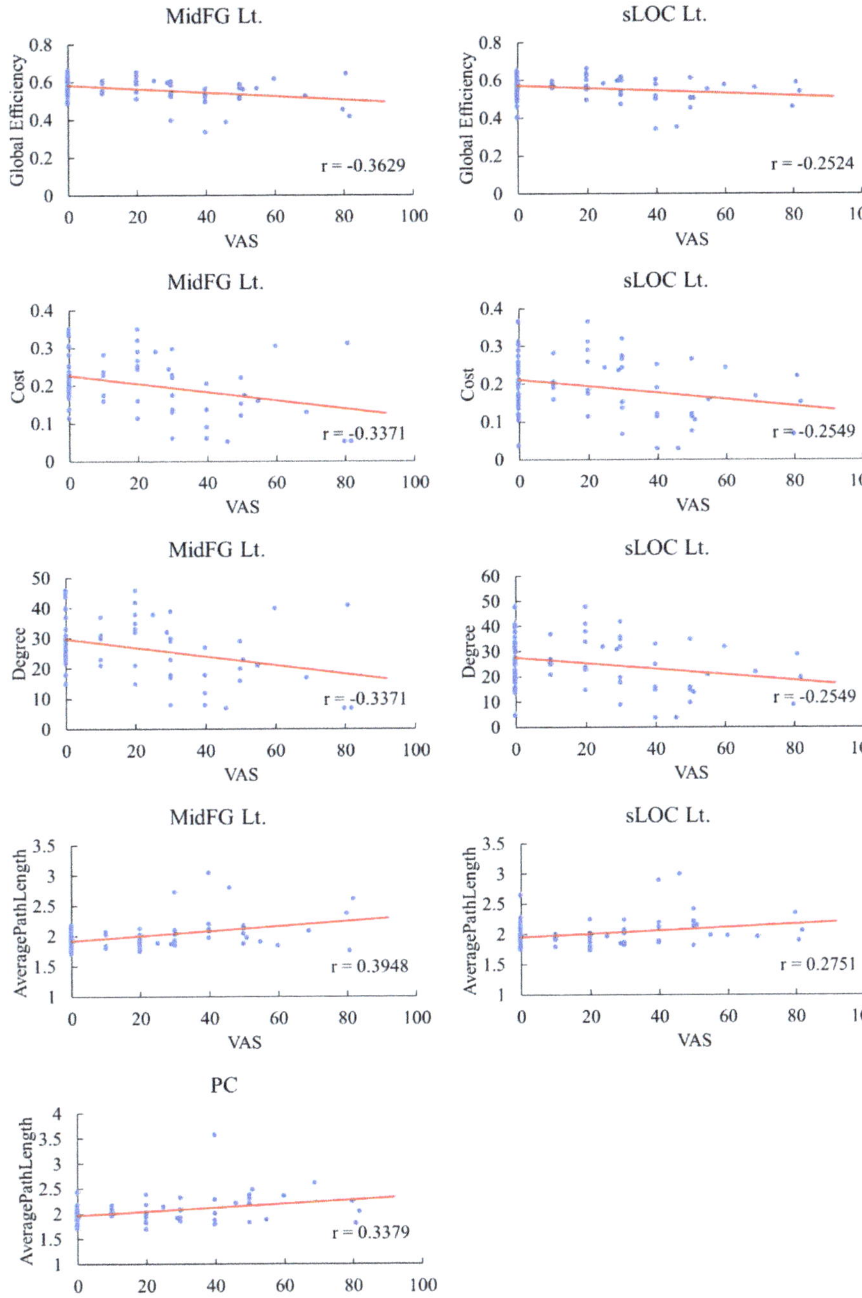

Figure 2. Correlation between the graph theory analysis values and VAS scores at nodes analyzed using the graph theory. Neuropathic pain, NP; visual analog scale, VAS; left, Lt.; middle frontal gyrus, MidFG; superior division of lateral occipital cortex, sLOC; posterior division of cingulate gyrus, PCC.

3.3. *fALFF Analysis According to the Severity of NP Following SCI*

Figure 3 and Tables 4–6 show the fALFF analysis results for three brain regions selected from the graph theory analysis results. Among the three regions selected through graph theory analysis, the regions showing significant differences among the three groups were the left MidFG (p = 0.004) and PCC (p = 0.041). No significant difference was observed among groups in other regions. For the left MidFG region, the fALFF values were ranked in the following order of groups: mild NP < control < moderate–severe NP. The fALFF value of the moderate–severe NP group differed significantly from those of the mild NP and control groups (no significant difference was observed between the mild NP and control groups). In the PCC region, the fALFF values were ranked in the following order of groups: control < mild NP < moderate–severe NP; the fALFF values differed significantly between the control and moderate–severe NP groups but not between the control and mild NP groups or between the mild NP and moderate–severe NP groups.

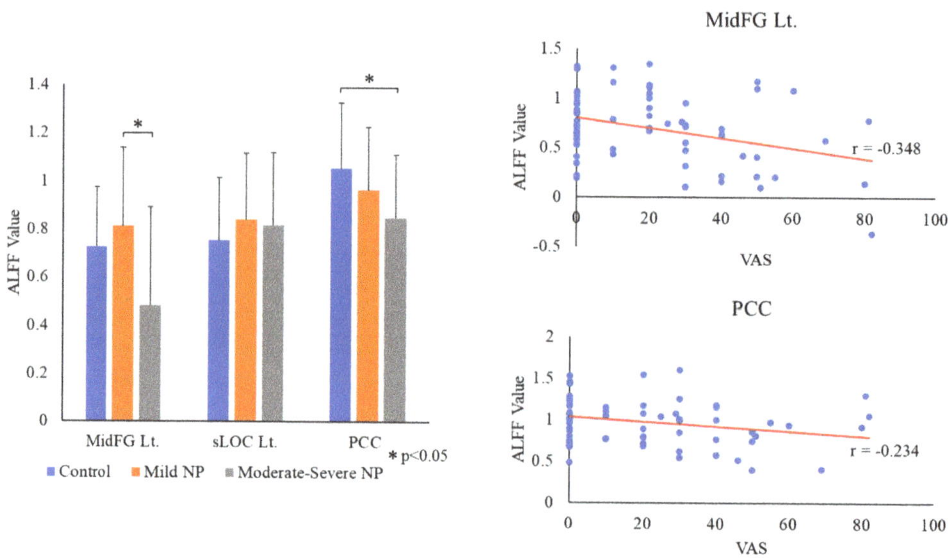

(a) (b)

Figure 3. (**a**) The graph on the left shows the fractional amplitude of low-frequency fluctuation (fALFF) analysis results for three regions selected from the graph theory analysis results. (**b**) The graph on the right shows two regions with fALFF values differing among the three groups and correlating with VAS scores. Neuropathic pain, NP; visual analog scale, VAS; left, Lt.; middle frontal gyrus, MidFG; superior division of lateral occipital cortex, sLOC; posterior division of cingulate gyrus, PCC.

Table 4. In the area of graph theory analysis result, the ANOVA result of ALFF analysis among three groups and the result of correlation with VAS.

ROI	ANOVA		Correlation with VAS	
	F	p-unc	r	p
MidFG Lt.	5.73	0.004	−0.348	0.002
sLOC Lt.	0.71	0.491	0.102	0.387
PCC	3.33	0.041	−0.234	0.045

Visual analog scale, VAS; left, Lt.; middle frontal gyrus, MidFG; superior division of lateral occipital cortex, sLOC; posterior division of cingulate gyrus, PCC.

Table 5. Post hoc analysis of ALFF results in the area where there is a statistical difference among the three groups and is correlated with VAS.

ROI	Control vs. Mild NP	Control vs. Moderate-Severe NP	Mild NP vs. Moderate-Severe NP
MidFG Lt.	0.619	0.096	0.027
PCC	0.451	0.043	0.409

Neuropathic pain, NP; visual analog scale, VAS; left, Lt.; middle frontal gyrus, MidFG; posterior division of cingulate gyrus, PCC.

Table 6. In the area of graph theory analysis result, the ALFF value for each group.

ROI	Control	Mild NP	Moderate-Severe NP
MidFG Lt.	0.726 ± 0.247	0.813 ± 0.326	0.48 ± 0.411
sLOC Lt.	0.755 ± 0.258	0.841 ± 0.276	0.816 ± 0.303
PCC	1.055 ± 0.274	0.963 ± 0.263	0.849 ± 0.264

Neuropathic pain, NP; left, Lt.; middle frontal gyrus, MidFG; superior division of lateral occipital cortex, sLOC; posterior division of cingulate gyrus, PCC.

The fALFF value of the left MidFG showed a high statistical level of negative correlation with the VAS score (less than $p < 0.01$, Figure 3b). Similarly, the fALFF value of the PCC showed a negative correlation with the VAS score, but the statistical level ($p < 0.05$) was lower than that of the left MidFG.

4. Discussion

This study found that changes in intrinsic BOLD oscillations in the MidFG and PCC were statistically significant in association with the severity of NP. Patients with moderate–severe NP showed reduced degree and fALFF in the MidFG compared with those with mild NP or healthy controls. Furthermore, patients with more severe NP had increased average path length and reduced fALFF in the PCC.

Neuroimaging studies of chronic pain have focused on nociceptive pain-processing-related brain regions. The brain regions most commonly activated by noxious stimuli are somatosensory (e.g., the S1, thalamus, and secondary somatosensory cortex), limbic (e.g., the anterior cingulate cortex and insula), and associative structures (e.g., the prefrontal cortex) [27]. Emotional and cognitive process regions, such as the dorsolateral prefrontal cortex, anterior cingulate cortex, and anterior insula [28,29], show increased activations under conditions of prolonged exposure to pain, including chronic pain induced by fibromyalgia, diabetes, and amputation. However, SCI-induced NP can be regarded as chronic pain if classified in terms of time and is not a nociceptive pain. Nevertheless, non-nociceptive pain is classified as a neurophysiological concept. A chronic lesion may cause chronic pain and continue to appear even after the injury has healed. It is known that the intensity and duration of pain is affected by adding other exacerbating factors surrounding the patient, such as stress, environment, emotions, occupation, chronic disease, and traumatic injury [30–32]. To the best of our knowledge, this study is the first to determine the brain regions that are highly associated with the severity of NP, affecting resting-state network efficiency and regional neural activity, after SCI.

In this study, we first analyzed the relationship between resting-state connectivity of whole-brain ROIs using graph theory and the severity of NP after SCI. Reduced global efficiency and cost values, which were interpreted in the weak potential for network integration in the left MidFG and left sLOC, were related to severe pain intensity. This finding indicates that MidFG and sLOC regions in patients with more severe NP demonstrate reduced communication with other distributed brain regions compared with healthy controls. Additionally, increased values of average path length in the left MidFG, left sLOC, and PCC were associated with increased NP severity. Higher average path length in the severe pain group was interpreted as lower efficiency in the network than that in the control group. Furthermore, reduced degree value, which was interpreted on the basis of decreased interaction with other nodes in the network in the left MidFG and sLOC, was associated with increased NP severity.

Compared with patients with mild NP or healthy controls, those with moderate–severe NP showed a reduced degree and fALFF in the MidFG. The dorsal attention network in resting-state neural networks comprises the intraparietal sulcus and the MidFG in the prefrontal cortex [33]. This network exhibits an increase in activity after presenting cues indicating where, when, or what subjects should receive attention [33,34]. Attention and cognitive function are known to modulate pain [35]. Animal studies have shown that pain stimuli interrupt attention; thus, distracting attention from that pain can reduce pain-related behaviors [36,37]. Human studies also showed that cognitive tasks induced a change in neural activity in brain regions related to pain processing, such as the thalamus, S1, anterior cingulate cortex, and insular cortex [38,39]. In addition, Wiech et al. reported that the prefrontal cortex mediated analgesic effects on expected or perceived pain [40]. In the present study, patients with moderate–severe NP showed a lower degree of connection in the MidFG than those with mild NP. A previous study proposed that a low degree of connections was linked to decreased glucose metabolism, as determined using 18F-FDG-PET [41]. Furthermore, reduced fALFF value in the MidFG might be related to reduced spontaneous neural activity in more severe pain. Prolonged NP following SCI might induce attention distraction in the MidFG to adjust to more severe pain. Further neuroimaging studies are warranted to confirm the relationship between attention-related cognitive task performance and the severity of NP in patients with SCI.

This study also showed an increased average path length and reduced fALFF in the PCC in patients with moderate–severe NP compared with those with mild NP. The average path length in graph metrics indicates brain network integration. The low efficiency of information transfer between any pair of neurons is equivalent to the increased average path length of connections between nodes [12,42]. The PCC area includes an intrinsic default mode network, and neural activity changes in the default mode network are related to chronic pain conditions, such as complex regional pain syndrome and fibromyalgia [43,44]. PCC has been closely associated with pain processing and was found to be more related to disgust than to the painful aspect of stimulation [43,44]. Particularly, the activation of ventral PCC is associated with catastrophizing and widespreadness of chronic pain [45]. Considering previous evidence, the results of our study suggest that inefficient path and reduced neuronal activity in PCC might be related to the modulation of severe NP.

Recent research demonstrates that physiotherapy may enhance the utilization of sensorimotor strategies in body representation for individuals suffering from SCI [46]. This finding implies that even when the body is affected by de-afferentation/de-efferantation, physiotherapy can modify the relationship between sensorimotor and visual aspects of body representation [46]. These alterations appear to be associated with changes in functional connectivity within and between frontal and parietal networks [47]. Furthermore, a recent study reported that alterations in intralimbic and limbostriatal connectivity were associated with the severity of neuropathic pain [48]. The study also found that the functional reorganization of cortico-cerebellar connectivity and subcortical areas varied depending on the onset time after SCI [49]. Based on these findings, the present study could contribute to the development of more comprehensive rehabilitation strategies to alleviate neuropathic pain. We propose that physiotherapy following SCI, which could affect functional neuroplastic changes, may help relieve neuropathic pain depending on the onset time of SCI. Additionally, further studies are needed to explore the relationship between spontaneous neural changes of functional reorganization after SCI, the severity of neuropathic pain, and the neuroplastic effects of post-SCI rehabilitation strategies.

This study has some limitations. First, the sample size was small; thus, this study is effectively considered a pilot. Further studies with a large population are required to conclude general outcomes in patients with NP after SCI. Second, although age- and sex-matched healthy subjects were enrolled as controls, further studies are needed to confirm the vulnerability of including patients with SCI without NP as a control group for investigating whether the current study findings are specific to individuals with NP following SCI. Finally, the interslice gap technique was not introduced in the current study. The gap technique is usually recommended for receiving good signal. However, using gap technique requires spatial interpolation when processing the imaging data, and thus, spatial uncertainty may have a chance to increase. Therefore, we decided not to use gap technique in the current study.

5. Conclusions

We investigated the changes in intrinsic BOLD oscillations associated with the severity of NP after incomplete SCI using graph theory and fALFF analyses. The results revealed that an increase in NP severity following SCI is related to less efficient connections, such as those represented by reduced degrees and increased average path length, and less spontaneous activity, such as those represented by reduced fALFF values, in the MidFG and PCC. Therefore, for possible therapeutic application, the findings of the current study could help determine the candidate brain regions to control severe and refractory NP using noninvasive brain stimulation.

Supplementary Materials: The following supporting information can be downloaded at: https://www.mdpi.com/article/10.3390/bioengineering10070860/s1, Table S1: Demographic and clinical characteristics of the incomplete spinal cord injury patients.

Author Contributions: Conceptualization, J.W.P.; methodology, J.W.P.; software, J.W.P.; validation, E.P. and H.J.L.; formal analysis, E.P. and J.W.P.; investigation, E.P. and Y.-S.M.; resources, E.K. and Y.C.; data curation, E.P.; writing—original draft, E.P. and J.W.P.; writing—review and editing, Y.C.; visualization, J.W.P.; supervision, T.-D.J.; project administration, T.-D.J. and Y.C.; funding acquisition, Y.C. All authors have read and agreed to the published version of the manuscript.

Funding: This study was supported by a grant of the Korea Institute of Radiological and Medical Sciences (KIRAMS), funded by Ministry of Science and ICT (MSIT), Republic of Korea. (No. 50539-2023) and Basic Science Research Program through the National Research Foundation of Korea (NRF) funded by the Ministry of Education (NRF-2020R1I1A3070636).

Institutional Review Board Statement: This study was approved by the Institutional Review Board of Kyungpook National University Chilgok Hospital (No. 2018-12-004).

Informed Consent Statement: Patient consent was waived because we submitted a consent exemption request form to the IRB as part of the retrospective study approval process.

Data Availability Statement: Not applicable.

Conflicts of Interest: The authors declare no conflict of interest.

References

1. Siddall, P.J.; Taylor, D.A.; McClelland, J.M.; Rutkowski, S.B.; Cousins, M.J. Pain Report and the Relationship of Pain to Physical Factors in the First 6 Months Following Spinal Cord Injury. *Pain* **1999**, *81*, 187–197. [CrossRef] [PubMed]
2. Finnerup, N.B.; Haroutounian, S.; Kamerman, P.; Baron, R.; Bennett, D.L.H.; Bouhassira, D.; Cruccu, G.; Freeman, R.; Hansson, P.; Nurmikko, T.; et al. Neuropathic Pain: An Updated Grading System for Research and Clinical Practice. *Pain* **2016**, *157*, 1599–1606. [CrossRef] [PubMed]
3. Jensen, M.P.; Chodroff, M.J.; Dworkin, R.H. The Impact of Neuropathic Pain on Health-Related Quality of Life: Review and Implications. *Neurology* **2007**, *68*, 1178–1182. [CrossRef] [PubMed]
4. Turner, J.A.; Cardenas, D.D.; Warms, C.A.; McClellan, C.B. Chronic Pain Associated with Spinal Cord Injuries: A Community Survey. *Arch. Phys. Med. Rehabil.* **2001**, *82*, 501–508. [CrossRef]
5. Siddall, P.J.; McClelland, J.M.; Rutkowski, S.B.; Cousins, M.J. A Longitudinal Study of the Prevalence and Characteristics of Pain in the First 5 Years Following Spinal Cord Injury. *Pain* **2003**, *103*, 249–257. [CrossRef]
6. Gustin, S.M.; Wrigley, P.J.; Siddall, P.J.; Henderson, L.A. Brain Anatomy Changes Associated with Persistent Neuropathic Pain Following Spinal Cord Injury. *Cereb. Cortex* **2010**, *20*, 1409–1419. [CrossRef]
7. Jutzeler, C.R.; Freund, P.; Huber, E.; Curt, A.; Kramer, J.L.K. Neuropathic Pain and Functional Reorganization in the Primary Sensorimotor Cortex After Spinal Cord Injury. *J. Pain* **2015**, *16*, 1256–1267. [CrossRef]

8. Seminowicz, D.A.; Jiang, L.; Ji, Y.; Xu, S.; Gullapalli, R.P.; Masri, R. Thalamocortical Asynchrony in Conditions of Spinal Cord Injury Pain in Rats. *J. Neurosci.* **2012**, *32*, 15843–15848. [CrossRef]
9. Park, E.; Cha, H.; Kim, E.; Min, Y.-S.; Kim, A.R.; Lee, H.J.; Jung, T.-D.; Chang, Y. Alterations in Power Spectral Density in Motor- and Pain-Related Networks on Neuropathic Pain after Spinal Cord Injury. *NeuroImage: Clin.* **2020**, *28*, 102342. [CrossRef]
10. Fransson, P. Spontaneous Low-Frequency BOLD Signal Fluctuations: An FMRI Investigation of the Resting-State Default Mode of Brain Function Hypothesis. *Hum. Brain Mapp.* **2005**, *26*, 15–29. [CrossRef]
11. Bullmore, E.; Sporns, O. Complex Brain Networks: Graph Theoretical Analysis of Structural and Functional Systems. *Nat Rev Neurosci* **2009**, *10*, 186–198. [CrossRef]
12. Van Den Heuvel, M.P.; Hulshoff Pol, H.E. Exploring the Brain Network: A Review on Resting-State FMRI Functional Connectivity. *Eur. Neuropsychopharmacol.* **2010**, *20*, 519–534. [CrossRef]
13. Lenoir, D.; Cagnie, B.; Verhelst, H.; De Pauw, R. Graph Measure Based Connectivity in Chronic Pain Patients: A Systematic Review. *Pain Physician* **2021**, *24*, E1037–E1058.
14. Duff, E.P.; Johnston, L.A.; Xiong, J.; Fox, P.T.; Mareels, I.; Egan, G.F. The Power of Spectral Density Analysis for Mapping Endogenous BOLD Signal Fluctuations. *Hum. Brain Mapp.* **2008**, *29*, 778–790. [CrossRef]
15. Zuo, X.-N.; Di Martino, A.; Kelly, C.; Shehzad, Z.E.; Gee, D.G.; Klein, D.F.; Castellanos, F.X.; Biswal, B.B.; Milham, M.P. The Oscillating Brain: Complex and Reliable. *NeuroImage* **2010**, *49*, 1432–1445. [CrossRef]
16. Aiello, M.; Salvatore, E.; Cachia, A.; Pappatà, S.; Cavaliere, C.; Prinster, A.; Nicolai, E.; Salvatore, M.; Baron, J.-C.; Quarantelli, M. Relationship between Simultaneously Acquired Resting-State Regional Cerebral Glucose Metabolism and Functional MRI: A PET/MR Hybrid Scanner Study. *NeuroImage* **2015**, *113*, 111–121. [CrossRef]
17. Kirshblum, S.C.; Waring, W.; Biering-Sorensen, F.; Burns, S.P.; Johansen, M.; Schmidt-Read, M.; Donovan, W.; Graves, D.E.; Jha, A.; Jones, L.; et al. Reference for the 2011 Revision of the International Standards for Neurological Classification of Spinal Cord Injury. *J. Spinal Cord Med.* **2011**, *34*, 547–554. [CrossRef]
18. Maynard, F.M.; Bracken, M.B.; Creasey, G.; Jr, J.F.D.; Donovan, W.H.; Ducker, T.B.; Garber, S.L.; Marino, R.J.; Stover, S.L.; Tator, C.H.; et al. International Standards for Neurological and Functional Classification of Spinal Cord Injury. *Spinal Cord* **1997**, *35*, 266–274. [CrossRef]
19. Siddall, P.J.; Taylor, D.A.; Cousins, M.J. Classification of Pain Following Spinal Cord Injury. *Spinal Cord* **1997**, *35*, 69–75. [CrossRef]
20. Bryce, T.N.; Biering-Sørensen, F.; Finnerup, N.B.; Cardenas, D.D.; Defrin, R.; Lundeberg, T.; Norrbrink, C.; Richards, J.S.; Siddall, P.; Stripling, T.; et al. International Spinal Cord Injury Pain Classification: Part I. Background and Description. *Spinal Cord* **2012**, *50*, 413–417. [CrossRef]
21. Celik, E.C.; Erhan, B.; Lakse, E. The Clinical Characteristics of Neuropathic Pain in Patients with Spinal Cord Injury. *Spinal Cord* **2012**, *50*, 585–589. [CrossRef] [PubMed]
22. Boonstra, A.M.; Schiphorst Preuper, H.R.; Balk, G.A.; Stewart, R.E. Cut-off Points for Mild, Moderate, and Severe Pain on the Visual Analogue Scale for Pain in Patients with Chronic Musculoskeletal Pain. *Pain* **2014**, *155*, 2545–2550. [CrossRef] [PubMed]
23. Hahn, H.; Yum, T.; Shin, Y.; Kim, K.; Yoon, D.; Chung, K. A Standardization Study of Beck Depression Inventory in Korea. *J. Korean Neuropsychiatr. Assoc.* **1982**, *25*, 487–502.
24. Whitfield-Gabrieli, S.; Nieto-Castanon, A. Conn: A Functional Connectivity Toolbox for Correlated and Anticorrelated Brain Networks. *Brain Connect* **2012**, *2*, 125–141. [CrossRef] [PubMed]
25. Tzourio-Mazoyer, N.; Landeau, B.; Papathanassiou, D.; Crivello, F.; Etard, O.; Delcroix, N.; Mazoyer, B.; Joliot, M. Automated Anatomical Labeling of Activations in SPM Using a Macroscopic Anatomical Parcellation of the MNI MRI Single-Subject Brain. *NeuroImage* **2002**, *15*, 273–289. [CrossRef]
26. Ferreira-Santos, F. Complex Network Analysis of Brain Connectivity: An Introduction (LabReport No. 5). Porto: Laboratory of Neuropsychophysiology (University of Porto). 2012. Available online: http://www.fpce.up.pt/labpsi/data_files/09labreports/LabReport_5.pdf (accessed on 15 March 2020).
27. Apkarian, A.V.; Bushnell, M.C.; Treede, R.-D.; Zubieta, J.-K. Human Brain Mechanisms of Pain Perception and Regulation in Health and Disease. *Eur. J. Pain* **2005**, *9*, 463. [CrossRef]
28. Garcia-Larrea, L.; Peyron, R. Pain Matrices and Neuropathic Pain Matrices: A Review. *Pain* **2013**, *154*, S29–S43. [CrossRef]
29. Kim, J.-Y.; Kim, S.-H.; Seo, J.; Kim, S.-H.; Han, S.W.; Nam, E.J.; Kim, S.-K.; Lee, H.J.; Lee, S.-J.; Kim, Y.-T.; et al. Increased Power Spectral Density in Resting-State Pain-Related Brain Networks in Fibromyalgia. *Pain* **2013**, *154*, 1792–1797. [CrossRef]
30. Barad, M.J.; Ueno, T.; Younger, J.; Chatterjee, N.; Mackey, S. Complex Regional Pain Syndrome Is Associated With Structural Abnormalities in Pain-Related Regions of the Human Brain. *J. Pain* **2014**, *15*, 197–203. [CrossRef]
31. Mackey, S.C.; Maeda, F. Functional Imaging and the Neural Systems of Chronic Pain. *Neurosurg. Clin. N. Am.* **2004**, *15*, 269–288. [CrossRef]
32. Bushnell, M.C.; Čeko, M.; Low, L.A. Cognitive and Emotional Control of Pain and Its Disruption in Chronic Pain. *Nat. Rev. Neurosci.* **2013**, *14*, 502–511. [CrossRef] [PubMed]
33. Fox, M.D.; Corbetta, M.; Snyder, A.Z.; Vincent, J.L.; Raichle, M.E. Spontaneous Neuronal Activity Distinguishes Human Dorsal and Ventral Attention Systems. *Proc. Natl. Acad. Sci. USA* **2006**, *103*, 10046–10051. [CrossRef] [PubMed]
34. Tekin, S.; Cummings, J.L. Frontal–Subcortical Neuronal Circuits and Clinical Neuropsychiatry. *J. Psychosom. Res.* **2002**, *53*, 647–654. [CrossRef] [PubMed]

35. Knudsen, L.; Petersen, G.L.; Nørskov, K.N.; Vase, L.; Finnerup, N.; Jensen, T.S.; Svensson, P. Review of Neuroimaging Studies Related to Pain Modulation. *Scand. J. Pain* **2011**, *2*, 108–120. [CrossRef]
36. Boyette-Davis, J.A.; Thompson, C.D.; Fuchs, P.N. Alterations in Attentional Mechanisms in Response to Acute Inflammatory Pain and Morphine Administration. *Neuroscience* **2008**, *151*, 558–563. [CrossRef]
37. Ford, G.K.; Moriarty, O.; McGuire, B.E.; Finn, D.P. Investigating the Effects of Distracting Stimuli on Nociceptive Behaviour and Associated Alterations in Brain Monoamines in Rats. *Eur. J. Pain* **2008**, *12*, 970–979. [CrossRef]
38. Dunckley, P.; Aziz, Q.; Wise, R.G.; Brooks, J.; Tracey, I.; Chang, L. Attentional Modulation of Visceral and Somatic Pain. *Neurogastroenterol Motil* **2007**, *19*, 569–577. [CrossRef]
39. Seminowicz, D.A.; Mikulis, D.J.; Davis, K.D. Cognitive Modulation of Pain-Related Brain Responses Depends on Behavioral Strategy. *Pain* **2004**, *112*, 48–58. [CrossRef]
40. Wiech, K.; Kalisch, R.; Weiskopf, N.; Pleger, B.; Stephan, K.E.; Dolan, R.J. Anterolateral Prefrontal Cortex Mediates the Analgesic Effect of Expected and Perceived Control over Pain. *J. Neurosci.* **2006**, *26*, 11501–11509. [CrossRef]
41. Tomasi, D.; Wang, G.-J.; Volkow, N.D. Energetic Cost of Brain Functional Connectivity. *Proc. Natl. Acad. Sci. USA* **2013**, *110*, 13642–13647. [CrossRef]
42. Bullmore, E.; Sporns, O. The Economy of Brain Network Organization. *Nat. Rev. Neurosci.* **2012**, *13*, 336–349. [CrossRef]
43. Freund, W.; Wunderlich, A.P.; Stuber, G.; Mayer, F.; Steffen, P.; Mentzel, M.; Weber, F.; Schmitz, B. Different Activation of Opercular and Posterior Cingulate Cortex (PCC) in Patients With Complex Regional Pain Syndrome (CRPS I) Compared With Healthy Controls During Perception of Electrically Induced Pain: A Functional MRI Study. *Clin. J. Pain* **2010**, *26*, 339–347. [CrossRef]
44. Ichesco, E.; Peltier, S.J.; Mawla, I.; Harper, D.E.; Pauer, L.; Harte, S.E.; Clauw, D.J.; Harris, R.E. Prediction of Differential Pharmacologic Response in Chronic Pain Using Functional Neuroimaging Biomarkers and a Support Vector Machine Algorithm: An Exploratory Study. *Arthritis Rheumatol.* **2021**, *73*, 2127–2137. [CrossRef]
45. Lee, J.; Protsenko, E.; Lazaridou, A.; Franceschelli, O.; Ellingsen, D.-M.; Mawla, I.; Isenburg, K.; Berry, M.P.; Galenkamp, L.; Loggia, M.L.; et al. Encoding of Self-Referential Pain Catastrophizing in the Posterior Cingulate Cortex in Fibromyalgia. *Arthritis Rheumatol.* **2018**, *70*, 1308–1318. [CrossRef]
46. Scandola, M.; Dodoni, L.; Lazzeri, G.; Arcangeli, C.A.; Avesani, R.; Moro, V.; Ionta, S. Neurocognitive benefits of physiotherapy for spinal cord injury. *J. Neurotrauma* **2019**, *36*, 2028–2035. [CrossRef]
47. Thomas, M. Age-related differences of neural connectivity during mental rotation. *Int. J. Psychophysiol.* **2016**, *101*, 33–42. [CrossRef]
48. Kowalski, J.L.; Morse, L.R.; Troy, K.; Nguyen, N.; Battaglino, R.A.; Falci, S.P.; Linnman, C. Resting state functional connectivity differentiation of neuropathic and nociceptive pain in individuals with chronic spinal cord injury. *NeuroImage Clin.* **2023**, *38*, 103414. [CrossRef]
49. Vallesi, V.; Richter, J.K.; Hunkeler, N.; Abramovic, M.; Hashagen, C.; Christiaanse, E.; Shetty, G.; Verma, R.K.; Berger, M.; Frotzler, A. Functional connectivity and amplitude of low-frequency fluctuations changes in people with complete subacute and chronic spinal cord injury. *Sci. Rep.* **2022**, *12*, 20874. [CrossRef]

bioengineering

Article

Near-Infrared Blood Vessel Image Segmentation Using Background Subtraction and Improved Mathematical Morphology

Ling Li [1], Haoting Liu [1,*], Qing Li [1,*], Zhen Tian [1], Yajie Li [1], Wenjia Geng [2] and Song Wang [3]

[1] Beijing Engineerin Research Center of Industrial Spectrum Imaging, School of Automation and Electrical Engineering, University of Science and Technology Beijing, Beijing 100083, China; s20200534@xs.ustb.edu.cn (Y.L.)
[2] Department of Traditional Chinese Medicine, Peking University People's Hospital, Beijing 100044, China
[3] Department of Nephrology, Peking University Third Hospital, Beijing 100191, China; songwang30@163.com
* Correspondence: liuhaoting@ustb.edu.cn (H.L.); liqing@ies.ustb.edu.cn (Q.L.)

Abstract: The precise display of blood vessel information for doctors is crucial. This is not only true for facilitating intravenous injections, but also for the diagnosis and analysis of diseases. Currently, infrared cameras can be used to capture images of superficial blood vessels. However, their imaging quality always has the problems of noises, breaks, and uneven vascular information. In order to overcome these problems, this paper proposes an image segmentation algorithm based on the background subtraction and improved mathematical morphology. The algorithm regards the image as a superposition of blood vessels into the background, removes the noise by calculating the size of connected domains, achieves uniform blood vessel width, and smooths edges that reflect the actual blood vessel state. The algorithm is evaluated subjectively and objectively in this paper to provide a basis for vascular image quality assessment. Extensive experimental results demonstrate that the proposed method can effectively extract accurate and clear vascular information.

Keywords: near-infrared image; blood vessel segmentation; noise reduction; image quality; mathematical morphology

Citation: Li, L.; Liu, H.; Li, Q.; Tian, Z.; Li, Y.; Geng, W.; Wang, S. Near-Infrared Blood Vessel Image Segmentation Using Background Subtraction and Improved Mathematical Morphology. *Bioengineering* **2023**, *10*, 726. https://doi.org/10.3390/bioengineering10060726

Academic Editors: Andrea Cataldo and Cuneyt M. Alper

Received: 3 May 2023
Revised: 31 May 2023
Accepted: 6 June 2023
Published: 15 June 2023

1. Introduction

Subcutaneous venipuncture is one of the common medical treatments encountered in daily life. As a category, it includes intravenous infusions and injections, and venous blood collection and transfusions [1]. In hospitals, most current methods of venipuncture require the use of a tourniquet fixed to a patient's limb and then request the patient to clench a fist in order to make the blood vessel more prominent before puncturing it. Clearly, this method relies on the experience and proficiency of doctors or nurses, and there are many drawbacks to this method: one is that it is affected by skin pigmentation, vessel depth and thickness, and fat thickness [2]. Second, some patients fail to make a hand-clenched fist in some cases, resulting in the blood vessels not being prominently displayed. These conditions may lead to venipuncture failure, which not only increases the patient's anxiety and pain but also brings psychological pressure to health care workers; sometimes, this also results in a risk of doctor-patient conflict. Therefore, the rapid and accurate acquisition of blood vessel location becomes a very important and urgent problem. The solution of this problem will help the evaluation of the quality of vessels, providing effective information to support nurses in their clinical care.

Many methods are currently proposed for blood vessel segmentation, and they are classified into two categories: the traditional methods and the machine learning methods. The traditional methods include the Gaussian function-based vessel segmentation, multiscale matched filtering, and threshold processing-based segmentation [3–5], etc. These methods usually need some preprocessing, such as the contrast limited adaptive histogram equalization (CLAHE) or Gabor filtering [6]. There is also a blood vessel feature extraction method which uses fuzzy theory to segment the veins [7]. The vessel tracking method, on the other hand, requires the initialization of a point on a vessel and then tracking of the localized vessel's centerline by the monitoring of the image's local information [8]. A vessel segmentation method based on the spatial feature point set of a rotating coordinate system clears off the error feature points, which has a good effect on peripheral blood vessels [9]. Some researchers have also proposed designing an active contour strategy coupled with Kalman filtering, which can be used to segment pulmonary vascular images [10].

The machine learning algorithm has two steps: the training step and image segmentation step. In the training step, the researchers use the segmented standard vascular images to train the classifiers according to certain learning methods. In the recognition segmentation phase, the unknown blood vessel images with certain features are fed into the trained classifier for segmentation. Support vector machine, convolutional neural networks, deep neural network [11–13], etc., are the representative methods. U-net is a typical deep learning network that can rely on data enhancement from a very small number of training images [14]. It has been used to segment human tissue images such as skeleton, muscle, and knee joints [15–17]. ResU-net focuses on solving problems such as tiny vessels at the end that are difficult to segment, and blurred edges caused by illumination [18]. R2U-net can enhance pixel correlation and increase the network depth effectively, having a good segmentation effect for lung CT images and fundus vascular images [19]. The LadderNet is essentially equivalent to the concatenation of two U-net networks [20]. The SMU-net proposes a saliency-guided morphology-aware U-net (SMU-net) for lesion segmentation, which tries to explore and exploit background-salient representations for assisting foreground segmentation [21].

Although many segmentation methods have been proposed, there are still many problems in their applications. The machine learning method has a high segmentation accuracy; however, its performance depends on the use of large amounts of data in the early stage to train the model. Due to the lack of unified calibration and tools, manual calibration by experts is still difficult. As a result, there is not enough standard dataset available for training currently. Therefore, it is not feasible to use the machine learning method to segment the near-infrared vascular images of forearm. Traditional segmentation methods do not need training, and their computation speeds are fast. However, traditional segmentation methods also have many shortcomings: first, because the edge of near-infrared vascular image is blurry, it is still hard to define the edge after enhancement, which requires higher criteria for the selection of threshold value in the traditional threshold segmentation. Second, the traditional segmentation is not effective at extracting some deep blood vessels and small blood vessels at the end. To address these problems of traditional segmentation algorithms, this paper proposes a vascular segmentation method based on background subtraction and improved mathematical morphology.

As shown in Figure 1, an improved vascular segmentation method using background subtraction and an improved mathematical morphology is proposed. Its computation can be roughly divided into the following steps: first, preprocessing, which improves the image brightness uniformity and prepares for image enhancement. Second, image background extraction, which uses the minimum filtering and median filtering to obtain image background information. Third, to obtain vascular information, the preprocessed image is used to subtract the extracted background information to obtain preliminary vascular information. Fourth, the method based on the connected domain is used to remove the isolated noises. Fifth, mathematical morphological processing is developed. This repairs some breakpoints in the preliminary results according to the state of blood vessel growth and the uniform blood vessel width is developed. The primary contributions of this paper include: (1) a vascular image segmentation method based on background subtraction is proposed, which ensures the accuracy of blood vessel assessments and simplifies the algorithm. (2) An improved mathematical morphology algorithm with a

T-type structural element is proposed to extract information from blood vessels and smooth the edges of blood vessels.

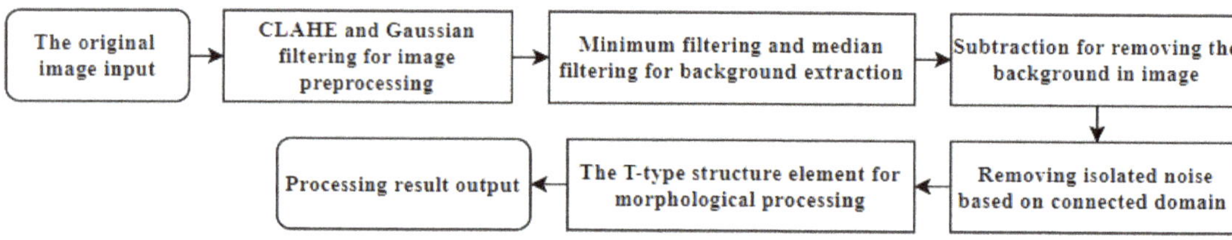

Figure 1. Proposed algorithm flow chart of near-infrared blood vessel image segmentation.

2. Primary Computational Methods

2.1. Dataset Preprocessing and Enhancement

In previous work [22], we built a near-infrared image acquisition system and collected a forearm near-infrared vascular image dataset containing 360 images. Figure 2a shows an image sample taken from this dataset. It can be seen that the vascular image of the forearm is characterized by low contrast and uneven brightness distribution in imaging area, and that the finer blood vessels cannot be clearly observed under the light source irradiation. This is because the background of the human vascular region is complex, the difference between muscle and vascular region is not clear, and the imaging effect is easily affected by factors such as shooting angle, light source distribution, and other factors. Figure 2b presents its local vascular area of the image. It is not difficult to find that the edges of blood vessels in the image are fuzzy. However, it is difficult to observe the thinner blood vessels, and that there are a lot of noises. Figure 2c is the gray histogram of the local image Figure 2b. The gray distribution in the figure is concentrated, which indicates that the image is brighter overall and the contrast between foreground and background of blood vessels is low. In order to improve image contrast and enhance its details, this paper applies CLAHE to improve image contrast [23].

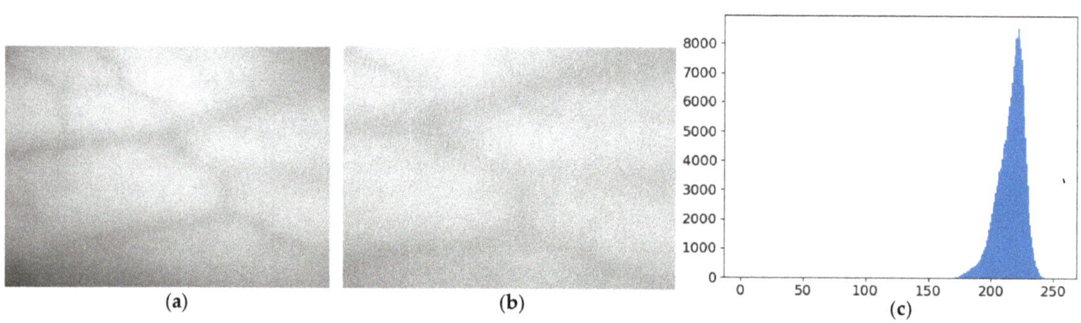

Figure 2. Image sample and its local vascular areas histogram. (**a**) An image sample from our dataset. (**b**) The local vascular areas image of (**a**). (**c**) The histogram of (**b**).

CLAHE is improved based on adaptive histogram equalization (AHE), which makes up for the noises brought by the AHE algorithm while enhancing image contrast [24]. The basic idea of CLAHE is to divide the image into a number of non-overlapping sub-regions and calculate the gray gradient value according to the statistics of gray level in each sub-region. When the local contrast of image is reduced or increased, CLAHE will automatically adjust, avoiding the problem of local excessive light or darkness. CLAHE's specific steps are as follows.

First, CLAHE segments the original low-contrast image into $M \times N$ non-overlapping sub-regions and then calculates their local histograms. Second, CLAHE constrains the grayscale height of each local histogram and calculates the average number of pixels assigned to, the calculation method is shown in Equation (1):

$$A_v = \frac{u_X u_Y}{N_{XY}} \tag{1}$$

where N_{XY} denotes the number of gray levels in the corresponding subregion, and u_X, and u_Y are the number of pixels in the X and Y directions of the subregion, respectively.

Third, CLAHE calculates the extreme values for which cropping is required, and calculate according to Equation (2):

$$L_C = N_{CLip} A_v \tag{2}$$

where N_{CLip} is the factor to limit the amplitude.

Fourth, let S mean the total number of pixels cropped, then the number of "cropped" pixels can be calculated as Equation (3):

$$a_v = \frac{S}{N_{XY}} \tag{3}$$

Finally, the step size for assigning the remaining pixels is calculated as Equation (4):

$$L = \frac{L_G}{S} \tag{4}$$

where L denotes the step size to be assigned to the remaining pixels and L_G is the length range of grayscale.

Due to the imperfections in the formation, transmission, and recording of digital image signals, these factors can cause the contamination of digital image signals with a variety of noises. This can disrupt the observation of useful information in images, degrade image quality severely, and affect the enhancement and edge segmentation of blood vessel images. Therefore, this is an indispensable step to suppressing the noises of blood vessel images while trying to keep that the image quality detail features are not affected. In this paper, Gaussian filtering is chosen to suppress the noise just because it can effectively remove spot noises while preserving the edge information of an image [25]. Gaussian filtering is a linear filter that can smooth an image and make it more homogeneous. It operates by using a Gaussian function to perform a convolution operation. The two-dimensional Gaussian function is given by:

$$G(x, y) = \frac{1}{2\pi\sigma^2} e^{-\frac{x^2+y^2}{2\sigma^2}} \tag{5}$$

where x and y denote the positions of the convolution kernel, respectively; additionally, σ is the standard deviation.

There are two reasons for using Gaussian filtering used in this paper. On the one hand, the Gaussian function can simulate the distribution of many phenomena in nature, such as sound, light, temperature [26], etc. The noise generated by the imaging sensor under many conditions belongs to Gaussian noise. On the other hand, the local image with relatively uniform brightness and non-vascular region in the near-infrared vascular image is captured and its gray histogram is calculated. As Figure 3b, from histogram, it can be found that its probability density function is similar to the density function of Gaussian noise distribution. Therefore, it can be concluded that the primary noise type in the near-infrared image is Gaussian noise. In this paper, the convolution kernel of 3×3 with $\sigma = 1.5$ is used to convolve the image, and the noises in the image will be removed smoothly after Gaussian filtering.

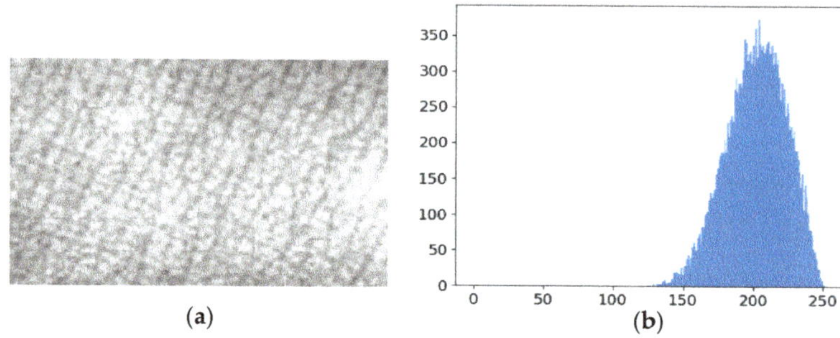

Figure 3. Local image of near-infrared blood vessel and its gray histogram. (**a**) Local image of near-infrared blood vessel. (**b**) The histogram of (**a**).

2.2. Vascular Information Extraction and Image Background Subtraction

It is not difficult to observe that the grayscale value of the vascular region is lower than that of the non-vascular region, showing that the blood vessel is darker compared to the non-vascular region. According to this feature, the near-infrared vascular image can be regarded as a combination of background and vascular information. The image can be expressed as follows:

$$I(x,y) = R(x,y) + G(x,y) \tag{6}$$

where $I(x,y)$ denotes the near-infrared vessel image information, $R(x,y)$ is the image information of vessel region, and $G(x,y)$ indicates the background image information.

According to Equation (6), it is not difficult to derive that:

$$R(x,y) = I(x,y) - G(x,y) \tag{7}$$

The function $G(x,y)$ can be obtained as a background by applying minimum filtering and median filtering methods. Minimum filtering is a relatively conservative method of image processing. The processing principle of it is that the surrounding pixels and the center pixel value are first sorted. Then, the center pixel value is compared with the minimum pixel values. If the center pixel is smaller than the minimum value, the center pixel is replaced by the minimum value. In this paper, small circular regions with different gray values appear in the image after minimum filtering with 3×3 template. In the image, the difference of gray values on both sides of the inner and outer edges of blood vessels increased, and the number of different gray values decreased, which is helpful for clarifying the edge of blood vessels.

After the processing, the image background can be obtained by using the median filtering algorithm, which is a nonlinear signal processing technique based on the statistical theory of sorting, which can effectively suppress noises [27]. The basic principle of median filtering is to replace the value of a point in a digital image or digital sequence with the median of the values of points in a neighborhood of the center point, so that the surrounding pixel values are closed to the true value. Clearly, median filtering can blur the blood vessel image effectively. In this paper, the median filtering uses a two-dimensional sliding template of some structure to sort the pixels in a plate according to the size of pixel values, generating a monotonically increasing (or decreasing) tendency for two-dimensional data sequences. The expressions are:

$$g(x,y) = \underset{(i,j)\in Sxy}{Med}[f(i,j)] \tag{8}$$

where S_{xy} is the set of neighborhoods centered on (x, y) and Med is a function to take the middle value of the set.

2.3. Image Denoising and Morphological Processing

After subtracting the background operation, there are still many noise points and breaks. Additionally, if these isolated noise points are not removed, they will affect the subsequent link to repair the edges of blood vessels. Therefore, a connectivity domain-based element screening method is used in this algorithm to eliminate the isolated noise points [28]. The connected-domain-based element screening method refers to the calculation of the size of each connected domain in an image to determine whether it is a noise point or not. Since most blood vessels have elongated structures, the isolated noise points are generally round and irregular in shape, and their areas are much smaller than those of the connected blood vessel areas, a threshold can be set. If the area of the connected domain is smaller than a specific threshold, the connected domain can be considered as a noise region [29]. Its mathematical expression is:

$$S_n = \begin{cases} S_n & S \geqslant T \\ 0 & S < T \end{cases} \quad (9)$$

where S_n represents the nth connected domain, S is the area of S_n, and T is the set threshold. In this algorithm, after a series of experiment tests, the T value can be set between 1500 and 1800 for our dataset, a value which can eliminate most of the noise points in blood vessel images.

After removing the isolated noise points, the edges of blood vessels may still have problems such as unevenness or even fractures. These problems do not correspond to the direction of normal blood vessels. To solve these problems, this paper develops an improved mathematical morphology algorithm. Mathematical morphology is a mathematical tool for image analysis based on morphological structural elements [30]. Its core idea is to use structures with certain morphologies to extract features such as the shape, structure, and spatial relationships of an image. When the template is continuously moved in an image, information about the interrelationships between various parts of the image can be collected to understand the image structure features. A novel mathematical morphology method of a T-type structural element is proposed in this paper. Using this method can allow researchers to simplify the image data, keep its basic shape features, and eliminate the skeleton of blood vessel.

The operations commonly used in mathematical morphology include dilation and erosion. In binary and grayscale images, each operation has its own unique characteristics. These basic operations can be further combined and derived to obtain various practical mathematical morphology algorithms. In dealing with the vascular edge problem, the dilation and erosion operations of morphological algorithm are used first. The dilation operation can expand the range of vessel edges and make them smoother, while the erosion operation can reduce the range of vessel edges and make them clearer. By properly combining these two operations, it can become more accurate and produce finer blood vessel images. In a binary image, the erosion operation is defined as:

$$(f \ominus e)(x,y) = min\{\begin{matrix} f(x+a, y+b) - e(a,b) \mid (x+a), \\ (y+b) \in D_f \; and \; (a,b) \in D_e \end{matrix}\} \quad (10)$$

where $f \ominus e$ indicates that e is used to erode f, f represents the original image, and D_f and D_e are the definition domains of f and e, respectively; $f(a,b)$ is outside the definition domain of f, assuming it is $+\infty$.

In the binary image, the dilation operation is defined as:

$$(f \oplus e)(x,y) = max\{\begin{matrix} f(x-a, y-b) + e(a,b) \mid (x-a), \\ (y-b) \in D_f \; and \; (a,b) \in D_e \end{matrix}\} \quad (11)$$

where $f \oplus e$ indicates that e is used to dilate, f represents the original image, and D_f and D_e are the definition domains of f and e, respectively; additionally, $f(a,b)$ is outside the definition domain of f, assuming it is $-\infty$.

The analysis of mathematical morphology theory shows that the selection of structural elements is crucial in the process of erosion and dilation, which directly determines the shape of processed image. The traditional structural elements are rectangles, circles, and crosses. However, in some cases, these structural elements do not fit well with the characteristics of an image and also do not perform well in the edge direction. Therefore, new structural elements need to be proposed to overcome this shortcoming. This paper proposes a novel T-type structural element. The advantage of this structural element is that it has high directionality, which can better process the line segmentations in different directions and smooth the protruding and depressed parts of image edges. As shown in Figure 4, the blue structure at the top is used to simulate a small segmentation of a blood vessel image, and the green structure at the bottom is the blood vessel structure corroded by a T-type structural element after background reduction and denoising. The vascular binary graph obtained initially has uneven edges, protrusions, or dents. The template of T-type structural element traverses all pixel points in image and retains the part that can completely overlap with this template, while the rest is corroded. As can be seen from the diagram, the T-type structural element effectively smooths some protruding parts, making the lines of the edge pixels smoother.

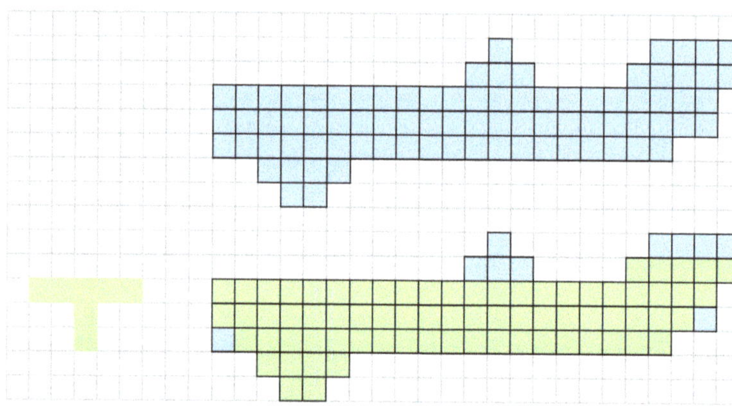

Figure 4. Proposed T-type structural element of mathematical morphology method.

The algorithm in this paper is based on the T-type structural element, which is rotated by 90° in sequence and processed several times. In the algorithm, the following four structural elements are used to erode images.

$$t_1 = \begin{bmatrix} 0 & 0 & 0 & 0 & 0 & 0 \\ 0 & 0 & 0 & 0 & 0 & 0 \\ 1 & 1 & 1 & 1 & 1 & 1 \\ 1 & 1 & 1 & 1 & 1 & 1 \\ 0 & 0 & 1 & 1 & 0 & 0 \\ 0 & 0 & 1 & 1 & 0 & 0 \end{bmatrix} \tag{12}$$

$$t_2 = \begin{bmatrix} 0 & 0 & 1 & 1 & 0 & 0 \\ 0 & 0 & 1 & 1 & 0 & 0 \\ 0 & 0 & 1 & 1 & 1 & 1 \\ 0 & 0 & 1 & 1 & 1 & 1 \\ 0 & 0 & 1 & 1 & 0 & 0 \\ 0 & 0 & 1 & 1 & 0 & 0 \end{bmatrix} \tag{13}$$

$$t_3 = \begin{bmatrix} 0 & 0 & 1 & 1 & 0 & 0 \\ 0 & 0 & 1 & 1 & 0 & 0 \\ 1 & 1 & 1 & 1 & 1 & 1 \\ 1 & 1 & 1 & 1 & 1 & 1 \\ 0 & 0 & 0 & 0 & 0 & 0 \\ 0 & 0 & 0 & 0 & 0 & 0 \end{bmatrix} \quad (14)$$

$$t_4 = \begin{bmatrix} 0 & 0 & 1 & 1 & 0 & 0 \\ 0 & 0 & 1 & 1 & 0 & 0 \\ 1 & 1 & 1 & 1 & 0 & 0 \\ 1 & 1 & 1 & 1 & 0 & 0 \\ 0 & 0 & 1 & 1 & 0 & 0 \\ 0 & 0 & 1 & 1 & 0 & 0 \end{bmatrix} \quad (15)$$

According to the definition of dilation and erosion, this paper defines f_n as the result of the nth processing. First, the t_1 structural element is used for erosion to remove small noise and fine lines on the edges; then, the circular template is also used to fill in some dotted lines and holes. The result of the first processing is shown below.

$$f_1 = (f \ominus t_1) \oplus b \quad (16)$$

where f is the original image and b represents the circular structural element.

For the second step, t_2 is used for corrosion and b is still used for dilation. The result of the second operation can be expressed as follows.

$$f_2 = (((f \ominus t_1) \oplus b) \ominus t_2) \oplus b) \quad (17)$$

Next, t_3 and t_4 are used for the same operation, and finally the processing result can be estimated by (13).

$$f_4 = (f_3 \ominus t_4) \oplus b \quad (18)$$

Four calculations are completed in this step and the image is fully processed to obtain the desired shape. By using the T-type structural element and multiple rotations, this method can better deal with the problems of lines and voids in the binarized image and achieve a better denoising effect. This process can effectively eliminate horizontal and vertical line segments and noises.

3. Experiment Results and Evaluations

This section validates the noise reduction algorithm, segmentation algorithm, and morphological processing algorithm. The comparison tests randomly select forearm infrared blood vessel images from our dataset and crop each image into a region of interest (ROI) with a size of 548 × 335. Python programming is used for simulation experiments on PC Intel(R) Core(TM) i5-1035G1 CPU @ 1.00 GH, 8 GB RAM, NVIDIA GeForce MX 250).

3.1. Evaluation of Image Enhancement Algorithm

To evaluate the enhancement effect further, five algorithms, i.e., CLAHE, histogram equalization (HE) algorithm [31], AHE algorithm [31], single-scale Retinex (SSR) algorithm [32], brightness compensation method for different brightness regions in the image (BCMDBR), and homomorphic filtering [33] are compared in this section. The results are shown in Figure 5. It can be observed from the results that, although the HE algorithm can improve image contrast, it can produce situations of local over dark or over bright images, and the effects of detail retention are not good. Although the above problems do not occur in AHE, noise is reflected and contrast enhancement is not as good as CLAHE. The SSR algorithm and BCMDBR appear to be over dark or over bright, and no vascular information can be observed. The homomorphic filtering algorithm is not effective for vascular edge processing, and its contrast is still not clear enough. Figure 5f shows the experimental results of the CLAHE. The forearm blood vessel image in ROI is investigated. The result shows that the CLAHE algorithm overcomes the problems of local overexposure and loss of details and can achieve better contrast enhancement while keeping the blood vessel images clearer.

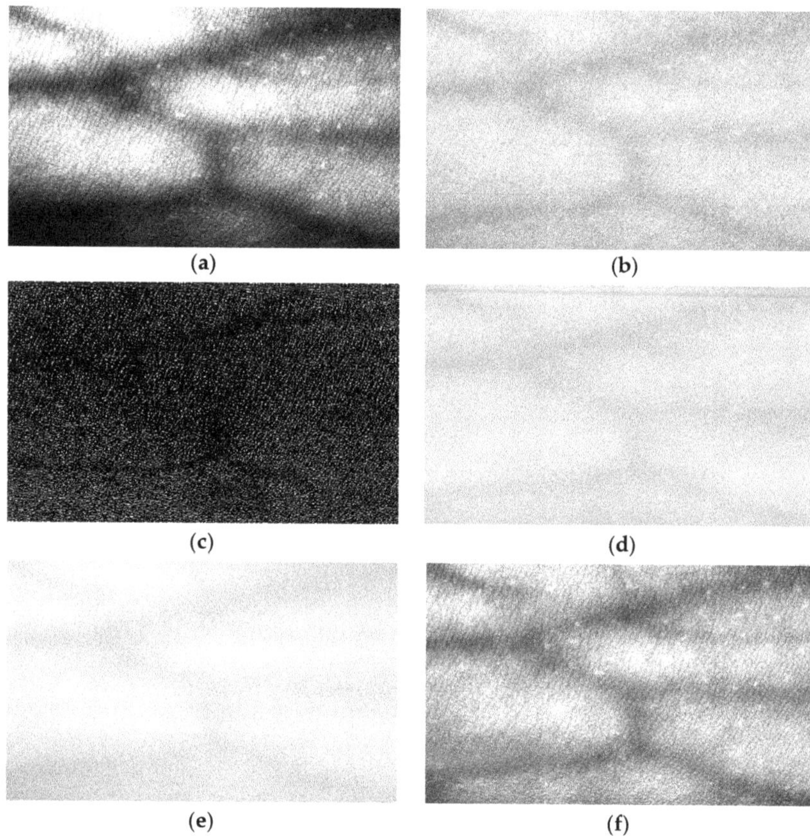

Figure 5. Experiment results of enhancement algorithms. (**a**) The result of HE. (**b**) The result of AHE. (**c**) The result of SSR method. (**d**) The result of BCMDBR. (**e**) The result of s homomorphic filtering. (**f**) The result of CLAHE.

3.2. Evaluation of Noise Reduction Algorithm

To compare the effect of noise reduction algorithms, the other traditional noise reduction methods like bilateral filtering [34], median filter algorithm [35], mean filter algorithm [36], non-local-mean filtering [37], and bandpass filtering are applied to the same noisy image for performance evaluation, respectively. The corresponding results are shown in Figure 6. From Figure 6, it can be observed that the bilateral filtering and mean filtering retain the edge information of images, but that the effects of noise removal are not as good as the effects of Gaussian filtering. The processing results of median filtering and non-local means denoising will cause the image to become blurred, and the bandpass filtering cannot remove the noise in the image well.

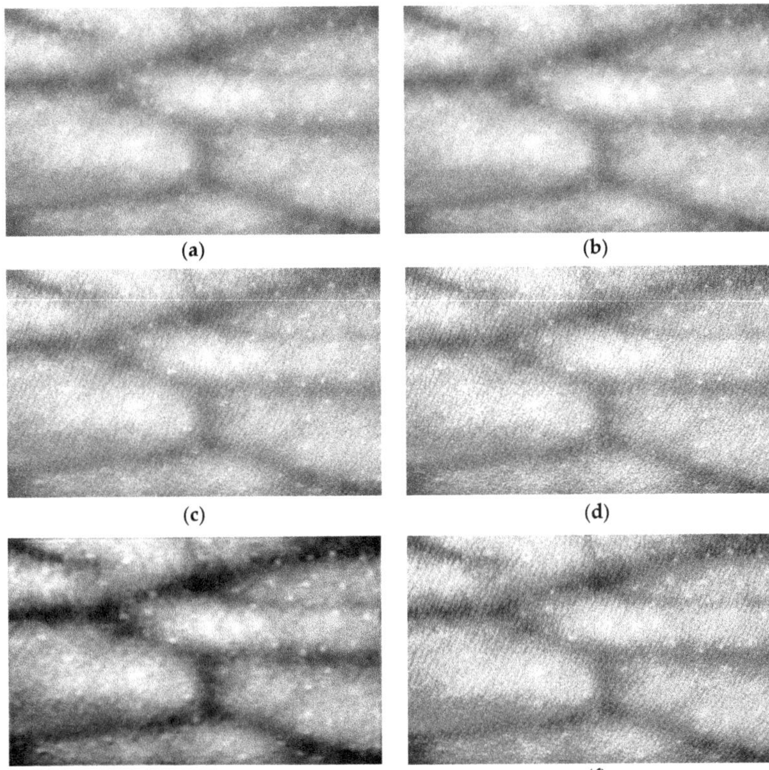

Figure 6. Experiment results of image noise reduction. (a) The result of bilateral filtering. (b) The result of median filtering. (c) The result of mean filtering. (d) The result of non-local-mean filtering. (e) The result of bandpass filtering. (f) The result of Gaussian filtering.

Since the image denoising evaluation is closely related to human vision and subjective perception, and human eye still has a certain deficiency in terms of detail discrimination, this paper introduces three metrics in order to further test the processing effect, i.e., the energy [38], Brenner [39], and visual information fidelity (VIF) [40]. This is carried out to evaluate image quality after denoising. The corresponding scores are shown in Table 1. Energy is an evaluation index used to measure the brightness value distribution, contrast, and other characteristics of an image. Generally, the higher the energy of an image is, the wider the distribution of luminance values and the stronger the contrast will be.

$$D(f) = \Sigma_y \Sigma_x (| f(x+1,y) - f(x,y) |^2 + | f(x,y+1) - f(x,y) |^2) \tag{19}$$

Table 1. Evaluation results of different noise reduction algorithms.

Name of Filtering Algorithm	Energy	Brenner	VIF
Bilateral filtering	0.111113	0.149405	0.0229357
Median filtering	0.142327	0.185371	0.0171459
Mean filtering	0.542328	0.717005	0.228890
Non-local means denoising	0.166525	0.200364	0.192109
Bandpass filtering	0	0	6.829285×10^{-5}
Gaussian filtering (Our method)	0.579129	0.788554	0.314532

Brenner is an evaluation metric based on image gradient information. This method is usually used to measure image sharpness. The Brenner metric is sensitive to grayscale changes and can effectively distinguish the sharpness of various parts of an image.

$$D(f) = \Sigma_y \Sigma_x \mid f(x+2,y) - f(x,y) \mid^2 \tag{20}$$

VIF is an objective metric used to evaluate the quality of an image or video. It is based on the human visual system's perception of information, and it also can quantify degrees of distortion. It takes into account several factors, including spatial frequency, brightness, or color, in order to more accurately simulate variation in perceived quality by the human eye. In the calculation of VIF function, multi-scale decomposition is firstly carried out, and then the information quantity is modeled and calculated at each scale. Finally, the VIF value is formed by combination. Higher VIF scores indicate better denoising effects. As shown in Table 1, the Gaussian filtering used in this paper has a good noise reduction effect on our dataset.

3.3. Evaluation of Image Segmentation Algorithm

Figure 7a shows the background of the near-infrared image obtained by minimum filtering processed by 3×3 template and the median filtering. The background image obtained can be considered as the luminance information without blood vessel information. By using Formula (7), the image information of the blood vessel region can be obtained without background information, as shown in Figure 7b. From the results, it can be observed that the background of near-infrared blood vessel image can be well extracted. Although there are some noises in the blood vessel information obtained by eliminating the background, the blood vessel information is well extracted without missing segmentation or wrong segmentation, and there are not many breakpoints in an image.

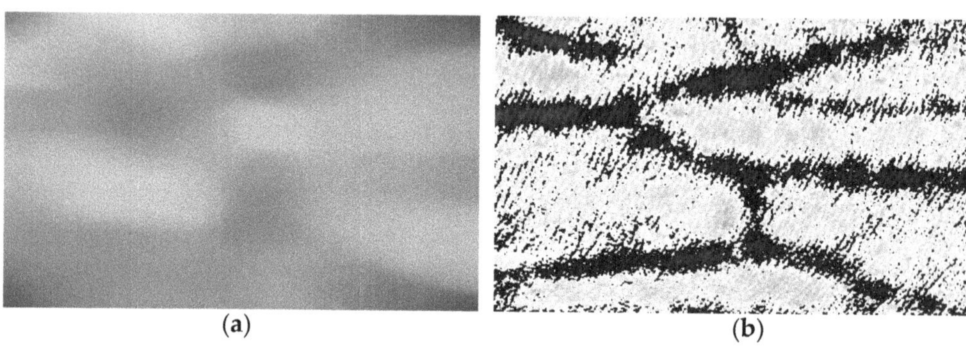

Figure 7. Processing result of background subtraction. (**a**) The background image extracted from the vascular image. (**b**) The extracted vascular image result.

To compare the experiment of segmentation algorithm performance, the global thresholding-based segmentation algorithm [41], region-growth-based thresholding method [42], entropy thresholding process [43], OTSU algorithm [44], and histogram-based segmentation algorithm [45] are applied to the same dataset to assess their processing effects, respectively. The results are shown in Figure 8. It can be observed that, although the segmentation algorithms based on global threshold, OTSU algorithm, and entropy threshold processing can segment the blood vessels, they cannot process the small-blood-vessel information well, and more details are lost. The segmentation effects of the threshold method assessed based on region growth, alongside the algorithms investigated based on histograms, are not good, respectively. Even for the primary blood vessels, the blood vessels cannot be clearly identified. In contrast, the algorithm in this paper does not omit details when considering the primary blood vessels, and its use can segment a relatively clear and complete vascular framework. This is because the traditional segmentation algorithms are based on the threshold value, which can be considered as the vascular feature. The segmentation algorithm based on the global threshold is to choose the manual threshold, and it needs to be adjusted continuously to obtain a more appropriate value, meaning that the algorithm is not robust. Based on the histogram technique and other algorithms, the appropriate threshold value can be obtained through calculation, but the easy threshold value will be affected by many factors in the calculation process, such as brightness, noise, etc. There are high requirements for using the pre-processing algorithm, and its use is not suitable for processing near infrared vascular images. However, our image reduction algorithm does not use threshold value and is less affected by brightness and other factors, and as a result it has a good effect.

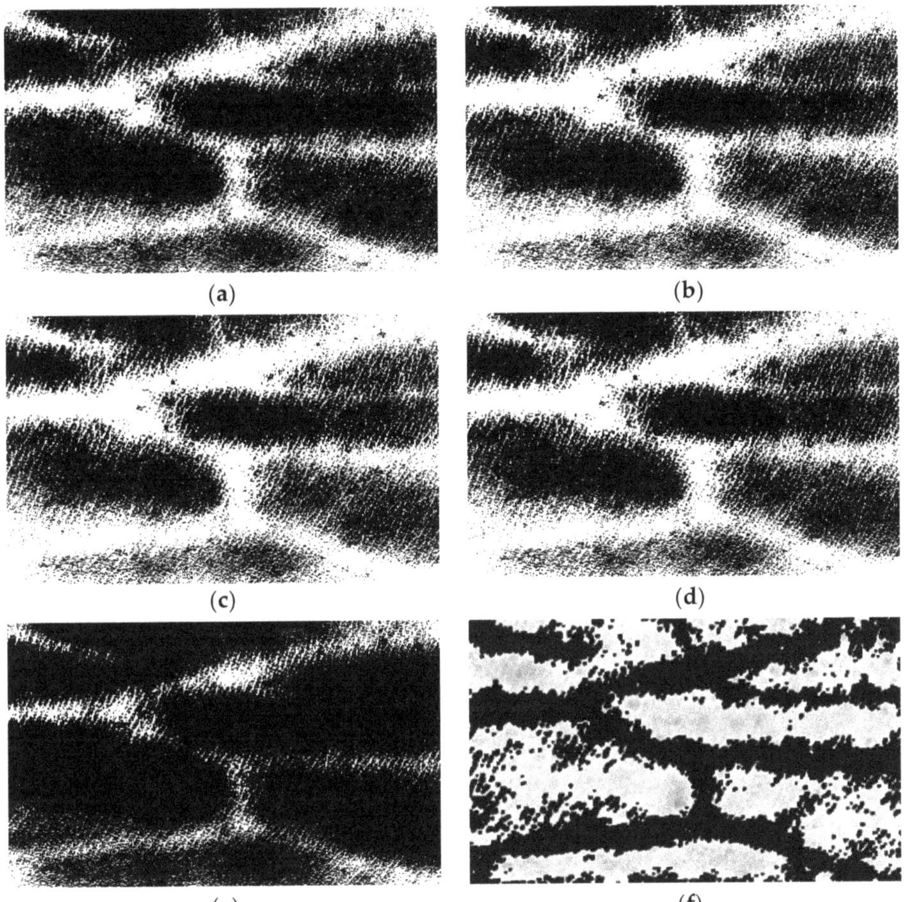

Figure 8. Experiment results of image segmentation. (**a**) The result of the global threshold-based segmentation. (**b**) The result of the region-growth-based thresholding method. (**c**) The result of the entropy thresholding method. (**d**) The result of the OTSU algorithm. (**e**) The result shows the histogram-based segmentation. (**f**) The result of our method.

Some objective quantitative analyses are also performed. Three indexes are considered in this paper, i.e., Dice, accuracy (Acc), and sensitivity (Sen) [46,47]. The Dice coefficient is a function to evaluate similarity. This is usually used to calculate the similarity or overlap of two samples. Its computation method is defined in (11). Acc indicates the proportion of correctly segmented pixel points in segmentation result to the total number of pixel points in the whole image. Its formula can be calculated by (12). Sen means the proportion of correct vessel points in the segmentation results to the total number of vessel points in the standard segmentation result [47]. Its calculation method is illustrated in (13).

$$Dice = 2\frac{V_{seg} \cap V_{gt}}{V_{seg} \cup V_{gt}} \tag{21}$$

$$Acc = (TP + TN)/(TP + TN + FP + FN) \tag{22}$$

$$Sen = TP/(TP + FN) \tag{23}$$

where V_{seg} denotes the segmentation result, V_{gt} means the expert segmentation result; $V_{seg} \cap V_{gt}$ represents the intersection of V_{seg} and V_{gt}, V_{seg}, and $\cup V_{gt}$ represents the union of V_{seg} and V_{gt}. True positive (TP) shows the correctly segmented vessel points; false positive (FP) indicates the incorrectly segmented vessel points; true negative (TN) indicates the correctly segmented background points; and false negative (FN) means the incorrectly segmented background points [47].

Table 2 presents the corresponding results. Table 2 shows that, although the Acc score of our method is not as good as those of other algorithms, the reason for the low Acc in the segmentation evaluation results in this paper is the class imbalance in the vascular images. Vascular segmentation is essentially a classification problem. In near-infrared vascular images, the area of non-vascular regions is much larger than the area of vascular regions, causing class imbalance and low Acc. In contrast, Dice and Sen focus more on the degree of overlap between prediction and ground truth. As a result, Dice and Sen are higher, which also reflects that the segmentation effects in this paper are closer to the manual segmentation results of experts.

Table 2. Evaluation results of different segmentation algorithms.

Name of Filtering Algorithm	Dice	Acc	Sen
Threshold segmentation method based on region growth	0.66151	0.570808	0.668102
Threshold method based on regional growth	0.670982	0.489470	0.811824
Entropy threshold method	0.662919	0.456471	0.8487629
OTSU threshold segmentation algorithm	0.670982	0.489471	0.811824
Histogram-based technique for threshold segmentation	0.457068	0.657206	0.319836
The algorithm in this paper	0.697933	0.443855	0.938922

3.4. Evaluation of Morphological Algorithm

This paper assesses the morphological processing algorithms. The traditional corroded and inflated structural elements, whether elliptical, rectangular, or cross-type, are compared with T-type structural element to obtain the same dataset of binarized images. In order to facilitate the understanding of the advantages of T-type structural elements, this paper firstly draws pictures to demonstrate the working process of ellipse, rectangular, cross-type and T-type structures. As can be seen from Figure 9, although the ellipse structure cannot handle the upper and lower edge structures, the gentle convex part of them cannot be processed well. The rectangular structure can remove the convex part of edge better, but the angles of shape details are lost. The cross-type structure has little effect on the sharp protrusions. The T-type structural element can retain as much shape information as possible, and it can also "smooth" the convex part of the upper and lower edges.

Figure 9. Processing effect examples of different shape structural elements.

Figure 10 shows the corresponding processing results of the blood vessel images using the T-type, elliptic, rectangular and cross-type structural elements. It can be seen from Figure 10 that the morphologic processing algorithm based on T-type structural element can remove the noise and holes better and extract the target region more accurately. Compared with the traditional ellipse type and cross-type, the T-type structural element smooths the upper and lower edges better. Although the rectangular structural element has a better smoothing effect, it displays more distortions compared with the real blood vessel region. Therefore, a T-type structural element is considered better able to adapt to the characteristics of blood vessel image, which can improve the effect of image processing to a certain extent.

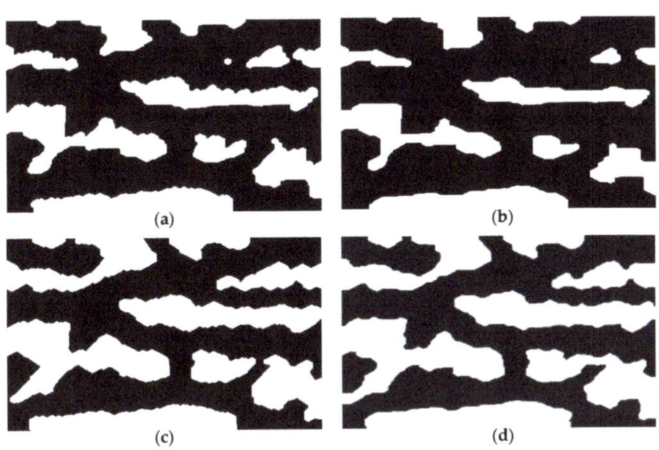

Figure 10. Experiment results of morphologic processing. (**a**) The result of elliptic structural elements. (**b**) The result of rectangular structural elements. (**c**) The result of cross-type structural elements. (**d**) The result of T-type structural elements.

4. Discussion

In this research, a method based on background subtraction and improved morphology is proposed to design an effective preprocessing and segmentation method for forearm near-infrared blood vessel images. Not only the gray characteristics, but also the morphological features, are considered when designing this algorithm. The target vascular region is accurately processed according to the relevant morphological characteristics in order to ensure segmentation accuracy. The algorithm has a better anti-noise performance than the traditional algorithm. The proposed algorithm aims to improve the extracted blood vessel information by applying several steps: firstly, the original image is preprocessed with CLAHE to achieve a highlighted vascular part. Then, Gaussian filtering is applied to reduce the noise in the image. Next, the background image is obtained by blurring the image with minimum filtering and median filtering. It is then subtracted from the enhanced image to obtain a preliminary binary image of a blood vessel. In the segmentation step, the isolated noises are removed using the element screening method based on the connected domain. Then, the image is dilated and eroded with T-type structural elements to render uniform the width of blood vessels and smooth edges. Through these steps, the vascular region in the forearm near-infrared venous vessel image can be effectively extracted with better performance and accuracy than other existing methods.

At present, there are limited datasets for near-infrared vascular images, and manual segmentation of vascular images cost lots of manpower; therefore, it is difficult to use supervised learning method to segment images. This paper also tries other techniques to segment the blood vessel image. For example, the K-means clustering algorithm [48], watershed algorithm [49], active contour method [50], and the method of implicit active contours, driven by local binary fitting energy (LBF) [51], are all considered. According to the results which are shown in Figure 11, the first three algorithms cannot segment the near-infrared vascular images with poor results, and cannot extract the blood vessel skeleton with more details lost. The segmentation result of LBF the method exhibit a serious over-segmentation phenomenon, and the blood vessel information cannot be extracted in the image. The primary reasons for these phenomena are as follows: the K-means segmentation is based on gray level segmentation in essence, which has high requirements on illumination changes, initial clustering center selection, and other factors. Additionally, the watershed segmentation easily produces segmentation deviation and has poor effects on images with low contrast. The algorithm model of active contour is dependent on parameters, and suitable parameters should be found for different images to achieve good segmentation effect. Additionally, the LBF method has high requirements for application to an the image, and when image noises are strong, its segmentation effects are not accurate.

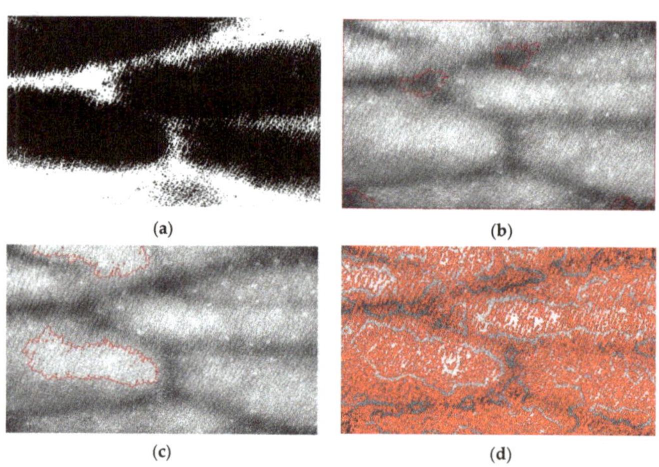

Figure 11. Comparison results of other segmentation algorithms. (**a**) The result of K-means. (**b**) The result of watershed algorithm. (**c**) The result of active contour method. (**d**) The result of LBF.

This study is of great significance for the field of medical image processing, which can improve the quality and readability of near-infrared vascular images and provide more accurate and effective information for clinical diagnosis and treatment. In addition, this study provides a new non-invasive detection method for the field of cardiovascular disease. It can be used to assist nurses in the evaluation of vascular quality and help doctors in the detection, localization and intervention to treat venous thrombosis, which is helpful in efforts to improve the quality of life and safety of patients. However, this study also has some limitations. These must be addressed in the future. For example, this study only verifies the algorithm on forearm near-infrared vascular images without considering other parts or other types of vascular images. The repair effect of this paper needs to be improved in relation to some fractures. Based on these limitations, future research can be carried out in the following aspects: first, the cooperation with hospitals to collect more data will be established for expanding the scope of the experiment, which will test the effect of the algorithm on other sites or other types of vessel images. Second, the ability of the algorithm to deal with more details of the blood vessels will be improved, and efforts will be constantly made to design a corresponding solution or improvement strategy will.

5. Conclusions

In this paper, background subtraction and improved mathematical morphology technique are used to extract blood vessel information for the task of forearm infrared blood vessel segmentation. Initially, image enhancement and spatial domain filtering are used to enhance near-infrared vascular images and highlight the target vascular region. Secondly, the fuzzy method is used to obtain the image background, and then the idea of removing the background from the image is used to obtain the blood vessel information. Finally, the element screening method based on the basic operation of connecting domain and enhanced morphology is used to remove noise and smooth edges in order to ensure the uniformity of the width of the target blood vessel region. Compared with traditional vascular detection technology, the method proposed in this work is simpler, more intuitive, efficient and accurate, and can provide valuable information and support for non-invasive medical diagnosis and treatment. In addition, it has a wide range of possible applications in medical image processing and is likely to make greater contributions to the field of medical diagnosis in the future.

Author Contributions: Conceptualization, L.L., H.L., Q.L., Z.T. and Y.L.; Data curation, L.L. and H.L.; Formal analysis, L.L., H.L., Q.L., Z.T., Y.L. and W.G.; Funding acquisition, H.L.; Investigation, L.L., H.L., Q.L. and Z.T.; Methodology, L.L., H.L., Q.L., Z.T., Y.L., W.G. and S.W.; Project administration, H.L.; Resources, L.L., H.L., W.G. and S.W.; Software, L.L., Z.T. and Y.L.; Supervision, H.L., Q.L., W.G. and S.W.; Validation, L.L., H.L., Q.L., Z.T., W.G. and S.W.; Visualization, L.L.; Writing—original draft, L.L.; Writing—review and editing, L.L. and H.L. All authors have read and agreed to the published version of the manuscript.

Funding: This work was funded by the National Natural Science Foundation of China under Grant 61975011, the Fund of Science and Technology on Near-Surface Detection Laboratory under Grant 6142414221403, the Fund of State Key Laboratory of Intense Pulsed Radiation Simulation and Effect under Grant SKLIPR2024, the Natural Science Foundation of Guangdong Province under Grant 2023A1515010275, and the Fundamental Research Fund for the China Central Universities of USTB under Grant FRF-BD-19-002A.

Institutional Review Board Statement: Not applicable.

Informed Consent Statement: Not applicable.

Data Availability Statement: The data presented in this study are available on request from the corresponding author Haoting Liu.

Conflicts of Interest: The authors declare no conflict of interest.

References

1. Waitt, C.; Waitt, P.; Pirmohamed, M. Intravenous therapy. *Postgrad. Med. J.* **2004**, *80*, 1–6. [CrossRef] [PubMed]
2. Lamperti, M.; Pittiruti, M., II. Difficult peripheral veins: Turn on the lights. *Br. J. Anaesth.* **2013**, *110*, 888–891. [CrossRef] [PubMed]
3. Chaudhuri, S.; Chatterjee, S.; Katz, N.; Nelson, M.; Goldbaum, M. Detection of blood vessels in retinal images using two-dimensional matched filters. *IEEE Trans. Med. Imaging* **1989**, *8*, 263–269. [CrossRef] [PubMed]
4. Li, Q.; You, J.; Zang, D. Vessel segmentation and width estimation in retinal images using multiscale production of matched filter responses. *Expert Syst. Appl.* **2012**, *39*, 7600–7610. [CrossRef]
5. Hoover, A.; Kouznetsova, V.; Goldbaum, M. Locating blood vessels in retinal images by piecewise threshold probing of a matched filter response. *IEEE Trans. Med. Imaging* **2000**, *19*, 203–210. [CrossRef]
6. Zhang, J.; Yang, J. Finger-vein image enhancement based on combination of gray-level grouping and circular Gabor filter. In Proceedings of the 2009 International Conference on Information Engineering and Computer Science, Wuhan, China, 19–21 December 2009; pp. 1–4.
7. Lee, E.C.; Park, K.R. Restoration method of skin scattering blurred vein image for finger vein recognition. *Electron. Lett.* **2009**, *45*, 1074–1076. [CrossRef]
8. Yin, Y.; Adel, M.; Bourennane, S. Retinal vessel segmentation using a probabilistic tracking method. *Pattern Recognit.* **2012**, *45*, 1235–1244. [CrossRef]
9. Liu, B.; Jiang, Q.; Liu, W.; Wang, M.; Zhang, S.; Zhang, X.; Zhang, B.; Yue, Z. A vessel segmentation method for serialized cerebral vascular DSA images based on spatial feature point set of rotating coordinate system. *Comput. Methods Programs Biomed.* **2018**, *161*, 55–72. [CrossRef]
10. Lee, S.H.; Lee, S. Adaptive Kalman snake for semiautonomous 3D vessel tracking. *Comput. Methods Programs Biomed.* **2015**, *122*, 56–75. [CrossRef]
11. Song, M.; Civco, D. Road extraction using SVM and image segmentation. *Photogramm. Eng. Rem.* **2004**, *70*, 1365–1371. [CrossRef]
12. Pathak, D.; Krähenbühl, P.; Darrell, T. Constrained convolutional neural networks for weakly supervised segmentation. *IEEE Int. Conf. Comput. Vis.* **2015**, *15*, 1796–1804.
13. Liskowski, P.; Krawiec, K. Segmenting retinal blood vessels with deep neural networks. *IEEE Trans. Med. Imaging* **2016**, *35*, 2369–2380. [CrossRef] [PubMed]
14. Ronneberger, O.; Fischer, P.; Brox, T. U-Net: Convolutional networks for biomedical image segmentation. *Lect. Notes Comput. Sci.* **2015**, *16*, 711–720.
15. Sánchez, J.C.G.; Magnusson, M.; Sandborg, M.; Carlsson; Tedgren, Å.; Malusek, A. Segmentation of bones in medical dual-energy computed tomography volumes using the 3D U-Net. *Phys. Med.* **2020**, *69*, 241–247. [CrossRef] [PubMed]
16. Hiasa, Y.; Otake, Y.; Takao, M.; Ogawa, T.; Sugano, N.; Sato, Y. Automated muscle segmentation from clinical CT using Bayesian U-Net for personalized musculoskeletal modeling. *IEEE Trans. Med. Imaging* **2020**, *39*, 1030–1040. [CrossRef] [PubMed]
17. Perslev, M.; Pai, A.; Runhaar, J.; Igel, C.; Dam, E.B. Cross-cohort automatic knee MRI segmentation with multi-planar U-Nets. *J. Magn. Reason. Imaging* **2022**, *55*, 1650–1663. [CrossRef]
18. Xiao, X.; Lian, S.; Luo, Z.; Li, S. Weighted Res-UNet for high-quality retina vessel segmentation. In Proceedings of the 2018 IEEE 9th International Conference on Information Technology in Medicine and Education, Hangzhou, China, 19–21 October 2018; pp. 327–331.
19. Alom, M.Z.; Yakopcic, C.; Hasan, M.; Taha, T.M.; Asari, V.K. Recurrent residual U-Net for medical image segmentation. *J. Med. Imaging* **2018**, *16*, 228–233. [CrossRef]
20. Zhao, P.; Zhang, Y.; Bian, K.; Tuo, H.; Song, L. LadderNet: Knowledge transfer based viewpoint prediction in 360° video. In Proceedings of the 2019 IEEE International Conference on Acoustics, Speech and Signal Processing, Brighton, UK, 12–17 May 2019; pp. 1657–1661.
21. Ning, Z.; Zhong, S.; Feng, Q.; Chen, W.; Zhang, Y. SMU-Net: Saliency-guided morphology-aware U-Net for breast lesion segmentation in ultrasound image. *IEEE Trans. Med. Imaging* **2021**, *41*, 476–490. [CrossRef]
22. Li, Y.; Liu, H.; Tian, Z.; Geng, W. Near-infrared vascular image segmentation using improved level set method. *Infrared Phys. Technol.* **2023**, *133*, 104678. [CrossRef]
23. Alwazzan, M.J.; Ismael, M.A.; Ahmed, A.N. A hybrid algorithm to enhance colour Retinal fundus images using a Wiener filter and CLAHE. *J. Digit. Imaging* **2021**, *34*, 750–759. [CrossRef]
24. Arpah, B.A.S.; Taib, M.N.; Khalid, N.E.; Taib, H. Variations of adaptive histogram equalization (AHE) analysis on intra-oral dental radiograph. In Proceedings of the 2015 IEEE 6th Control and System Graduate Research Colloquium, Shah Alam, Malaysia, 10–11 August 2015; pp. 87–92.
25. Liu, X.W.; Liu, C.Y. An optional Gauss filter image denoising method based on difference image fast fuzzy clustering. *Appl. Mech. Mater.* **2013**, *414*, 1348–1352. [CrossRef]
26. Portilla, J.; Strela, V.; Wainwright, M.J.; Simoncelli, E.P. Image denoising using scale mixtures of Gaussians in the wavelet domain. *IEEE Trans. Image Process.* **2003**, *12*, 1338–1351. [CrossRef] [PubMed]
27. Tang, J.; Wang, Y.; Cao, W.; Yang, J. Improved adaptive median filtering for structured light image denoising. In Proceedings of the 2019 7th International Conference on Information, Communication and Networks, Macao, China, 24–26 April 2019; pp. 146–149.
28. Cheng, H.; Jiang, X.; Sun, Y.; Wang, J. Color image segmentation: Advances and prospects. *Pattern Recognit.* **2001**, *34*, 2259–2281. [CrossRef]

29. Xiaosong, S.; Yongjie, H.; Yongge, L. Text on Oracle rubbing segmentation method based on connected domain. In Proceedings of the 2016 IEEE Advanced Information Management, Communicates, Electronic and Automation Control Conference, Xi'an, China, 3–5 October 2016; pp. 414–418.
30. Haralick, R.M.; Sternberg, S.R.; Zhuang, X. Image analysis using mathematical morphology. *IEEE Trans. Pattern Anal. Mach. Intell.* **1987**, *9*, 532–550. [CrossRef] [PubMed]
31. Sargun; Rana, S.B. Performance evaluation of HE, AHE and fuzzy image enhancement. *Int. J. Comput. Appl.* **2015**, *122*, 14–19. [CrossRef]
32. Zhang, G.; Sun, D.; Yan, P.; Zhao, H.; Li, Z. A LDCT image contrast enhancement algorithm based on single-scale Retinex theory. In Proceedings of the 2008 International Conference on Computational Intelligence for Modelling Control & Automation, Vienna, Austria, 10–12 December 2008; pp. 1282–1287.
33. Yu, H.; Li, X.; Lou, Q.; Yan, L. Underwater image enhancement based on color-line model and homomorphic filtering. *Signal Image Video P* **2021**, *16*, 83–91. [CrossRef]
34. Chen, B.; Tseng, Y.; Yin, J. Gaussian-adaptive bilateral filter. *IEEE Signal Process. Lett.* **2020**, *27*, 1670–1674. [CrossRef]
35. Erkan, U.; Gökrem, L.; Enginoğlu, S. Different applied median filter in salt and pepper noise. *Comput. Electr. Eng.* **2018**, *70*, 789–798. [CrossRef]
36. Fu, B.; Xiong, X.; Sun, G. An efficient mean filter algorithm. In Proceedings of the 2011 IEEE/ICME International Conference on Complex Medical Engineering, Harbin, China, 22–25 May 2011; pp. 466–470.
37. Yang., J.; Fan, J.; Ai, D.; Zheng, Y.; Tang, S.; Wang, Y. Local statistics and non-local mean filter for speckle noise reduction in medical ultrasound image. *Neural Comput.* **2016**, *195*, 88–95. [CrossRef]
38. Malladi, S.R.; Ram, S.; Rodríguez, J.J. Image denoising using superpixel-based PCA. *IEEE Trans. Multimedia* **2021**, *23*, 2297–2309. [CrossRef]
39. Yan, Z.; Chen, G.; Xu, W.; Yang, C.; Lu, Y. Study of an image autofocus method based on power threshold function wavelet reconstruction and a quality evaluation algorithm. *Appl. Opt.* **2018**, *33*, 9714–9721. [CrossRef] [PubMed]
40. Peng, C.; Wu, M.; Liu, K. Multiple levels perceptual noise backed visual information fidelity for picture quality assessment. In Proceedings of the 2022 International Symposium on Intelligent Signal Processing and Communication Systems, Penang, Malaysia, 22–25 November 2022; pp. 1–4.
41. Kim, D.; Lee, S.; Hong, W.; Lee, H.; Jeon, S.; Han, S.; Nam, J. Image segmentation for FIB-SEM serial sectioning of a Si/C–graphite composite anode microstructure based on preprocessing and global thresholding. *Microsc. Microanal.* **2019**, *25*, 1139–1154. [CrossRef] [PubMed]
42. Hasikin, K.; Isa, N.A.; Mohamed, M.; Suandi, S.A.; Kadri, N.A. A new region-based adaptive thresholding for sperm motility segmentation. *Malays. J. Comput. Sci.* **2017**, *29*, 272–286. [CrossRef]
43. Wong, A.K.C.; Sahoo, P.K. A gray-level threshold selection method based on maximum entropy principle. *IEEE Trans. Syst. Man Cybern. Syst.* **1989**, *19*, 866–871. [CrossRef]
44. Fu, Z.L. Image thresholding method—A generalization of Otsu's method. *Comput. Appl.* **2000**, *20*, 37–39.
45. Otsu, N. A threshold selection method from gray-level histograms. *IEEE Trans. Syst. Man Cybern. Syst.* **2007**, *9*, 62–66. [CrossRef]
46. Eelbode, T.; Bertels, J.; Berman, M.; Vandermeulen, D.; Maes, F.; Bisschops, R.; Blaschko, M.B. Optimization for medical image segmentation: Theory and practice when evaluating with Dice score or Jaccard index. *IEEE Trans. Med. Imaging* **2020**, *39*, 3679–3690. [CrossRef]
47. Moccia, S.; Momi, E.D.; Hadji, S.E.; Mattos, L.S. Blood vessel segmentation algorithms—Review of methods, datasets and evaluation metrics. *Comput. Methods Programs Biomed.* **2018**, *158*, 71–91. [CrossRef]
48. Ng, H.P.; Ong, S.H.; Foong, K.W.; Goh, P.; Nowinski, W.L. Medical image segmentation using K-Means clustering and improved watershed algorithm. In Proceedings of the 2006 IEEE Southwest Symposium on Image Analysis and Interpretation, Denver, CO, USA, 26–28 March 2006; pp. 61–65.
49. Roerdink, J.B.T.M.; Meijster, A. The watershed transform: Definitions, algorithms and parallelization strategies. *Fund. Inform.* **2000**, *41*, 187–228. [CrossRef]
50. Chan, T.F.; Vese, L.A. Active contours without edges. *IEEE Trans. Med. Imaging* **2001**, *10*, 266–277. [CrossRef]
51. Li, C.; Kao, C.-Y.; Gore, J.C.; Ding, Z. Implicit active contours driven by local binary fitting energy. In Proceedings of the 2007 IEEE Conference on Computer Vision and Pattern Recognition, Minneapolis, MN, USA, 17–22 June 2007; pp. 1–7.

bioengineering

MDPI

Article

Reference Standards for Digital Infrared Thermography Measuring Surface Temperature of the Upper Limbs

Seong Son [1], Byung Rhae Yoo [1,*] and Ho Yeol Zhang [2]

[1] Department of Neurosurgery, Gil Medical Center, Gachon University College of Medicine, Incheon 21565, Republic of Korea; sonseong44@gmail.com

[2] Department of Neurosurgery, National Health Insurance Service Ilsan Hospital, Yonsei University College of Medicine, Ilsan 10444, Republic of Korea

* Correspondence: byungryoo@gilhospital.com; Tel.: +82-32-460-3304

Abstract: (1) Background: although digital infrared thermographic imaging (DITI) is used for diverse medical conditions of the upper limbs, no reference standards have been established. This study aims to establish reference standards by analyzing DITI results of the upper limbs. (2) Methods: we recruited 905 healthy Korean adults and conducted thermography on six regions (dorsal arm, ventral arm, lateral arm, medial arm, dorsal hand, and ventral hand region). We analyzed the data based on the proximity of regions of interest (ROIs), sex, and age. (3) Results: the average temperature (°C) and temperature discrepancy between the right and the left sides (ΔT) of each ROI varied significantly ($p < 0.001$), ranging from 28.45 ± 5.71 to 29.74 ± 5.14 and from 0.01 ± 0.49 to 0.15 ± 0.62, respectively. The temperature decreased towards the distal ROIs compared to proximal ROIs. The average temperatures of the same ROIs were significantly higher for men than women in all regions ($p < 0.001$). Across all regions, except the dorsal hand region, average temperatures tended to increase with age, particularly in individuals in their 30s and older ($p < 0.001$). (4) Conclusions: these data could be used as DITI reference standards to identify skin temperature abnormalities of the upper limbs. However, it is important to consider various confounding factors, and further research is required to validate the accuracy of our results under pathological conditions.

Keywords: infrared rays; reference standard; skin temperature; thermography; upper limb

Citation: Son, S.; Yoo, B.R.; Zhang, H.Y. Reference Standards for Digital Infrared Thermography Measuring Surface Temperature of the Upper Limbs. *Bioengineering* **2023**, *10*, 671. https://doi.org/10.3390/bioengineering10060671

Academic Editor: Cuneyt M. Alper

Received: 28 April 2023
Revised: 27 May 2023
Accepted: 30 May 2023
Published: 1 June 2023

1. Introduction

Digital infrared thermographic imaging (DITI) has been utilized as an ancillary diagnostic method for various medical conditions related to the upper extremities. For instance, cervical radiculopathy; peripheral nerve entrapment syndrome (e.g., carpal tunnel syndrome); rheumatic disease (e.g., Raynaud's disease); complex regional pain syndrome; tendinopathy; hand arthritis (e.g., psoriatic arthritis); and skin cancer can be confirmed by DITI [1–13]. The hypo-radiant (hypothermia) or hyper-radiant (hyperthermia) regions can be identified by comparing the temperature between the right and left arms or with empirical normal ranges. However, the significant temperature differences between both sides and deviations from normal ranges have not been established. Furthermore, body surface temperatures are influenced by the measurement environment, ana-tomical area, and subject characteristics, which presents a challenge in deriving reference values from the existing literature [14,15].

Consequently, previous attempts to establish reference standards for DITI using systematic reviews based on meta-analysis or machine learning methods have been limited in providing detailed standard DITI values [10,16–18]. Although a recent study suggested a correct differential diagnosis process for Raynaud's phenomenon in the hand using a deep convolutional neural network, a type of deep learning method, its small data set had limitations [19]. To address this gap and establish scientific reference standards for DITI, we investigated the DITI results of several body regions, including the lower limbs, upper limbs, trunk, and face, in a large cohort of subjects under a controlled protocol. We previously reported DITI reference standards for the lower limbs [16]. In the present study, we provide DITI reference standards for the upper limbs, including the hands.

2. Materials and Methods

2.1. Study Design, Ethics, and Sample Size

This multi-center, single-arm, open label study was conducted in compliance with global/local ethics. The Institutional Review Board of three hospitals approved the study protocol. In addition, this clinical trial was registered with the Clinical Research Information Service of Korea (CRIS, http://cris.nih.go.kr, accessed on 1 April 2023, number KCT0006880) [20].

The sample size for this study was determined to be 922 participants, with a dropout rate of 15%, based on the equation provided in our previous report [20]. The equation used was as follows:

$$n = \frac{\theta(1 - \theta)^2 z_{\alpha/2}^2}{d^2}$$

(Population proportion [θ] = 0.85; margin of error [d] = 0.025; and significance level of 5% with confidence level of 95%).

2.2. Subjects Recruitment

Healthy Korean adults were recruited through open announcements at three hospitals between March 2018 and December 2020.

The general qualifications for participation in the DITI examination of whole body, including the upper extremities, lower extremities, trunk, and face, were as follows: (1) age between 20 and 69; (2) the absence of any potential contraindication for DITI (e.g., pregnancy or claustrophobia); (3) the ability to maintain posture during the test; and (4) the absence of any other issue considered by the investigator to adversely influence DITI results [20].

To minimize potential confounders, the following criteria were adopted: (1) no specific medical history, such as cervical spine disease, diabetes mellitus, peripheral neuropathy or entrapment, joint disorder of the arm or hand, previous history of cervical spine or upper extremity surgery; and (2) currently no pain or skin abnormality of the upper extremity.

Initially, 922 healthy Korean adults were enrolled. However, 17 patients were excluded due to test failure or withdrawal of consent. Therefore, the results from 905 participants were included in the analysis [20]. All testing was performed under controlled conditions after obtaining informed consent.

2.3. Equipment and Examination Protocol

DITI testing was performed using the same equipment and methodology as we previously described [20]. The Iris-XP Digital infrared imaging system (Medicore, Seoul, Republic of Korea) was employed to obtain the images. Prior to the test, subjects were required to spend 20 min in the test room to acclimatize to the controlled room conditions, which maintained a temperature of 20.0–23.0 °C and humidity of 30–75%. The distance between the subject and the equipment was set at 1.5 m.

During the DITI imaging, upper limbs were divided into six sections based on empirical classification, which facilitated classification into various two-dimensional planes, including the front, back, and lateral sides. The upper limbs were tested in a neutral standing position, encompassing six regions, including the dorsal, ventral, lateral, and medial arm regions, as well as the dorsal and ventral hand regions. Regions of interest (ROIs) were manually marked on both sides and comprised 18 ROIs in the dorsal arm region, 19 ROIs in the ventral arm region, 19 ROIs in the lateral arm region, 7 ROIs in the medial arm region, 26 ROIs in the dorsal hand region, and 26 ROIs in the ventral hand region (Figure 1). Five certified examiners performed the testing using a diagram that provided details of measurements and ROI locations.

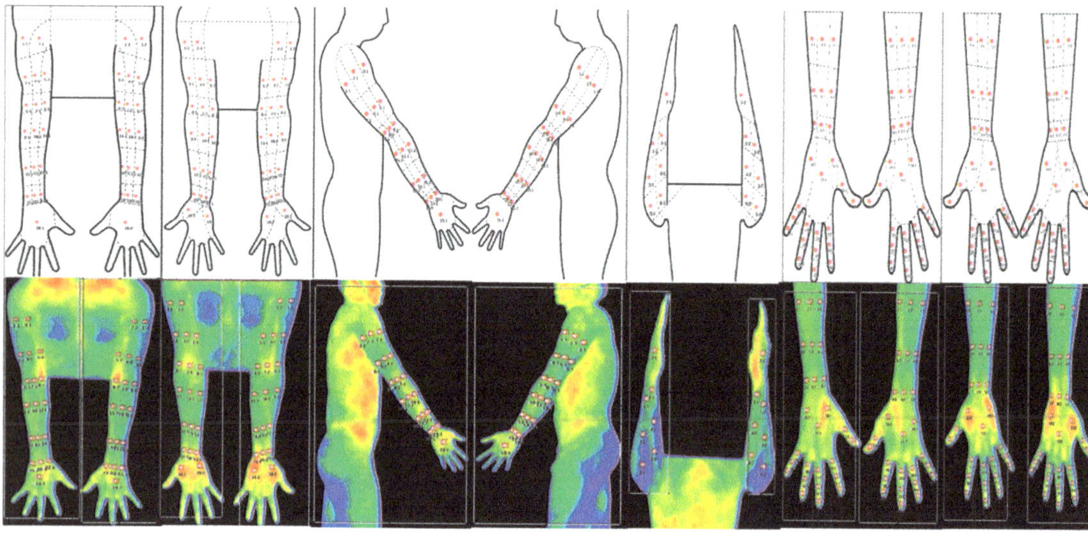

Figure 1. Schematics and actual photographs showing the regions of interest in the dorsal arm, ventral arm, lateral arm, medial arm, dorsal hand, and ventral hand of the upper limbs. The red dots represent each region of interest, and the related numbers are the serial numbers assigned to each regions of interest.

2.4. Statistical Analysis

SPSS version 27.0 (IBM Corporation, Armonk, NY, USA) was used for statistical analysis. Data were checked for normality using the Kolmogorov–Smirnov test, and the results were expressed as mean ± standard deviation (SD), mean with 95% confidence intervals (CI), or median with range. One-way analysis of variance (ANOVA), paired *t*-test, or linear regression analysis were applied according to the purpose of analysis. A *p* value of less than 0.05 was considered statistically significant.

3. Results

3.1. Basic Demographic Data

The average age of all subjects (n = 905) was 42.86 ± 12.87 years, and the male-to-female ratio was 411:405. The sample consisted of 183 individuals (97 males and 86 females) in their 20s, 213 (108 males and 105 females) in their 30s; 228 (109 males and 119 females) in their 40s; 177 (65 males and 112 females) in their 50s; and 104 individuals (32 males and 72 females) in their 60s [20].

3.2. Average Temperatures and Temperature Discrepancies between Right and Left Sides for Each ROI (°C)

The average temperatures and average temperature discrepancies between the right and left (ΔT, temperature of right side−temperature of the left side) of each ROI were as follows: 29.22 ± 5.71 (range, 28.56 ± 5.42–30.24 ± 5.90) and 0.06 ± 0.49 (range, −0.08 ± 0.64–0.25 ± 0.48) in the dorsal arm region; 29.63 ± 5.74 (range, 29.08 ± 5.69–30.51 ± 6.00) and 0.01 ± 0.49 (range, −0.18 ± 0.49–0.18 ± 0.43) in the ventral arm region; 28.45 ± 5.71 (range, 27.59 ± 5.39–29.53 ± 5.90) and 0.15 ± 0. 62 (range,–0.20 ± 0.76–0.52 ± 0.78) in the lateral arm region; 28.84 ± 5.57 (range, 28.00 ± 5.23–29.46 ± 5.79) and 0.04 ± 0.48 (range, −0.10 ± 0.69–0.11 ± 0.42) in the medial arm region; 29.74 ± 5.14 (range, 28.08 ± 4.89–31.27 ± 5.53) and 0.06 ± 0.59 (range, −0.17 ± 0.57–0.26 ± 0.49) in the dorsal hand region; and 28.62 ± 5.44 (range, 27.08 ± 5.14–29.98 ± 5.60) and 0.07 ± 0.69 (range, −0.15 ± 0.57–0.20 ± 0.88) in the ventral hand region (Tables 1–6).

Table 1. The average temperature of each region of interest in the dorsal arm region.

ROI	Mean (°C)	SD	Extended Uncertainty	Difference (ΔT, Right Side–Left Side)				
				Mean (°C)	SD	95% CI Lower	95% CI Upper	Extended Uncertainty
1_1	29.63	5.88	11.78	0.00				
1_2	29.74	5.93	11.88	−0.10	0.55	−0.22	0.01	1.35
2_1	29.96	5.93	11.88	0.00				
2_2	29.99	5.97	11.96	−0.03	0.54	−0.14	0.08	1.33
3_1	28.78	5.76	11.54	0.00				
3_2	28.82	5.77	11.56	−0.04	0.52	−0.15	0.06	1.30
4_1	29.18	5.80	11.62	0.00				
4_2	29.24	5.80	11.63	−0.06	0.49	−0.16	0.04	1.24
5_1	29.59	5.81	11.63	0.00				
5_2	29.62	5.82	11.67	−0.02	0.45	−0.12	0.07	1.17
6_1	29.26	5.81	11.64	0.00				
6_2	29.16	5.80	11.62	0.10	0.51	0.00	0.21	1.26
7_1	29.64	5.87	11.76	0.00				
7_2	29.72	5.94	11.90	−0.08	0.64	−0.21	0.06	1.48
8_1	30.24	5.90	11.82	0.00				
8_2	30.20	5.86	11.74	0.04	0.65	−0.09	0.17	1.49
9_1	29.66	5.86	11.74	0.00				
9_2	29.43	5.82	11.66	0.23	0.50	0.12	0.33	1.26
10_1	29.47	5.81	11.65	0.00				
10_2	29.21	5.77	11.56	0.25	0.48	0.15	0.35	1.22
11_1	29.22	5.73	11.48	0.00				
11_2	29.09	5.71	11.45	0.14	0.41	0.05	0.22	1.10
12_1	28.87	5.61	11.25	0.00				
12_2	28.77	5.61	11.24	0.11	0.52	0.00	0.21	1.28
13_1	28.97	5.62	11.26	0.00				
13_2	28.80	5.61	11.23	0.17	0.43	0.08	0.26	1.14
14_1	28.96	5.58	11.19	0.00				
14_2	28.81	5.56	11.14	0.15	0.43	0.06	0.23	1.13
15_1	28.56	5.42	10.87	0.00				
15_2	28.59	5.43	10.89	−0.03	0.42	−0.11	0.06	1.12
16_1	28.75	5.44	10.90	0.00				
16_2	28.69	5.43	10.89	0.06	0.44	−0.03	0.15	1.14
17_1	28.85	5.44	10.90	0.00				
17_2	28.72	5.43	10.89	0.13	0.44	0.03	0.22	1.15
18_1	28.88	5.49	11.01	0.00				
18_2	28.90	5.51	11.05	−0.01	0.47	−0.11	0.08	1.20
Mean	29.22	5.71	11.44	0.06	0.49	−0.05	0.16	1.24

CI, confidence interval; ROI, region of interest; SD, standard deviation.

Table 2. The average temperature of each region of interest in the ventral arm region.

ROI	Mean (°C)	SD	Extended Uncertainty	Difference (ΔT, Right Side–Left Side)				
				Mean (°C)	SD	95% CI Lower	95% CI Upper	Extended Uncertainty
1_1	29.81	5.90	11.82					
1_2	29.80	5.91	11.83	0.01	0.43	−0.08	0.10	1.13
2_1	29.88	5.91	11.84					
2_2	29.96	5.94	11.90	−0.08	0.44	−0.17	0.01	1.14
3_1	29.14	5.74	11.51					
3_2	29.16	5.73	11.48	−0.03	0.49	−0.13	0.07	1.23

Table 2. *Cont.*

ROI	Mean (°C)	SD	Extended Uncertainty	Difference (ΔT, Right Side–Left Side)				
				Mean (°C)	SD	95% CI Lower	95% CI Upper	Extended Uncertainty
4_1	29.60	5.77	11.57					
4_2	29.64	5.80	11.62	−0.04	0.46	−0.13	0.06	1.18
5_1	29.84	5.84	11.69					
5_2	29.86	5.88	11.78	−0.01	0.49	−0.11	0.09	1.24
6_1	29.76	5.83	11.67					
6_2	29.85	5.87	11.75	−0.08	0.54	−0.19	0.03	1.31
7_1	30.51	6.00	12.03					
7_2	30.47	5.96	11.94	0.04	0.51	−0.06	0.15	1.26
8_1	30.11	5.93	11.89					
8_2	29.92	5.84	11.71	0.19	0.54	0.07	0.30	1.32
9_1	29.12	5.70	11.42					
9_2	29.08	5.69	11.41	0.04	0.39	−0.04	0.12	1.07
10_1	29.46	5.70	11.43					
10_2	29.54	5.77	11.57	−0.07	0.42	−0.16	0.01	1.12
11_1	29.80	5.83	11.68					
11_2	29.73	5.83	11.67	0.08	0.41	−0.01	0.16	1.11
12_1	29.24	5.57	11.16					
12_2	29.42	5.65	11.33	−0.18	0.49	−0.29	−0.08	1.24
13_1	29.81	5.75	11.52					
13_2	29.84	5.77	11.57	−0.02	0.42	−0.11	0.07	1.13
14_1	29.69	5.76	11.55					
14_2	29.50	5.70	11.43	0.18	0.43	0.10	0.27	1.14
15_1	29.37	5.61	11.24					
15_2	29.53	5.65	11.32	−0.16	0.54	−0.27	−0.05	1.30
16_1	29.69	5.66	11.34					
16_2	29.59	5.63	11.28	0.10	0.52	−0.01	0.21	1.28
17_1	29.45	5.58	11.18					
17_2	29.27	5.49	11.00	0.18	0.59	0.06	0.30	1.39
18_1	29.45	5.51	11.04					
18_2	29.43	5.51	11.05	0.02	0.53	−0.09	0.13	1.30
19_1	29.42	5.50	11.02					
19_2	29.32	5.47	10.96	0.10	0.66	−0.03	0.24	1.51
Mean	29.63	5.74	11.51	0.01	0.49	−0.09	0.12	1.23

CI, confidence interval; ROI, region of interest; SD, standard deviation.

Table 3. The average temperature of each region of interest in the lateral arm region.

ROI	Mean (°C)	SD	Extended Uncertainty	Difference (ΔT, Right Side–Left Side)				
				Mean (°C)	SD	95% CI Lower	95% CI Upper	Extended Uncertainty
1_1	28.68	5.76	11.55					
1_2	28.88	5.88	11.79	−0.20	0.76	−0.36	−0.04	1.69
2_1	28.73	5.85	11.72					
2_2	28.62	5.86	11.75	0.12	0.56	0.00	0.23	1.34
3_1	28.63	5.84	11.70					
3_2	28.11	5.73	11.49	0.52	0.78	0.36	0.68	1.72
4_1	28.13	5.70	11.42					
4_2	28.19	5.74	11.51	−0.06	0.67	−0.19	0.08	1.52
5_1	28.28	5.73	11.48					
5_2	28.16	5.72	11.47	0.11	0.42	0.02	0.19	1.12
6_1	28.16	5.74	11.51					
6_2	27.88	5.67	11.37	0.28	0.56	0.16	0.40	1.35

Table 3. *Cont.*

ROI	Mean (°C)	SD	Extended Uncertainty	Difference (ΔT, Right Side–Left Side)				
				Mean (°C)	SD	95% CI Lower	95% CI Upper	Extended Uncertainty
7_1	28.64	5.81	11.63					
7_2	28.36	5.73	11.48	0.27	0.76	0.12	0.43	1.69
8_1	29.02	5.89	11.79					
8_2	28.73	5.83	11.68	0.29	0.63	0.16	0.42	1.46
9_1	28.46	5.79	11.60					
9_2	28.22	5.77	11.56	0.23	0.79	0.07	0.39	1.75
10_1	29.14	5.86	11.74					
10_2	28.71	5.79	11.61	0.43	0.67	0.29	0.57	1.54
11_1	29.53	5.90	11.82					
11_2	29.27	5.89	11.80	0.26	0.42	0.17	0.35	1.12
12_1	29.19	5.88	11.79					
12_2	29.11	5.90	11.82	0.07	0.59	−0.05	0.19	1.39
13_1	28.35	5.64	11.31					
13_2	28.12	5.61	11.25	0.22	0.71	0.07	0.37	1.59
14_1	28.76	5.71	11.44					
14_2	28.66	5.71	11.45	0.10	0.60	−0.03	0.22	1.41
15_1	28.47	5.65	11.32					
15_2	28.56	5.72	11.47	−0.10	0.70	−0.24	0.05	1.58
16_1	27.73	5.40	10.82					
16_2	27.59	5.39	10.80	0.14	0.65	0.01	0.28	1.49
17_1	28.01	5.44	10.90					
17_2	27.91	5.42	10.86	0.10	0.46	0.00	0.20	1.19
18_1	27.70	5.47	10.96					
18_2	27.63	5.44	10.91	0.06	0.69	−0.08	0.21	1.57
19_1	28.43	5.50	11.03					
19_2	28.37	5.50	11.03	0.05	0.45	−0.05	0.14	1.17
Mean	28.45	5.71	11.44	0.15	0.62	0.02	0.28	1.46

CI, confidence interval; ROI, region of interest; SD, standard deviation.

Table 4. The average temperature of each region of interest in the medial arm region.

ROI	Mean (°C)	SD	Extended Uncertainty	Difference (ΔT, Right Side–Left Side)				
				Mean (°C)	SD	95% CI Lower	95% CI Upper	Extended Uncertainty
1_1	28.60	5.48	10.98					
1_2	28.57	5.47	10.97	0.03	0.49	−0.07	0.13	1.24
2_1	28.68	5.43	10.88					
2_2	28.70	5.45	10.93	−0.01	0.45	−0.11	0.08	1.17
3_1	29.46	5.79	11.60					
3_2	29.36	5.76	11.54	0.10	0.45	0.01	0.20	1.17
4_1	28.85	5.57	11.16					
4_2	28.77	5.56	11.14	0.08	0.40	0.00	0.17	1.09
5_1	29.39	5.84	11.71					
5_2	29.31	5.80	11.63	0.08	0.43	−0.02	0.17	1.14
6_1	29.02	5.70	11.42					
6_2	28.91	5.66	11.34	0.11	0.42	0.02	0.20	1.12
7_1	28.00	5.23	10.50					
7_2	28.10	5.28	10.58	−0.10	0.69	−0.25	0.04	1.57
Mean	28.84	5.57	11.17	0.04	0.48	−0.06	0.14	1.21

CI, confidence interval; ROI, region of interest; SD, standard deviation.

Table 5. The average temperature of each region of interest in the dorsal hand region.

ROI	Mean (°C)	SD	Extended Uncertainty	Difference (ΔT, Right Side–Left Side)				
				Mean (°C)	SD	95% CI Lower	95% CI Upper	Extended Uncertainty
1_1	31.25	5.51	11.05					
1_2	31.11	5.53	11.08	0.13	0.44	0.04	0.22	1.15
2_1	31.27	5.53	11.08					
2_2	31.01	5.49	11.00	0.26	0.49	0.15	0.36	1.22
3_1	30.85	5.38	10.79					
3_2	30.73	5.37	10.76	0.11	0.59	−0.01	0.23	1.41
4_1	30.56	5.29	10.60					
4_2	30.60	5.31	10.65	−0.04	0.48	−0.14	0.06	1.21
5_1	30.76	5.32	10.67					
5_2	30.65	5.33	10.68	0.11	0.48	0.01	0.21	1.22
6_1	30.54	5.25	10.52					
6_2	30.36	5.21	10.46	0.18	0.52	0.07	0.29	1.28
7_1	29.93	5.06	10.15					
7_2	30.04	5.13	10.29	−0.11	0.45	−0.20	−0.01	1.17
8_1	30.34	5.13	10.29					
8_2	30.29	5.13	10.29	0.06	0.44	−0.03	0.15	1.16
9_1	30.37	5.14	10.32					
9_2	30.14	5.11	10.24	0.23	0.48	0.13	0.33	1.21
10_1	30.27	5.16	10.35					
10_2	30.31	5.22	10.46	−0.04	0.53	−0.15	0.07	1.29
11_1	30.47	5.08	10.20					
11_2	30.62	5.13	10.29	−0.16	0.50	−0.26	−0.05	1.25
12_1	30.64	5.52	11.07					
12_2	30.57	5.50	11.03	0.07	0.72	−0.07	0.22	1.62
13_1	30.05	5.13	10.29					
13_2	29.84	5.10	10.23	0.21	0.71	0.06	0.36	1.61
14_1	29.76	4.96	9.95					
14_2	29.69	4.97	9.98	0.06	0.64	−0.07	0.20	1.49
15_1	28.73	5.19	10.40					
15_2	28.77	5.23	10.48	−0.04	0.66	−0.18	0.10	1.51
16_1	28.33	5.05	10.13					
16_2	28.27	5.07	10.16	0.06	0.63	−0.07	0.19	1.47
17_1	28.15	4.91	9.84					
17_2	28.08	4.89	9.82	0.07	0.75	−0.09	0.23	1.67
18_1	29.26	4.96	9.94					
18_2	29.43	4.95	9.92	−0.17	0.57	−0.29	−0.06	1.36
19_1	28.89	4.96	9.95					
19_2	28.88	4.97	9.96	0.01	0.61	−0.11	0.14	1.43
20_1	28.77	5.13	10.29					
20_2	28.74	5.15	10.32	0.02	0.80	−0.14	0.18	1.76
21_1	29.28	4.96	9.95					
21_2	29.25	4.90	9.83	0.03	0.52	−0.08	0.14	1.28
22_1	28.99	5.07	10.17					
22_2	28.98	4.90	9.84	0.00	0.69	−0.14	0.14	1.56
23_1	29.11	5.12	10.28					
23_2	28.97	5.03	10.10	0.14	0.74	−0.02	0.29	1.65
24_1	29.29	4.87	9.77					
24_2	29.11	4.82	9.67	0.17	0.55	0.06	0.29	1.32
25_1	29.10	4.95	9.94					
25_2	29.02	4.86	9.74	0.08	0.64	−0.05	0.22	1.48
26_1	29.10	5.14	10.32					
26_2	28.88	5.05	10.13	0.21	0.79	0.05	0.38	1.75
Mean	29.74	5.14	10.30	0.06	0.59	−0.06	0.19	1.40

CI, confidence interval; ROI, region of interest; SD, standard deviation.

Table 6. The average temperature of each region of interest in the ventral hand region.

ROI	Mean (°C)	SD	Extended Uncertainty	Difference (ΔT, Right Side–Left Side)				
				Mean (°C)	SD	95% CI Lower	95% CI Upper	Extended Uncertainty
1_1	29.10	5.66	11.34					
1_2	29.09	5.67	11.36	0.01	0.40	−0.07	0.10	1.09
2_1	29.63	5.73	11.47					
2_2	29.71	5.77	11.57	−0.08	0.41	−0.17	0.00	1.11
3_1	29.73	5.80	11.62					
3_2	29.57	5.76	11.54	0.17	0.38	0.08	0.25	1.06
4_1	29.01	5.53	11.08					
4_2	29.13	5.59	11.20	−0.11	0.51	−0.22	−0.01	1.27
5_1	29.82	5.74	11.49					
5_2	29.85	5.76	11.55	−0.03	0.44	−0.13	0.06	1.16
6_1	29.53	5.70	11.43					
6_2	29.38	5.65	11.32	0.16	0.45	0.07	0.25	1.17
7_1	29.19	5.52	11.06					
7_2	29.33	5.60	11.23	−0.15	0.57	−0.26	−0.03	1.37
8_1	29.69	5.60	11.22					
8_2	29.62	5.59	11.21	0.07	0.57	−0.05	0.18	1.38
9_1	29.31	5.52	11.06					
9_2	29.12	5.50	11.02	0.19	0.68	0.05	0.33	1.55
10_1	28.16	5.13	10.29					
10_2	28.17	5.14	10.30	0.00	0.63	−0.14	0.13	1.47
11_1	28.70	5.33	10.68					
11_2	28.80	5.36	10.74	−0.10	0.83	−0.27	0.08	1.81
12_1	29.32	5.51	11.05					
12_2	29.30	5.57	11.16	0.02	0.73	−0.13	0.17	1.64
13_1	29.98	5.60	11.22					
13_2	29.88	5.60	11.22	0.10	0.63	−0.03	0.23	1.48
14_1	29.15	5.51	11.03					
14_2	29.09	5.47	10.96	0.06	0.86	−0.12	0.24	1.88
15_1	27.55	5.26	10.55					
15_2	27.49	5.32	10.67	0.06	0.87	−0.12	0.24	1.89
16_1	28.16	5.23	10.48					
16_2	28.05	5.31	10.65	0.11	0.84	−0.06	0.29	1.84
17_1	28.41	5.25	10.52					
17_2	28.37	5.27	10.57	0.04	0.69	−0.10	0.18	1.56
18_1	27.71	5.25	10.52					
18_2	27.52	5.21	10.44	0.19	0.82	0.02	0.36	1.80
19_1	28.16	5.31	10.64					
19_2	27.97	5.34	10.70	0.20	0.88	0.02	0.38	1.91
20_1	28.41	5.35	10.72					
20_2	28.30	5.36	10.74	0.11	0.74	−0.04	0.26	1.66
21_1	27.35	5.29	10.61					
21_2	27.26	5.23	10.49	0.09	0.84	−0.09	0.26	1.85
22_1	28.00	5.36	10.75					
22_2	27.88	5.25	10.54	0.12	0.85	−0.06	0.29	1.86
23_1	28.40	5.36	10.75					
23_2	28.20	5.30	10.63	0.20	0.76	0.04	0.36	1.70
24_1	27.25	5.16	10.35					
24_2	27.08	5.14	10.31	0.17	0.77	0.01	0.33	1.71
25_1	27.59	5.34	10.71					
25_2	27.41	5.33	10.69	0.19	0.79	0.02	0.35	1.75
26_1	27.84	5.48	10.98					
26_2	27.65	5.48	10.98	0.19	0.93	−0.01	0.38	2.01
Mean	28.62	5.44	10.91	0.07	0.69	−0.07	0.22	1.58

CI, confidence interval; ROI, region of interest; SD, standard deviation.

The average temperatures of the ROIs differed significantly between the six regions ($p < 0.001$, ANOVA) (Figure 2).

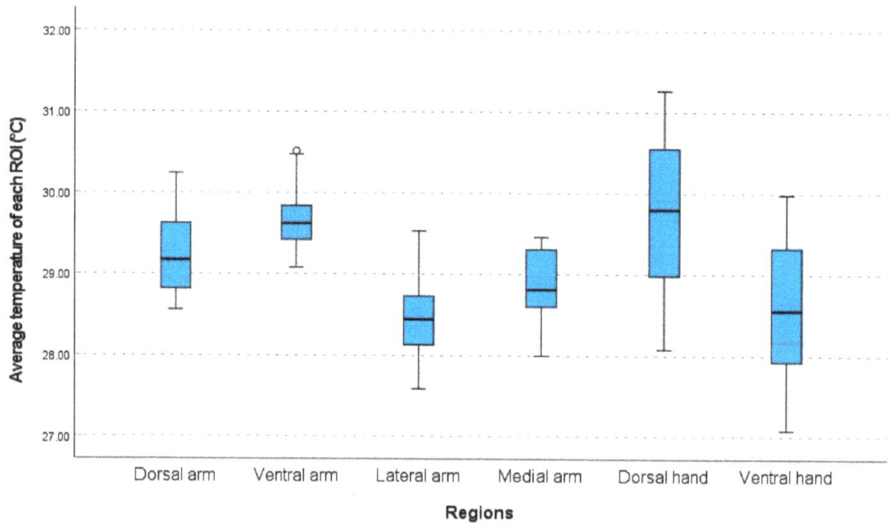

Figure 2. Comparison of the average temperatures of each ROI in the six regions. ROI: region of interest; and ° indicates a label out of range.

The overall median of the average of the absolute value of the ΔT (|ΔT|, $\sqrt{[\text{temperature of the right side} - \text{left side}]^2}$) was 0.10 (range, 0.00–0.52). Additionally, |ΔT| values differed significantly between the six regions ($p < 0.001$, ANOVA). In particular, the |ΔT| in the lateral arm region was significantly larger (median 0.14 [range, 0.05–0.52]) than that of any other regions (Figure 3).

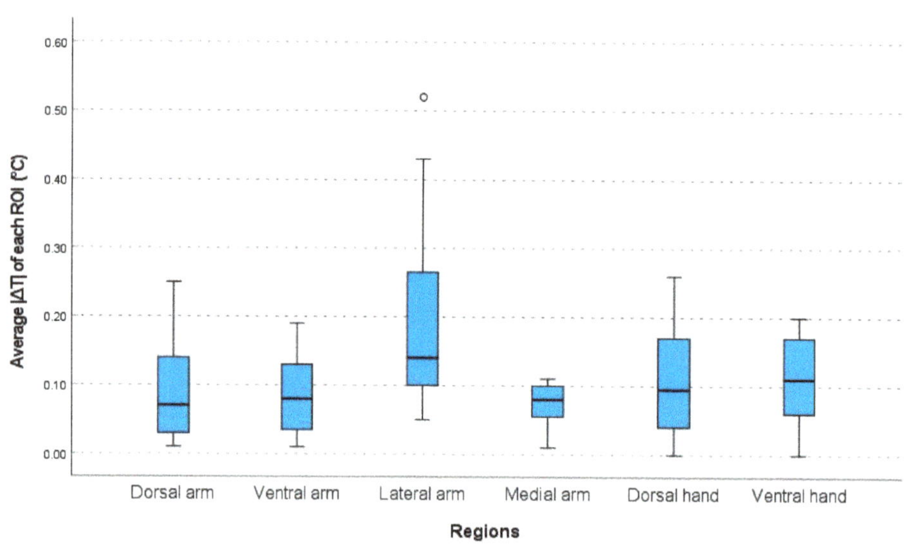

Figure 3. Comparison of the absolute value of the temperature discrepancies between the right and left side (|ΔT|) of each ROI in the six regions. ROI: region of interests; and ° indicates a label out of range.

3.3. Subgroup Analysis Based on ROI Location

Distally located ROIs had significantly lower average surface temperatures, except for the lateral arm region, as indicated by the regression analysis. The relationship between the average temperature of each ROI and the proximity of ROIs were as follows: dorsal arm region temperature = 29.916 − (0.181 × ROIs) ($p < 0.001$, $R^2 = 0.480$); ventral arm region temperature = 29.915 − (0.070 × ROIs) ($p = 0.012$, $R^2 = 0.162$); lateral arm region temperature = 28.684 − (0.063 × ROIs) ($p = 0.132$, $R^2 = 0.062$); medial arm region temperature = 29.591 − (0.188 × ROIs) ($p < 0.001$, $R^2 = 0.750$); dorsal hand region temperature = 31.474 − (0.399 × ROIs) ($p < 0.001$, $R^2 = 0.799$); and ventral hand region temperature = 29.855 − (0.284 × ROIs) ($p < 0.001$, $R^2 = 0.429$) (Figure 4).

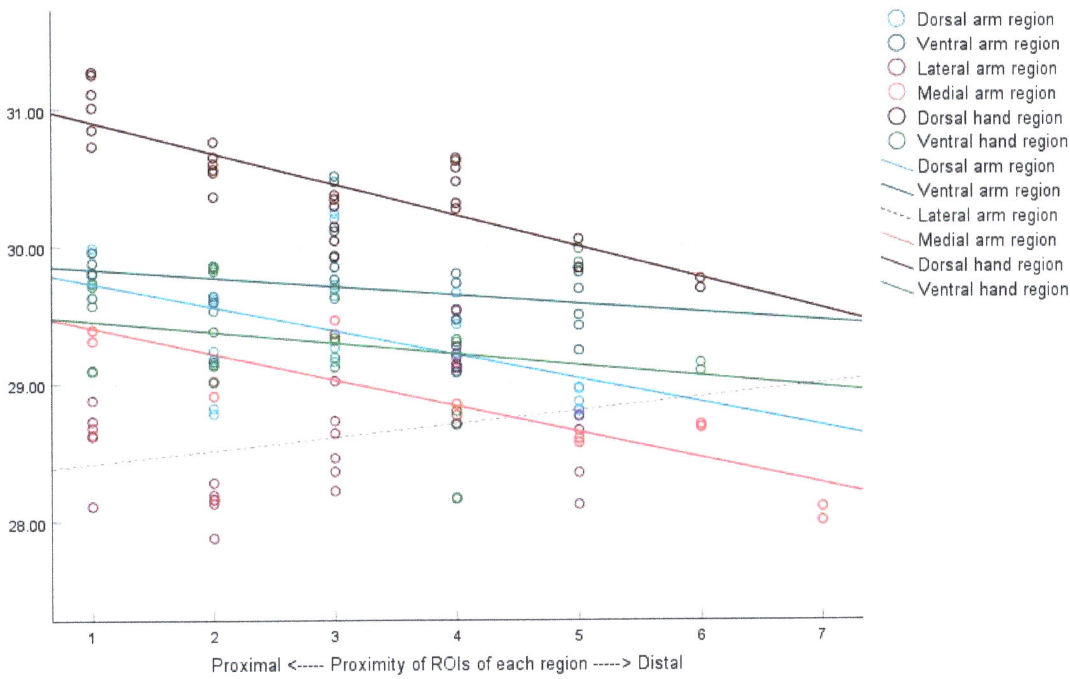

Figure 4. Correlation between the average temperatures of each ROI and the proximity of ROIs to the body core of the six regions, as determined by linear regression analysis. The trend observed in the figure demonstrates a gradual decrease in surface temperature from proximal to distal for each ROI, except for the lateral arm region. ROIs, regions of interest.

The average $|\Delta T|$ values of the ROIs did not show any specific trend. The relationship between the average $|\Delta T|$ of each ROI and the proximity of ROIs were as follows: dorsal arm region $|\Delta T|$ = 0.074 + (0.006 × ROIs) ($p = 0.535$, $R^2 = 0.024$); ventral arm region $|\Delta T|$ = 0.031 + (0.013 × ROIs) ($p = 0.090$, $R^2 = 0.160$); lateral arm region $|\Delta T|$ = 0.299 − (0.030 × ROIs) ($p = 0.066$, $R^2 = 0.185$); medial arm region $|\Delta T|$ = 0.090 − (0.001 × ROIs) ($p = 0.855$, $R^2 = 0.007$); dorsal hand region $|\Delta T|$ = 0.157 − (0.012× ROIs) ($p = 0.117$, $R^2 = 0.099$); and ventral hand region $|\Delta T|$ = 0.078 + (0.008 × ROIs) ($p = 0.225$, $R^2 = 0.061$) (Figure 5).

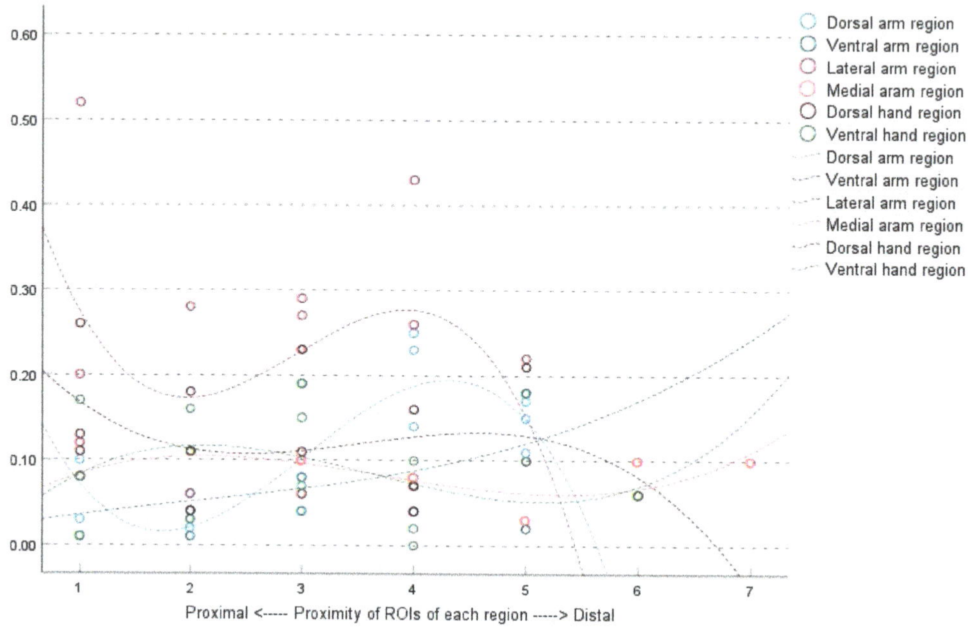

Figure 5. Correlation between right to left temperature discrepancies and the proximity of ROIs to the body core of the six regions, as determined by linear regression analysis. No perceptible correlation was observed between the absolute temperature difference (|ΔT|) and the proximity of each ROI. ROIs, regions of interest.

3.4. Subgroup Analysis Based on Sex

The average temperatures of each same ROI were significantly higher in male subjects compared to female subjects across all regions and age groups ($p < 0.001$, paired t-test). The average temperatures for each region were as follows: 30.31 ± 5.43 (range, 29.45 ± 5.83–32.51 ± 4.59) for men and 28.32 ± 5.94 (range, 26.06 ± 5.85–29.59 ± 5.82) for women in the dorsal arm region; 30.72 ± 5.47 (range, 29.9 ± 5.93–32.89 ± 4.55) for men and 28.73 ± 5.97 (range, 26.51 ± 5.92–30.05 ± 5.86) for women in the ventral arm region; 29.66 ± 5.45 (range, 28.76 ± 5.89–31.67 ± 4.60) for men and 27.44 ± 5.92 (range, 25.13 ± 5.85–28.81 ± 5.73) for women in the lateral arm region; 29.84 ± 5.33 (range, 29.11 ± 5.72–31.84 ± 4.56) for men and 28.00 ± 5.78 (range, 25.80 ± 5.64–29.34 ± 5.72) for women in the medial arm region; 30.51 ± 4.93 (range, 29.67 ± 5.33–32.38 ± 3.83) for men and 28.81 ± 5.38 (range, 27.34 ± 5.21–29.38 ± 5.25) for women in the dorsal hand region; and 29.71 ± 5.22 (range, 29.12 ± 5.14–31.30 ± 4.55) for men and 27.72 ± 5.63 (range, 25.48 ± 5.43–29.14 ± 5.67) for women in the ventral hand region (Table 7).

Table 7. Comparison of the average temperature in the same region of interest according to sex and age group.

Characteristics	Male (°C)	Female (°C)	Mean (°C)	Difference between Sexes (°C)	p Value
Dorsal arm region					<0.001 [a]
20s (n = 183)	29.82 ± 5.37	28.99 ± 6.05	29.43 ± 5.69	0.83 (95% CI, 0.57–1.09)	<0.001 [b]
30s (n = 213)	29.45 ± 5.83	26.06 ± 5.85	27.79 ± 5.84	3.39 (95% CI, 3.19–3.59)	<0.001 [b]
40s (n = 228)	30.44 ± 5.55	28.58 ± 6.41	29.47 ± 6.00	1.86 (95% CI, 1.61–2.12)	<0.001 [b]
50s (n = 177)	31.15 ± 5.04	28.82 ± 5.53	29.68 ± 5.35	2.33 (95% CI, 2.12–2.54)	<0.001 [b]
60s (n = 104)	32.51 ± 4.59	29.59 ± 5.82	30.49 ± 5.44	2.92 (95% CI, 2.65–3.18)	<0.001 [b]
Sum	30.31 ± 5.43	28.32 ± 5.94	29.22 ± 5.71	1.99 (95% CI, 1.76–2.22	<0.001 [b]

Table 7. *Cont.*

Characteristics	Male (°C)	Female (°C)	Mean (°C)	Difference between Sexes (°C)	*p* Value
Ventral arm region					<0.001 [a]
20s (n = 183)	30.18 ± 5.41	29.40 ± 6.12	29.81 ± 5.74	0.78 (95% CI, 0.60–0.96)	<0.001 [b]
30s (n = 213)	29.95 ± 5.93	26.51 ± 5.92	28.25 ± 5.93	3.44 (95% CI, 3.31–3.58)	<0.001 [b]
40s (n = 228)	30.82 ± 5.59	28.90 ± 6.35	29.82 ± 5.99	1.92 (95% CI, 1.73–2.10)	<0.001 [b]
50s (n = 177)	31.59 ± 5.05	29.25 ± 5.56	30.11 ± 5.37	2.34 (95% CI, 2.19–2.49)	<0.001 [b]
60s (n = 104)	32.89 ± 4.55	30.05 ± 5.86	30.92 ± 5.46	2.84 (95% CI, 2.64–3.05)	<0.001 [b]
Sum	30.72 ± 5.47	28.73 ± 5.97	29.63 ± 5.74	2.00 (95% CI, 1.84–2.16)	<0.001 [b]
Lateral arm region					<0.001 [a]
20s (n = 183)	29.18 ± 5.40	28.29 ± 6.16	28.76 ± 5.76	0.90 (95% CI, 0.66–1.14)	<0.001 [b]
30s (n = 213)	28.76 ± 5.89	25.13 ± 5.85	26.97 ± 5.87	3.63 (95% CI, 3.43–3.83)	<0.001 [b]
40s (n = 228)	29.85 ± 5.52	27.62 ± 6.31	28.69 ± 5.93	2.23 (95% CI, 1.99–2.48)	<0.001 [b]
50s (n = 177)	30.57 ± 5.07	27.88 ± 5.52	28.87 ± 5.35	2.69 (95% CI, 2.48–2.90)	<0.001 [b]
60s (n = 104)	31.67 ± 4.60	28.81 ± 5.73	29.69 ± 5.38	2.85 (95% CI, 2.58–3.13)	<0.001 [b]
Sum	29.66 ± 5.45	27.44 ± 5.92	28.62 ± 5.67	2.22 (95% CI, 2.00–2.45)	<0.001 [b]
Medial arm region					<0.001 [a]
20s (n = 183)	29.25 ± 5.25	28.62 ± 5.97	28.94 ± 5.60	0.63 (95% CI, 0.24–1.01)	0.003 [b]
30s (n = 213)	29.11 ± 5.72	25.80 ± 5.64	27.38 ± 5.68	3.15 (95% CI, 2.90–3.40)	<0.001 [b]
40s (n = 228)	29.96 ± 5.44	28.21 ± 6.18	28.85 ± 5.91	1.74 (95% CI, 1.29–2.20)	<0.001 [b]
50s (n = 177)	30.77 ± 4.98	28.50 ± 5.37	29.20 ± 5.25	2.27 (95% CI, 1.96–2.58)	<0.001 [b]
60s (n = 104)	31.84 ± 4.56	29.34 ± 5.72	30.48 ± 5.19	2.51 (95% CI, 2.06–2.95)	<0.001 [b]
Sum	29.84 ± 5.33	28.00 ± 5.78	28.84 ± 5.57	1.84 (95% CI, 1.49–2.20)	<0.001 [b]
Dorsal hand region					<0.001 [a]
20s (n = 183)	29.86 ± 5.31	29.04 ± 5.81	29.46 ± 5.56	0.82 (95% CI, 0.45–1.18)	<0.001 [b]
30s (n = 213)	32.38 ± 3.83	29.38 ± 5.25	30.81 ± 4.57	3.00 (95% CI, 2.63–3.38)	<0.001 [b]
40s (n = 228)	30.70 ± 4.92	28.93 ± 5.36	29.87 ± 5.13	1.77 (95% CI, 1.44–2.11)	<0.001 [b]
50s (n = 177)	29.67 ± 5.33	27.34 ± 5.21	28.52 ± 5.27	2.33 (95% CI, 2.01–2.66)	<0.001 [b]
60s (n = 104)	29.58 ± 5.51	28.76 ± 4.98	29.15 ± 5.23	0.83 (95% CI, 0.51–1.14)	<0.001 [b]
Sum	30.51 ± 4.93	28.81 ± 5.38	29.74 ± 5.14	1.70 (95% CI, 1.36–2.05)	<0.001 [b]
Ventral hand region					<0.001 [a]
20s (n = 183)	29.12 ± 5.14	28.07 ± 5.79	28.39 ± 5.59	1.05 (95% CI, 0.70–1.40)	<0.001 [b]
30s (n = 213)	29.28 ± 5.78	25.48 ± 5.43	27.21 ± 5.59	3.80 (95% CI, 3.49–4.10)	<0.001 [b]
40s (n = 228)	29.52 ± 5.21	27.88 ± 5.80	28.75 ± 5.49	1.64 (95% CI, 1.27–2.02)	<0.001 [b]
50s (n = 177)	30.86 ± 4.75	28.45 ± 5.49	29.67 ± 5.11	2.41 (95% CI, 2.12–2.70)	<0.001 [b]
60s (n = 104)	31.30 ± 4.55	29.14 ± 5.67	30.17 ± 5.13	2.16 (95% CI, 1.74–2.59)	<0.001 [b]
Sum	29.71 ± 5.22	27.72 ± 5.63	28.62 ± 5.44	2.00 (95% CI, 1.67–2.33)	<0.001 [b]

[a] analysis of variance, [b] paired t-test.

The temperature difference between both sexes ranged from 0 °C 63 (95% CI, 0.24–1.01) to 3.80 °C (95% CI, 3.49–4.10), depending on region and age group (Figure 6).

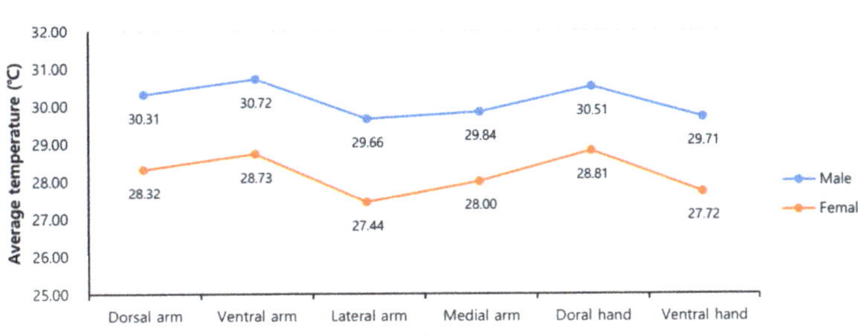

Figure 6. The average regional temperatures in men and women.

3.5. Subgroup Analysis Based on Age Group

The average temperatures of the ROIs significantly increased with age, particularly in individuals in their 30s and older in all regions, except for the dorsal hand region ($p < 0.001$, ANOVA). However, subjects in their 20s had a higher surface temperature than those in their 30s ($p < 0.001$, paired t-test). Consequently, surface temperatures were lowest for those in their 30s for all regions, except for the dorsal hand region ($p < 0.001$, ANOVA post hoc analysis) (Table 7 and Figure 7).

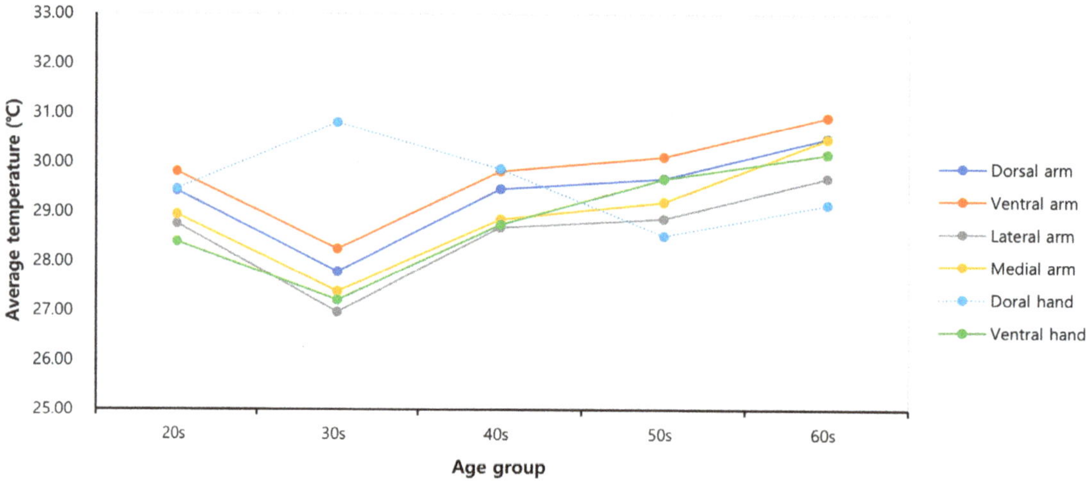

Figure 7. Relationship between the average regional temperatures and subject age.

4. Discussion

4.1. Average Temperature and |ΔT| of Each ROI

The average surface temperature of the upper limbs varied depending on regions and ROIs, ranging from 27.08 ± 5.14 °C to 31.27 ± 5.53 °C, which is consistent with the results found in our previous study on the lower limbs. However, the average temperatures of the upper limbs were higher than those of the lower limbs (range, 24.60 ± 5.06 °C–27.75 ± 5.76 °C) [20]. We believe this difference is due to the different distance from the heart [21–23].

The average of |ΔT| of each ROI also varied by regions and ROIs, ranging from 0.00 to 0.52 °C. However, these values are smaller compared to the reference standards of the lower limbs, where |ΔT| reached 0.76 °C [20]. Nevertheless, the practical value of |ΔT| of the upper limbs was not within 0.1–0.3 °C, which is considered the normal range, based on the previous consensus [15,24,25].

These data for each ROI provide reference standards for DITI of the upper limbs. Clinically significant cold/hot areas, or specific areas with a significant difference between both sides, can be detected based on comparative analysis between the practical patient's image and these data. Additionally, the detailed ROIs in this study can be modified simply during the processing of DITI capture in actual clinical practice. However, accurate diagnosis requires a comparison of each ROI, not just the averages of whole regions, as normal ranges of surface temperature and ΔT values vary depending on region or ROI.

Various intrinsic factors, such as sex, age, fat percentage, and menstrual cycle stage, and extrinsic factors, such as test environment, testing time, and season, can influence the results [26–29]. As a result, actual DITI measurements may fall outside the reference standard values. Therefore, further clinical studies are necessary to validate these data by comparing the DITI results of patients with specific diseases with those of healthy subjects.

4.2. Subgroup Analysis Based on Proximity of ROIs, Sex, and Age Group

A significant correlation was found between the proximity of ROIs to the body core and surface temperature in all regions, except for the lateral arm region, i.e., surface temperatures decreased from the proximal regions to the distal ends. These findings are consistent with a previous suggestion that the surface temperatures of peripheral regions (e.g., hand or foot) tend to be cooler than central regions (e.g., trunk or proximal arm), due to the greater distances from the main thermal organs, such as the heart, large vessels, or viscera [20–23]. This tendency was found to be greater for the upper limbs than lower limbs, which can be attributed to the fact that the upper limbs are closer to the heart than the lower limbs, resulting in the proximal region being distinctly warmer than the distal region [20].

However, the $|\Delta T|$ values showed no specific trend based on proximity of ROIs to the body core, which is different from a previous suggestion that the $|\Delta T|$ values are higher in the distal regions compared to the proximal regions [2,15]. This was also found in the lower limbs in a previous study [20].

Surface temperatures were found to be higher for men in all regions, which is consistent with suggestions that surface temperatures are lower for women due to the insulating effect of thicker subcutaneous fat [30]. This trend was also observed for the lower limbs [20].

Surface temperatures increased significantly with age from the 30s, except for the dorsal hand region, which is also consistent with previous research on the lower extremities [20]. These trends may be caused by age-related diminished vasoconstriction by the sympathetic nervous system [31–33]. The paradoxical higher temperatures for those in their 20s can be explained by a higher basal metabolic rate and less subcutaneous fat [34,35].

4.3. Limitations and Significance

Establishing reference standards for DITI is a complex task due to the influence of various confounding factors on skin temperature and DITI measurements [36]. Therefore, it is important to acknowledge several limitations of our study. First, body fat percentage is a significant confounding factor in measuring body surface temperature using DITI [37] However, we did not collect data on body fat percentage or calculate the body mass index based on body weight and height. Secondly, daily biorhythms, such as menstrual cycle status, menopause, sleep patterns, and emotional stress, can also impact DITI results [38,39]. Unfortunately, we did not account for these factors during the examination. Thirdly, various extrinsic factors, including diurnal testing time, season, and ethnicity were not considered [39]. Moreover, even though we recruited healthy adult volunteers based on questionnaire responses, the process does not guarantee the exclusion of specific diseases or conditions that may have affected DITI results.

Nonetheless, the study holds value due to its large sample size and the use of a consistent protocol for measurements. It represents the first attempt to establish reference standards for DITI measurements of the upper limbs.

5. Conclusions

The findings of this study provide a basis for establishing reference standards for DITI measurements of surface temperatures in the upper extremities. These standards can aid physicians in making objective diagnoses by comparing patient DITI results with the provided data. However, it is crucial to consider the influence of various confounding factors in surface temperature measurements. Moreover, before confirming the results, it is important to consider several parameters, including the specific location of the ROIs, sex, and age.

Author Contributions: Conceptualization: S.S.; methodology: S.S. and B.R.Y.; software: S.S. and B.R.Y.; validation: S.S. and B.R.Y.; formal analysis: S.S.; investigation: H.Y.Z.; resources: H.Y.Z.; data curation: S.S. and B.R.Y.; writing—original draft preparation: S.S. and B.R.Y.; writing—review and editing: S.S. and B.R.Y.; visualization: S.S.; supervision: H.Y.Z.; project administration: H.Y.Z.; funding acquisition: H.Y.Z. All authors have read and agreed to the published version of the manuscript.

Funding: This research was supported by the Korea Evaluation Institute of Industrial Technology (Grant no. 20016363) and the National Research Foundation of Korea (NRF) funded by the Korean government (MIST) (Grant no. 2021M3I2A1077405).

Institutional Review Board Statement: The study was conducted according to the guidelines of the Declaration of Helsinki and approved by the Institutional Review Board of Ilsan Hospital (IRB No. NHIMC 2018–01-018, 20 January 2018), Gangnam Severance Hospital (IRB No. 3-2018-0021, 5 January 2018), and Ajou University Hospital (IRB No, AJIRB-MED-MDB-18-032, 2 February 2018).

Informed Consent Statement: Informed consent was obtained from all subjects involved in the study.

Data Availability Statement: The data presented in this study are available from the corresponding author on receipt of reasonable request and are not publicly available.

Acknowledgments: We express our gratitude to the statistician, Tae Mi Youk, and surveyors, Bo Eun Lee, Ko Woon Yoon, Ho Gun Lee, Ji Eun Lee, and Ji Hye Lee, for conducting the DITI examinations and analysis.

Conflicts of Interest: The authors declare no conflict of interest.

References

1. Verstockt, J.; Verspeek, S.; Thiessen, F.; Tjalma, W.A.; Brochez, L.; Steenackers, G. Skin Cancer Detection Using Infrared Thermography: Measurement Setup, Procedure and Equipment. *Sensors* **2022**, *22*, 3327. [CrossRef] [PubMed]
2. Park, T.Y.; Son, S.; Lim, T.G.; Jeong, T. Hyperthermia associated with spinal radiculopathy as determined by digital infrared thermographic imaging. *Medicine* **2020**, *99*, e19483. [CrossRef] [PubMed]
3. Lim, M.J.; Kwon, S.R.; Jung, K.H.; Joo, K.; Park, S.G.; Park, W. Digital thermography of the fingers and toes in Raynaud's phenomenon. *J. Korean Med. Sci.* **2014**, *29*, 502–506. [CrossRef] [PubMed]
4. García Becerra, A.; Olguín Tiznado, J.E.; García Alcaraz, J.L.; Camargo Wilson, C.; López Barreras, J.A.; Cano Gutiérrez, J.C.; Garcia-Rivera, R.B. Temperature Asymmetry Analysis between Left and Right Wrist with Sensory and Infrared Thermography. *Int. J. Environ. Res. Public Health* **2022**, *19*, 10240. [CrossRef] [PubMed]
5. Dias de Lacerda, A.P.; Rodrigues de Andrade, P.; Kamonseki, D.H.; Parizotto, N.A.; Alves da Silva, A.S.; Bernardo de Medeiros, L.; de Almeida Ferreira, J.J. Accuracy of infrared thermography in detecting tendinopathy: A systematic review with meta-analysis. *Phys. Ther. Sport Off. J. Assoc. Chart. Physiother. Sport. Med.* **2022**, *58*, 117–125. [CrossRef]
6. de Almeida, A.N.S.; de Souza Ferreira, S.L.; Balata, P.M.M.; da Cunha, D.A.; Pernambuco, L.; da Silva, H.J. Thermography in complementary assessments of head and neck muscles: A scoping review. *J. Oral Rehabil.* **2022**, *49*, 1188–1196. [CrossRef]
7. Ćurković, S.; Antabak, A.; Halužan, D.; Luetić, T.; Prlić, I.; Šiško, J. Medical thermography (digital infrared thermal imaging—DITI) in paediatric forearm fractures—A pilot study. *Injury* **2015**, *46* (Suppl. S6), S36–S39. [CrossRef]
8. Chlebicka, I.; Matusiak, Ł.; Maj, J.; Baran, E.; Szepietowski, J.C. Freezing fingers syndrome, primary and secondary Raynaud's phenomenon: Characteristic features with hand thermography. *Acta Derm. Venereol.* **2013**, *93*, 428–432. [CrossRef]
9. Branco, J.H.L.; Branco, R.L.L.; Siqueira, T.C.; de Souza, L.C.; Dalago, K.M.S.; Andrade, A. Clinical applicability of infrared thermography in rheumatic diseases: A systematic review. *J. Therm. Biol.* **2022**, *104*, 103172. [CrossRef]
10. Albuquerque, N.F.; Lopes, B.S. Musculoskeletal applications of infrared thermography on back and neck syndromes: A systematic review. *Eur. J. Phys. Rehabil. Med.* **2021**, *57*, 386–396. [CrossRef]
11. Capo, A.; Ismail, E.; Cardone, D.; Celletti, E.; Auriemma, M.; Sabatini, E.; Merla, A.; Amerio, P. Joint functional impairment and thermal alterations in patients with Psoriatic Arthritis: A thermal imaging study. *Microvasc. Res.* **2015**, *102*, 86–91. [CrossRef] [PubMed]
12. Park, Y.E.; Lee, S.E.; Eom, Y.S.; Cho, J.M.; Yang, J.W.; Kim, M.S.; Kwon, H.D.; Lee, J.W.; Park, D. Infrared thermographic changes after decompression surgery in patients with carpal tunnel syndrome. *BMC Musculoskelet. Disord.* **2023**, *24*, 79. [CrossRef] [PubMed]
13. Park, D.; Kim, B.H.; Lee, S.E.; Kim, D.Y.; Eom, Y.S.; Cho, J.M.; Yang, J.W.; Kim, M.; Kwon, H.D.; Lee, J.W. Application of digital infrared thermography for carpal tunnel syndrome evaluation. *Sci. Rep.* **2021**, *11*, 21963. [CrossRef]
14. Edeiken, J.; Shaber, G. Thermography: A reevaluation. *Skelet. Radiol.* **1986**, *15*, 545–548. [CrossRef]
15. Zhang, H.Y.; Kim, Y.S.; Cho, Y.E. Thermatomal changes in cervical disc herniations. *Yonsei Med. J.* **1999**, *40*, 401–412. [CrossRef] [PubMed]
16. Hoffman, R.M.; Kent, D.L.; Deyo, R.A. Diagnostic accuracy and clinical utility of thermography for lumbar radiculopathy. A meta-analysis. *Spine* **1991**, *16*, 623–628. [CrossRef]

17. Berner, J.E.; Pereira, N.; Troisi, L.; Will, P.; Nanchahal, J.; Jain, A. Accuracy of infrared thermography for perforator mapping: A systematic review and meta-analysis of diagnostic studies. *J. Plast. Reconstr. Aesthetic Surg.* **2021**, *74*, 1173–1179. [CrossRef]
18. Kim, G.N.; Zhang, H.Y.; Cho, Y.E.; Ryu, S.J. Differential Screening of Herniated Lumbar Discs Based on Bag of Visual Words Image Classification Using Digital Infrared Thermographic Images. *Healthcare* **2022**, *10*, 1094. [CrossRef]
19. Filippini, C.; Cardone, D.; Perpetuini, D.; Chiarelli, A.M.; Gualdi, G.; Amerio, P.; Merla, A. Convolutional Neural Networks for Differential Diagnosis of Raynaud's Phenomenon Based on Hands Thermal Patterns. *Appl. Sci.* **2021**, *11*, 3614. [CrossRef]
20. Zhang, H.Y.; Son, S.; Yoo, B.R.; Youk, T.-M. Reference Standard for Digital Infrared Thermography of the Surface Temperature of the Lower Limbs. *Bioengineering* **2023**, *10*, 283. [CrossRef]
21. Abe, N.; Kodama, H. Distal-proximal skin temperature gradient prior to sleep onset in infants for clinical use. *Pediatr. Int. Off. J. Jpn. Pediatr. Soc.* **2015**, *57*, 227–233. [CrossRef] [PubMed]
22. Johnson, J.M.; Kellogg, D.L., Jr. Thermoregulatory and thermal control in the human cutaneous circulation. *Front. Biosci. (Sch. Ed.)* **2010**, *2*, 825–853. [CrossRef]
23. Braverman, I.M. The cutaneous microcirculation: Ultrastructure and microanatomical organization. *Microcirculation* **1997**, *4*, 329–340. [CrossRef] [PubMed]
24. Uematsu, S.; Edwin, D.H.; Jankel, W.R.; Kozikowski, J.; Trattner, M. Quantification of thermal asymmetry. Part 1: Normal values and reproducibility. *J. Neurosurg.* **1988**, *69*, 552–555. [CrossRef]
25. Uematsu, S.; Jankel, W.R.; Edwin, D.H.; Kim, W.; Kozikowski, J.; Rosenbaum, A.; Long, D.M. Quantification of thermal asymmetry. Part 2: Application in low-back pain and sciatica. *J. Neurosurg.* **1988**, *69*, 556–561. [CrossRef]
26. Neves, E.B.; Salamunes, A.C.C.; de Oliveira, R.M.; Stadnik, A.M.W. Effect of body fat and gender on body temperature distribution. *J. Therm. Biol.* **2017**, *70*, 1–8. [CrossRef]
27. Chudecka, M.; Lubkowska, A. Thermal Imaging of Body Surface Temperature Distribution in Women with Anorexia Nervosa. *Eur. Eat. Disord. Rev. J. Eat. Disord. Assoc.* **2016**, *24*, 57–61. [CrossRef]
28. Te Lindert, B.H.W.; Van Someren, E.J.W. Skin temperature, sleep, and vigilance. *Handb. Clin. Neurol.* **2018**, *156*, 353–365. [CrossRef]
29. Van Someren, E.J.; Raymann, R.J.; Scherder, E.J.; Daanen, H.A.; Swaab, D.F. Circadian and age-related modulation of thermoreception and temperature regulation: Mechanisms and functional implications. *Ageing Res. Rev.* **2002**, *1*, 721–778. [CrossRef]
30. Savastano, D.M.; Gorbach, A.M.; Eden, H.S.; Brady, S.M.; Reynolds, J.C.; Yanovski, J.A. Adiposity and human regional body temperature. *Am. J. Clin. Nutr.* **2009**, *90*, 1124–1131. [CrossRef]
31. Kenney, W.L.; Munce, T.A. Invited review: Aging and human temperature regulation. *J. Appl. Physiol.* **2003**, *95*, 2598–2603. [CrossRef]
32. Kenney, W.L.; Armstrong, C.G. Reflex peripheral vasoconstriction is diminished in older men. *J. Appl. Physiol.* **1996**, *80*, 512–515. [CrossRef]
33. Borisov, V.V.; Lin, D.C. Temperature fluctuations in the lower limbs of young and elderly individuals during activities of daily living. *Exp. Gerontol.* **2014**, *57*, 243–249. [CrossRef]
34. Marion, G.S.; McGann, K.P.; Camp, D.L. Core body temperature in the elderly and factors which influence its measurement. *Gerontology* **1991**, *37*, 225–232. [CrossRef] [PubMed]
35. Fox, R.H.; Woodward, P.M.; Exton-Smith, A.N.; Green, M.F.; Donnison, D.V.; Wicks, M.H. Body temperatures in the elderly: A national study of physiological, social, and environmental conditions. *Br. Med. J.* **1973**, *1*, 200–206. [CrossRef] [PubMed]
36. Sivanandam, S.; Anburajan, M.; Venkatraman, B.; Menaka, M.; Sharath, D. Medical thermography: A diagnostic approach for type 2 diabetes based on non-contact infrared thermal imaging. *Endocrine* **2012**, *42*, 343–351. [CrossRef] [PubMed]
37. Morrissey, M.C.; Wu, Y.; Zuk, E.F.; Livingston, J.; Casa, D.J.; Pescatello, L.S. The impact of body fat on thermoregulation during exercise in the heat: A systematic review and meta-analysis. *J. Sci. Med. Sport* **2021**, *24*, 843–850. [CrossRef]
38. Miyake, T.; Doi, M. Roles of the Circadian Clock Mechanism in the Regulation of Daily Rhythms of Body Temperature. *Brain Nerve = Shinkei Kenkyu No Shinpo* **2022**, *74*, 159–166. [CrossRef]
39. Costa, C.M.A.; Moreira, D.G.; Sillero-Quintana, M.; Brito, C.J.; de Azambuja Pussieldi, G.; de Andrade Fernandes, A.; Cano, S.P.; Bouzas Marins, J.C. Daily rhythm of skin temperature of women evaluated by infrared thermal imaging. *J. Therm. Biol.* **2018**, *72*, 1–9. [CrossRef]

bioengineering

Article

Comparing Performance of Spectral Image Analysis Approaches for Detection of Cellular Signals in Time-Lapse Hyperspectral Imaging Fluorescence Excitation-Scanning Microscopy

Marina Parker [1,2,†], Naga S. Annamdevula [3,4,†], Donald Pleshinger [3,4], Zara Ijaz [5], Josephine Jalkh [5], Raymond Penn [6], Deepak Deshpande [6], Thomas C. Rich [3,4] and Silas J. Leavesley [1,2,3,4,*]

1 Department of Chemical and Biomolecular Engineering, University of South Alabama, 150 Student Services Dr., Mobile, AL 36688, USA
2 Department of Systems Engineering, University of South Alabama, 150 Student Services Dr., Mobile, AL 36688, USA
3 Department of Pharmacology, University of South Alabama, 5851 USA Drive N., Mobile, AL 36688, USA; annamdevula@southalabama.edu (N.S.A.)
4 Center for Lung Biology, University of South Alabama, 5851 USA Drive N., Mobile, AL 36688, USA
5 College of Medicine, University of South Alabama, 5851 USA Drive N., Mobile, AL 36688, USA
6 College of Medicine, Thomas Jefferson University, Philadelphia, PA 19107, USA
* Correspondence: leavesley@southalabama.edu
† These authors contributed equally to this work.

Abstract: Hyperspectral imaging (HSI) technology has been applied in a range of fields for target detection and mixture analysis. While HSI was originally developed for remote sensing applications, modern uses include agriculture, historical document authentication, and medicine. HSI has also shown great utility in fluorescence microscopy. However, traditional fluorescence microscopy HSI systems have suffered from limited signal strength due to the need to filter or disperse the emitted light across many spectral bands. We have previously demonstrated that sampling the fluorescence excitation spectrum may provide an alternative approach with improved signal strength. Here, we report on the use of excitation-scanning HSI for dynamic cell signaling studies—in this case, the study of the second messenger Ca^{2+}. Time-lapse excitation-scanning HSI data of Ca^{2+} signals in human airway smooth muscle cells (HASMCs) were acquired and analyzed using four spectral analysis algorithms: linear unmixing (LU), spectral angle mapper (SAM), constrained energy minimization (CEM), and matched filter (MF), and the performances were compared. Results indicate that LU and MF provided similar linear responses to increasing Ca^{2+} and could both be effectively used for excitation-scanning HSI. A theoretical sensitivity framework was used to enable the filtering of analyzed images to reject pixels with signals below a minimum detectable limit. The results indicated that subtle kinetic features might be revealed through pixel filtering. Overall, the results suggest that excitation-scanning HSI can be employed for kinetic measurements of cell signals or other dynamic cellular events and that the selection of an appropriate analysis algorithm and pixel filtering may aid in the extraction of quantitative signal traces. These approaches may be especially helpful for cases where the signal of interest is masked by strong cellular autofluorescence or other competing signals.

Keywords: hyperspectral imaging; fluorescence microscopy; theoretical sensitivity; cellular autofluorescence

Citation: Parker, M.; Annamdevula, N.S.; Pleshinger, D.; Ijaz, Z.; Jalkh, J.; Penn, R.; Deshpande, D.; Rich, T.C.; Leavesley, S.J. Comparing Performance of Spectral Image Analysis Approaches for Detection of Cellular Signals in Time-Lapse Hyperspectral Imaging Fluorescence Excitation-Scanning Microscopy. *Bioengineering* **2023**, *10*, 642. https://doi.org/10.3390/bioengineering10060642

Academic Editor: Cuneyt M. Alper

Received: 24 April 2023
Revised: 13 May 2023
Accepted: 16 May 2023
Published: 25 May 2023

1. Introduction

Hyperspectral imaging (HSI) technologies were originally developed by NASA and the United States Department of Defense for Earth resource monitoring and military applications [1,2]. More recently, HSI technologies have been applied to a wide range of fields, including document preservation [3,4], agriculture monitoring [5], and medicine [6]. Within the field of biological cellular imaging, HSI technologies have displayed great potential for use with fluorescence microscopy [2,7,8]. In specific, HSI has enabled the detection of many fluorescent labels simultaneously [9–11]; the separation of fluorescence signals from cellular and tissue autofluorescence [11–13]; and the quantitation of signals, such as for measurement of ratiometric Förster resonance energy transfer (FRET) reporters [14–16].

A range of technologies and approaches have been developed in adapting HSI to clinical and biomedical research applications. If the imaging application presents a strong optical signal, high contrast, and does not require high-speed or time-sensitive imaging, a straight-forward approach may be to apply a pre-existing or semi-customized HSI camera or spectral filtering module, along with customized analysis algorithms. This approach is well-suited for transmitted or reflected light HSI, where broad-band illumination may be used to enable a wide spectral scanning range. For example, imaging of fixed and stained histology slides, which present a bright transmitted light signal with negligible photobleaching, can be imaged using a spectral detector consisting of a camera and tunable filter, such as a liquid-crystal tunable filter (LCTF) [17–19], acousto-optic tunable filter (AOTF) [20,21], or thin-film tunable filter (TFTF) [22]. Li and colleagues have implemented this approach on an upright microscope platform with illumination provided by a standard microscope lamp (presumed to be a halogen bulb) and spectral detection provided by an AOTF module coupled with a charge-coupled device (CCD) camera [23]. This system provided an HSI range of 550–1000 nm and allowed segmentation and spectral analysis of red blood cell (RBC) smears. The same system was later utilized by the group in a manuscript by Wang et al. for HSI-based studies of melanoma pathology slides, where a convolutional neural network (CNN) analysis was developed to allow segmentation of pathology images as well as automated detection of melanomas, with an overall accuracy of 92% [24]. Finally, HSI datasets of pathology slides of bile duct tissues were acquired using this same system, analyzed with deep neural network (DNN) approaches for detection of cholangiocarcinoma, and found to achieve a patch-level classification accuracy of 94% using the ResNet50 algorithm [25].

In other scenarios, a sample may present a sensitive fluorescence signal that is prone to photobleaching or a live cell, tissue, or in vivo sample may require dynamic imaging to measure a range of cell signaling or physiologic processes. In these cases, there are inherent trade-offs or limitations when implementing HSI for advanced microscopy applications [26,27]. These limitations primarily occur due to the need to divide the photon budget—the fluorescence emission signal—among several dimensions: spatial (X, Y, and in some instances, Z), temporal for time-dependent studies, and spectral (λ). In many cases, the selection of a fluorescence microscope platform and corresponding spectral filtering is determined so as to optimize the imaging performance and shift the weight of the trade-off for a specific application. For example, if high-speed imaging is required for kinetic cellular signaling studies, a widefield or spinning disk confocal microscope may be used where only one or several axial images are acquired, and only one or several wavelength bands are sampled [28,29]. By sampling only one or a minimal number of axial slices and wavelength bands, the temporal sampling may be increased at the expense of axial or spectral sampling. However, there are many biological imaging scenarios where multiple requirements must be considered, such as performing cell signaling studies in cell lines or tissues that are inherently autofluorescent, where both temporal sampling and spectral sampling are needed [30,31]. In these cases, the selection of an appropriate imaging platform and acquisition settings must be performed carefully in order to achieve optimal microscope settings for the study. Post-acquisition image processing may also be applied in an effort to mitigate imaging system trade-offs [32]. Often there is no perfect solution.

We, and others, have previously demonstrated that an alternative approach for hyperspectral imaging—where spectral filtering is implemented in the excitation optics rather than the emission optics, as shown in the optical light path in Figure 1—may provide an increased signal for some studies [33–35]. In traditional emission-scanning HSI fluorescence microscope systems (Figure 1A), illumination is typically provided by a broad-band source and a band-pass excitation filter, and the fluorescence emission spectrum is scanned using a tunable filter and camera or a dispersion-based spectral detector based on a grating [36] or prism [37]. A typical emission-scanning experiment would utilize spectral settings so as to excite one or several fluorescence labels near the corresponding excitation spectral peak wavelength(s) and acquire images of fluorescence emission sampled across many spectral bands using a narrow wavelength band for each (Figure 1B). By contrast, in excitation-scanning HSI fluorescence microscope systems (Figure 1C), narrow-band illumination is provided that can be tuned over many different excitation wavelengths while the fluorescence emission is detected using a broad-band or long-pass emission filter and camera. A typical excitation-scanning experiment would utilize spectral settings so as to scan the fluorescence excitation over the range of excitation spectral peak wavelengths while sequentially acquiring an image of fluorescence emission for each excitation wavelength scanned (Figure 1D). Hence, the excitation-scanning HSI approach utilizes the wavelength-dependent properties of the excitation spectrum of each fluorophore in order to perform spectral separation. Because the fluorescence emission is minimally processed in the spectral dimension, the signal reaching the detector is stronger, enabling increased imaging speeds and/or decreased photobleaching.

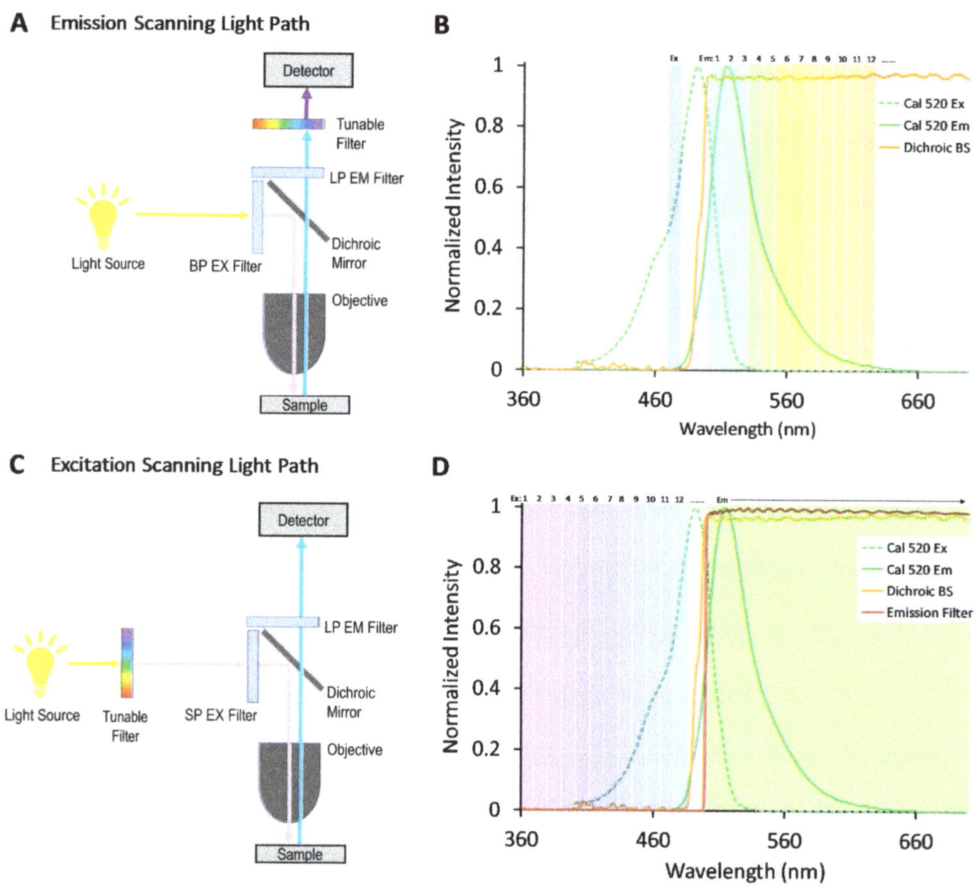

Figure 1. Light path schematics and corresponding spectral scan range illustrations for emission-scanning and excitation-scanning HSI microscope systems. (**A**) A typical emission-scanning HSI

microscope system utilizes one or several narrow bandwidths of illumination for excitation while acquiring fluorescence emission over many narrow spectral bands using a tunable filter or other wavelength sampling device (prism, grating, etc.). (**B**) An illustration of the spectral scan range for an emission-scanning HSI system. The fluorescence excitation and emission spectra of a hypothetical fluorescence label are shown as dotted and solid green lines. Excitation is provided at one narrow band, illustrated by the solid green bar placed at the peak excitation wavelength. Fluorescence emission is sampled across many narrow bands, illustrated by the many individual bars across the fluorescence emission spectrum. (**C**) The excitation-scanning HSI microscope system utilizes a tunable filter to sequentially select between many narrow excitation wavelength bands while acquiring fluorescence emission using a broad-band or long-pass emission filter (LP EM Filter). (**D**) An illustration of the spectral scan range for an excitation-scanning HSI system. The fluorescence excitation spectrum is sequentially sampled using many narrow excitation bands, as indicated by narrow bars placed over the excitation spectrum. The emitted fluorescence is detected in bulk using a broad-band or long-pass emission filter.

The overall goal of this study is to compare the performance of common spectral analysis algorithms for use in kinetic cell signaling experiments where excitation-scanning HSI is implemented to enable the separation of the fluorescent label of interest from cellular autofluorescence. Specifically, the goal is to accurately identify a fluorescent label for intracellular Ca^{2+} and to separate this signal from autofluorescence in time-lapse excitation-scanning HSI image data. Recent studies highlight the significance of signal localization in differentiating contractile from non-contractile agonists [38–42]. However, non-uniform cellular autofluorescence can complicate these studies, and HSI approaches offer the potential to separate autofluorescence from Ca^{2+} and other labels. Here, we compare four spectral analysis algorithms: linear unmixing (LU) [43,44], constrained energy minimization (CEM) [45], matched filtering (MF) [46,47], and spectral angle mapper (SAM) [48]. In addition, the effects of using signal intensity thresholding and pixel filtering are evaluated, which enables the analysis of selected pixels above a defined minimum detectable limit. The results indicate that utilizing an optimal spectral analysis algorithm, as well as pixel filtering, can provide an improved ability to quantify weak transient signals, or signals from small regions of the image or localized subcellular regions.

2. Materials and Methods

2.1. Cell Culture

Human airway smooth muscle cells (HASMCs) were selected as a model system because of the critical role that localized Ca^{2+} signals play in maintaining a contractile state. Recent studies highlight the significance of signal localization in segregating contractile from non-contractile agonists [38–42]. HASMCs were isolated and cultured as described previously [49]. Briefly, HASMCs were grown in 100 mm culture dishes and were maintained in Dulbecco's Modified Eagles Medium (DMEM, GIBCO) supplemented with 5% fetal bovine serum (Gemini), basic fibroblast growth factor (SIGMA), epidermal growth factor (Invitrogen), 100 U/mL penicillin, and 100 μg/mL streptomycin (Gibco) at pH 7.0. HASMCs were seeded onto 20 mm laminin-coated round glass coverslips and were incubated 37 °C and 5% CO_2 for 48 h (or until the cells were grown to 70–80% confluency).

Samples with a single label of either NucBlue (nuclear label, ThermoFisher Scientific, Inc., Waltham, MA, USA), Cal 520-AM (Ca^{+2} indicator, AAT Bioquest, Inc., Pleasanton, CA, USA), or unlabeled cells for cellular autofluorescence validation were prepared according to manufacturer labeling specifications. Briefly, single-label NucBlue samples were prepared using 2 drops of NucBlue followed by incubation at 37 °C for 20 min. Single-label Cal 520-AM samples were prepared using 5 μM Cal 520-AM followed by incubation at 37 °C for 30 min. Mixed-label samples were prepared for time-lapse measurements using both Cal 520-AM and NucBlue labeling, using identical labeling concentrations and incubation times as the single-label control samples. After labeling, coverslips were transferred to an attoflour holder (ThermoFisher) for imaging and covered with an extracellular buffer containing (mM): 145 NaCl, 4 KCl, 10 HEPES, 10 D-glucose, 1 $MgCl_2$, 1 $CaCl_2$, pH 7.4.

2.2. Hyperspectral Imaging Acquisition

Excitation-scanning hyperspectral images were acquired utilizing a custom TE-2000 inverted widefield fluorescence microscope (Nikon Instruments, Melville, NY, USA) equipped with 20X objective (Plan Apo 20X, N/A 0.75, Nikon Instruments), Titan 300 Xenon arc lamp (Sunoptics Surgical, Jacksonville, FL, USA), and Prime 95B sCMOS camera (Teledyne Photometrics, Tucson, AZ, USA) as previously described [22,33]. The image acquisition parameters are described in Table 1. A set of thin-film tunable filters (VersaChrome, Semrock, IDEX Health & Science LLC, Rochester, NY, USA) were mounted in a custom tiltable filter wheel (VF-5, Sutter Instrument Co., Novato, CA, USA) that allowed switching between individual TFTFs and tuning of TFTFs by adjusting the angle of the filter wheel. Hyperspectral image data were acquired by sequentially scanning excitation wavelengths from 360 to 480 nm in 5 nm increments while detecting the fluorescence emission at each excitation wavelength using a long-pass dichroic beamsplitter (FF495-Di03, Semrock) and corresponding long-pass emission filter (FF01-496/LP, Semrock). For single-label control samples, only a single hyperspectral image of each sample was acquired. For dynamic Ca^{2+} studies, time-lapse hyperspectral images were acquired by acquiring an excitation-scanning hyperspectral image every 30 s for 15 min. After 5 min of baseline acquisition, cells were treated with either 50 μM carbachol or vehicle control. Spectral image stacks were exported as individual tiff images using NIS Elements software (Nikon Instruments) for subsequent image analysis.

Table 1. Experimental spectral microscope acquisition parameters.

Microscope	Light Source	Objective	Detector
TE2000-U inverted epifluorescence widefield microscope (Nikon Instruments)	Titan 300 Xe arc lamp (Sunoptics Surgical) TFTF array: scan range 360–480 nm in 5 nm increments	Plan Apo 20X, N/A 0.75 (Nikon Instruments)	Prime 95B sCMOS camera (Teledyne Photometrics) 30 ms exposure time 0 averaging 2 × 2 binning 0 gain or offset 200 MHz readout speed Overall imaging speed = ~15 s/spectral image stack

2.3. Hyperspectral Image Data Preprocessing

To account for background and wavelength-dependent attenuation, spectral image data were corrected to a NIST-traceable spectral response, as described previously [33]. In brief, a microscope slide-configured laser power meter (ArgoPower, Argolight SA, Pessac, France) was used to measure the illumination power at each excitation wavelength, and the resulting illumination power vs. wavelength data were used as the excitation spectral power and saved as an array, $\vec{\Phi}$. The inverse of the illumination power vs. wavelength data series was normalized to a peak value of unity and used as a spectral correction coefficient.

$$\vec{cc} = \frac{\vec{\Phi}^{-1}}{\max\vec{\Phi}^{-1}} \tag{1}$$

Spectral images were corrected by background subtraction and multiplying by the spectral correction coefficient.

2.4. Reference Spectra and Spectral Library

Reference spectra were identified from single-label samples for each fluorescent species: Cal 520, NucBlue, and cellular autofluorescence (unlabeled cells). Spectral images were acquired for each single-label sample, and images were corrected to a flat spectral response, as described above. Regions of interest were then selected corresponding to areas of intense, but not oversaturated, signals using the freehand selection region tool in ImageJ software [50]. The pixel-averaged spectrum from each region was extracted using the Image → Stacks → Plot Z-axis Profile command in ImageJ. This served as the reference spectrum for each label. Reference spectra were compiled in Excel (Microsoft Corporation, Redmond, WA, USA) and subsequently transferred to MATLAB (The MathWorks, Inc., Natick, MA, USA) as a spectral library for use with the analysis algorithms described below.

2.5. Spectral Analysis Algorithms

2.5.1. Linear Unmixing

Linear unmixing (LU) is a standard approach used by the fluorescence microscopy spectral imaging community for estimating endmember abundances [43,51]. LU enables estimation of the abundance of each endmember present in each pixel, typically using least-squares regression—for fluorescence microscopy, this corresponds to an estimation of the relative signal from each fluorescent label. To utilize linear unmixing, control samples were prepared and imaged to build a spectral library a priori. Linear unmixing was implemented using a modified version of the "lsqnonneg" function in MATLAB, which utilizes a non-negatively constrained least-squares unmixing approach [43,44] (Equation (2)).

$$\vec{x} = \sum_{i=1}^{m} a_i \vec{r_i} + \vec{w} \tag{2}$$

where $\vec{x}$ is the detected pixel spectrum vector, a_i is the endmember, $\vec{r_i}$ is the respective endmember spectrum and $\vec{w}$ is the additive observation noise vector.

2.5.2. Spectral Angle Mapper

Spectral angle mapper (SAM) is an algorithm that measures the spectral similarities by finding the angle between the spectral signatures of two pixel vectors and has been widely used in remote sensing due to its advantage of only requiring the target endmember spectrum to be known [48,51]. The angle between pixels and the reference spectrum is then used to classify the image, usually by applying a global threshold. SAM was implemented using a custom MATLAB script to calculate the arccosine of the dot product of the spectra between the measured pixel spectrum and the *i*th endmember spectrum (Equation (3))

$$\theta_i = \cos^{-1}\left(\frac{\vec{r} \cdot \vec{x}}{\| \vec{r_i} \| \cdot \| \vec{x} \|}\right) \tag{3}$$

where $\vec{r}$ is the test spectrum and $\vec{x}$ is the reference spectrum.

2.5.3. Constrained Energy Minimization

Constrained energy minimization (CEM) is an algorithm that applies a mathematical filter to amplify the desired spectral signature while minimizing the output energy (e.g., background noise) resulting from target sources other than the desired target [45,52]. CEM utilizes an optical linear operator, L_{CEM} (Equation (4)), to minimize the intensity of undesired signatures while intensifying desired signatures. When using CEM, only the target endmember signature needs to be known. However, CEM presents limitations as it is a single-target detection method and therefore does not allow the preprocessing of known undesired targets for detection enhancement.

$$L_{CEM} = R_{corr}^{-1} \vec{r_i} \left(\vec{r_i}^{T} R_{corr}^{-1} \vec{r_i} \right)^{-1} \tag{4}$$

where the inverse of the sample correlation matrix of the observation pixel vectors (R_{corr}^{-1}) is calculated as:

$$R_{corr}^{-1} = \frac{\sum_{j=1}^{kl}\left(\vec{x}_j \vec{x}_j^{T}\right)}{kl} \quad (5)$$

2.5.4. Matched Filter

Matched filter (MF) is an algorithm used to quantify the abundance of a known spectral signature by partially unmixing the desired endmember from the image and maximizing the intensity while minimizing the remaining signatures [46,47]. The MF algorithm increases the signal-to-noise ratio for the desired endmember while minimizing the background signals by calculating a rejection operator (Equation (6))

$$P = I - UU^{\#} \quad (6)$$

where U represents all signatures in the spectral library acquired a priori while excluding the target endmember, and $U^{\#}$ is the pseudo-inverse of U and is calculated by:

$$U^{\#} = \left(U^{T}U\right)^{-1}U^{T} \quad (7)$$

The rejection operator (Equation (8)) is then used to calculate the filter operator, a_i, (Equation (9)) used for calculating the abundance via the MF method.

$$\vec{q} = \vec{r_i}^{T}P \quad (8)$$

$$a_i = \vec{q}\left(\vec{x_i}\right) \quad (9)$$

2.6. Theoretical Sensitivity Analysis Approaches

A theoretical sensitivity analysis (TSA) framework that was previously described [34] was implemented to estimate the performance of various spectral image analysis algorithms for the detection of the Cal 520 signal in excitation-scanning spectral image data. To accomplish this, a region of interest (ROI) containing all spectral components (except for the endmember spectra) was selected from each control sample to perform the TSA. Next, a binary mask corresponding to the selected ROI was created using ImageJ (Equation (6)), and the target endmember was multiplied by a scalar (a) and added to each pixel in the region.

$$\vec{x_i} = \vec{x_i} + a\vec{r_i} \quad (10)$$

The value of a was varied across a range resulting in a series of images, each with a different level of the target signature added to a selected region of the spectral autofluorescence image. This enabled the creation of a specified region within the image with a known simulated ground truth for the target signature—in this case, Cal 520.

The resulting spectral images were analyzed to produce three curves: a theoretical sensitivity curve (TSC, see example in Figure 2A), a thresholded positive pixel curve (TPPC, Figure 2B), and a receiver operator characteristic (ROC, Figure 2C) curve. The TSC was generated by plotting the unmixed intensity of each pixel in the selected ROI for different levels of the added endmember signal. The black squares represent unmixed pixel intensities, while the red error bars represent the standard deviation in pixel intensities. The mean pixel intensity was also calculated for each level of the added endmember signal and fit to a linear response. The TSC can be used to visualize the linearity of the detected signal response.

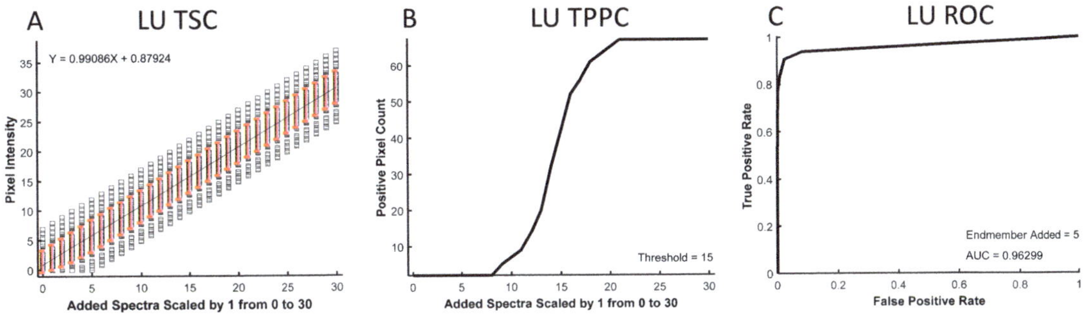

Figure 2. An example theoretical sensitivity analysis, as performed for non-negatively constrained linear unmixing (LU). (**A**) The theoretical sensitivity curve (TSC) demonstrates a linear response of the LU algorithm to varying levels of the added target endmember, Cal 520. (**B**) The thresholded positive pixel curve (TPPC) indicates a sharp slope in detection accuracy for a threshold of 15 (unmixed intensity units). Importantly, using the threshold of 15, no false-positive pixels were detected when the 0 endmember signal was added. (**C**) The receiver operator characteristic (ROC) curve demonstrates a high predicted performance for LU for this application of Cal 520 detection, with an area under the curve (AUC) of 0.96. The TSC analysis was performed for all 4 spectral analysis algorithms, as described in the Results.

The TPPC was generated by plotting the number of pixels within the entire image that were detected as "positive", as defined by an intensity above a specific threshold—in this example, the threshold = 15 A.U. The TPPC can be used to visualize the number of unmixed "positive" pixels above this fixed threshold value as a function of the endmember signal added to pixels in the ROI. The TPPC also allowed visualization of the detection response slope and the false-positive rate. A more vertical response (increased slope) signifies an increased ability to discriminate negative and positive pixels.

The ROC curve [53] was generated by counting the number of positive pixels that were detected within the ROI as true-positive detections while counting the number of positive pixels counted outside of the ROI as false-positive detections. The number of true-positive and false-positive detections was measured for a range of detection threshold levels at a fixed level of target endmember signal that was added to the ROI. The ROC curve provides an overall visualization of spectral analysis algorithm performance, with ideal target detection represented by an area under the curve equal to 1.

2.7. Pixel Threshold Analysis Approach

To improve the reliability of kinetic measurements and to remove bias from non-cell background pixels, the detection threshold can be utilized to allow quantification of only pixels above a minimum detectable limit within a field of view or single-cell region. Utilizing a pixel filtering technique may help to identify subtle changes in time course dynamics that would not be visible when averaging signals from all pixels within a field or large region. These signals, essentially, are hidden or diluted by signals from background pixels or those with weak labeling. To implement the pixel filtering approach, a binary mask was created and used to filter the Cal 520 signal using three different threshold levels: 0 (no threshold), 20 (threshold calculated using the theoretical sensitivity analysis), and 24 (threshold calculated using Otsu, as implemented in ImageJ software [50]). Time-dependent Cal 520 signal levels were quantified for the entire field. In addition, single-cell regions were defined, and Cal 520 signal traces were measured for three different cells using each pixel-filtering threshold level.

3. Results and Discussion

Hyperspectral imaging approaches have been applied to microscope systems using a variety of spectral technologies: LCTFs [18,19], AOTFs [13,20], TFTFs [22], gratings [36], prisms [37], interferometers [54], spectral snapshot cameras [55], and light-emitting diode (LED) arrays [56,57]. Each technology has associated advantages and disadvantages, and in some cases, may be implemented in more than one way, such as for scanning of the fluorescence emission or excitation spectrum. The focus of this study was to utilize HSI microscopy for a particularly challenging application—measurement of time-dependent fluorescence in the midst of cellular autofluorescence for dynamic, live-cell, cell signaling experiments. The spectral imaging requirements for this application include the need for:

1. High sensitivity—sufficient to detect subtle differences in the Cal 520 signal due to temporal or spatial changes in Ca^{2+} concentration.
2. High speed—sufficient to detect dynamic changes in the Cal 520 signal on the order of 20–30 s consistent with slower Ca^{2+} signals or waves.
3. High photon efficiency—necessary to minimize photobleaching effects during time-lapse experiments.
4. High spatial resolution—sufficient to allow localization of Ca^{2+} signals to spatial regions within the cell, such as near the cell membrane, near the nucleus, etc.
5. Moderate field of view—sufficient to allow simultaneous imaging of multiple (~20–100) cells.
6. Moderate spectral resolution—sufficient to allow discrimination of spectral peaks from Cal 520, NucBlue, and cellular autofluorescence—a spectral resolution of 10–20 nm is likely sufficient for this study as Cal 520 and NucBlue excitation wavelength peaks are ~150 nm in separation, and autofluorescence presents a broad intensity across this excitation spectral range.
7. A moderate and compromised spectral bandwidth—sufficiently narrow to achieve the needed spectral resolution (the TFTF system provides a bandwidth of ~18–20 nm), but sufficiently broad as to provide a strong excitation intensity for detecting weak signals.
8. A clean spectral band with high out-of-band rejection—sufficient to prevent excitation-emission spectral cross-talk and to minimize cross-talk between spectral bands.

The excitation-scanning HSI microscope system and corresponding acquisition settings were selected for this study to satisfy the range of spectral imaging requirements described above. In specific, the use of a TFTF array provides ~95% optical transmission, a bandwidth of between 18 and 20 nm depending on center wavelength, and a clean spectral band with out-of-band rejection of OD 5+. The excitation-scanning approach provides high overall photon efficiency for the system and helps to minimize photobleaching. These capabilities could be further enhanced by the development of very-high-speed rotation mechanisms to allow rapid tuning of the TFTF array, which could reduce the temporal sampling speed to 1–2 s per spectral band and allow the sampling of rapid Ca^{2+} transient signals.

3.1. Experimental Results and Discussion

Agonist-induced Ca^{2+} responses were measured using excitation-scanning HSI and analysis approaches. To provide the a priori information required for linear unmixing, as well as other spectral analysis approaches, images of single-label control samples were acquired for NucBlue (Figure 3A), unlabeled cells for cellular autofluorescence (Figure 3B), and Cal 520 (Figure 3C). The magnitude of signals within spectral images was visualized by summing all spectral bands. A region was then selected in each single-label image that corresponded to an area of high signal intensity. This region was transferred to the corresponding spectral image stack, and the pixel-averaged spectrum was extracted, yielding the measured reference spectrum (also known as the pure endmember spectrum). Pure spectra were combined into a spectral library (Figure 3D) for use with spectral analysis algorithms. Spectra were also validated against manufacturer-supplied and previously reported excitation spectra, as measured using a spectrofluorometer (Figure 3E). In general, there was high agreement between the spectral shape of the measured pure spectra and previously reported spectra.

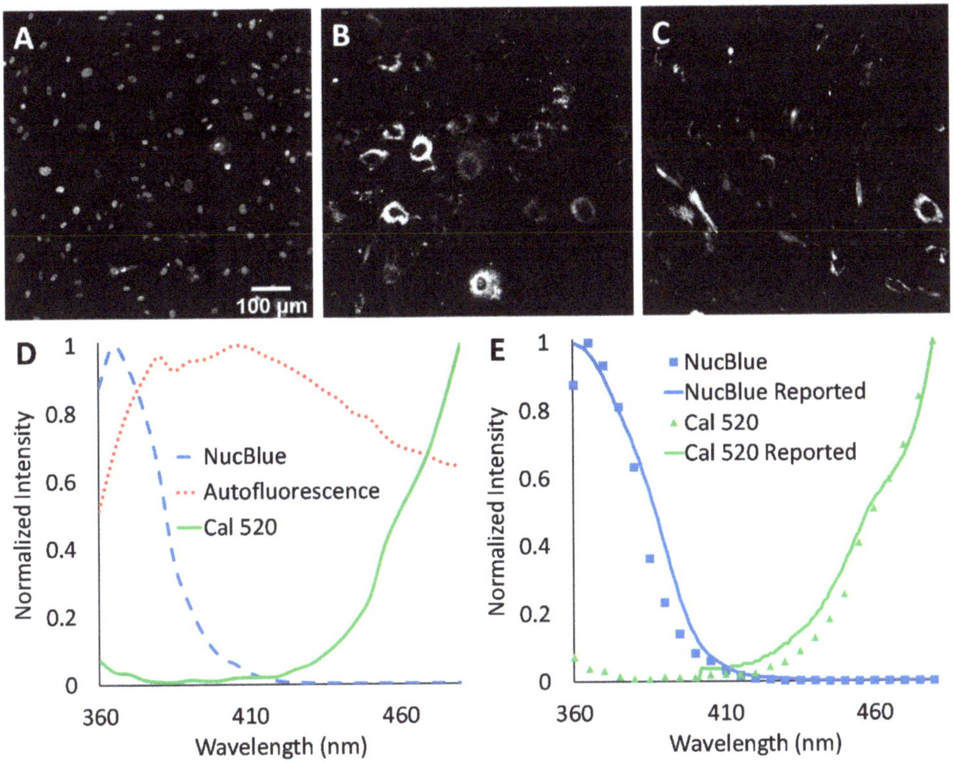

Figure 3. Spectral image data from three single-label control samples were analyzed to construct a spectral library. (**A**) Representation of HSI image data from HASMCs labeled with the nuclear label, NucBlue. For visualization purposes, all wavelength bands have been summed to represent a total or summed fluorescence intensity image, and the intensity range linearly adjusted from 0 to 6300 A.U. for display. (**B**) A summed fluorescence intensity representation of HSI image data from a separate sample of unlabeled HASMCs with intensity range linearly adjusted from 0 to 315 A.U. for display. (**C**) A summed fluorescence intensity representation of HSI image data from a separate sample of HASMCs labeled with Cal 520 with intensity range linearly adjusted from 0 to 550 A.U. (**D**) A spectral library was generated by selecting a region of high signal intensity within each of the single-label control spectral images (**A**–**C**), extracting the pixel-averaged spectrum, and normalizing to a peak value of unity. (**E**) Comparison of measured spectra for NucBlue (blue squares) and Cal 520 (green triangles) to reported spectra. The Cal 520 spectrum was supplied from AAT Bioquest, while the NucBlue spectrum was approximated as that of DAPI and obtained using the Semrock Searchlight spectral plotting tool [58].

Fluorescence excitation-scanning HSI data of Ca^{2+} signaling in HASMCs were then analyzed to detect each endmember. For visualization in a print format, single time points have been selected for viewing time-lapse microscopy data (Figure 4). The full time-lapse image data set for this study is available in Supplemental Video S1. Results from non-negatively constrained linear unmixing (LU) demonstrate the ability to separate the Ca^{2+} label signal (Cal 520) from the NucBlue and Autofluorescence. Results indicate that all three signals are identifiable and that the residual root mean square (RMS) (Figure 4) error is less than any of the unmixed endmembers. Increased Ca^{2+} signaling activity is clearly visible at time point 340 s after the addition of 50 μM carbachol, and dynamics can be seen in Supplemental Video S1.

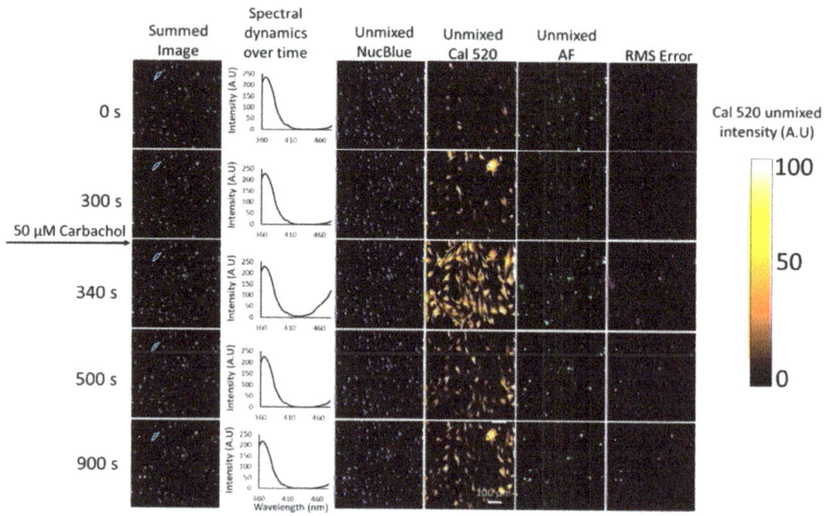

Figure 4. Linear spectral unmixing of spectral images from Cal 520 labeled HASMCs. All wavelength bands were summed to show the total intensity of raw spectral images (column 1) at different time points. A region of interest corresponding to a single cell was selected to illustrate the change in excitation spectrum over time (Column 2). The primary contributors to the mixed spectra shown in Column 2 were NucBlue (excitation peak at 360 nm) and Cal 520 (excitation peak at 520 nm). Note that the excitation peak of Cal 520 is beyond the scan range used for the current study, and hence the highest wavelength response for Cal 520 occurs at the last wavelength scanned of 480 nm—see Figure 3E for a comparison of measured and manufacturer excitation spectra. Note that the Cal 520 contribution is increased at time point 340 s, which corresponds to 40 s after agonist addition and a strong Ca^{2+} release response visualized in the unmixed Cal 520 images (Column 4). Columns 3–5 show false-colored unmixed endmember images of NucBlue, Cal 520, and AF signals. Column 6 shows the RMS error associated with linear spectral unmixing. The color bar at the right represents the color look-up table used to visualize Ca^{2+} signal intensity in Column 4.

Four spectral analysis algorithms—LU, SAM, CEM, and MF—were utilized to evaluate the ability to separate and identify endmember signals in fluorescence excitation-scanning time-lapse HSI data (Figure 5). While these algorithms have been previously used for the analysis of remote sensing HSI data, only the LU algorithm has been applied to fluorescence excitation-scanning HSI microscopy data. Hence, it is important to evaluate the performance of these algorithms with respect to the ability to detect Cal 520 and NucBlue signals and the ability to accurately differentiate these signals from confounding cellular autofluorescence. This is especially important when considering that signal levels and the mixture of spectral signatures will dynamically change for time-lapse live-cell imaging studies.

In a side-by-side comparison, all four algorithms were able to identify dynamic changes in Cal 520 due to changes in the Ca^{2+} concentration. However, the different algorithms presented different abilities to separate endmember signals. For example, SAM is a spectral angle measure and not an unmixing algorithm and, as such, provided spectral angle measurements that, when viewed as images, displayed contributions from endmembers other than the target endmember. Most notable is that the AF signal is visible in the SAM channel for Cal 520. All the other three analysis algorithms (CEM, LU, and MF) are subpixel analysis techniques and provide endmember images with visibly lower contributions from endmembers other than the target endmember. However, subtle analysis artifacts can still be visualized, especially when viewing time-dependent analyzed image data (Supplemental Video S1). For example, time-dependent changes in NucBlue can be seen, which are likely artifactual. The nuclei signal should be relatively static over the time span of these experiments, and rapid changes in the nuclear signal are likely due to the Cal 520 signal being incorrectly identified as NucBlue. This incorrect signal assignment is more apparent in CEM than in LU and MF and is also present in the AF channel.

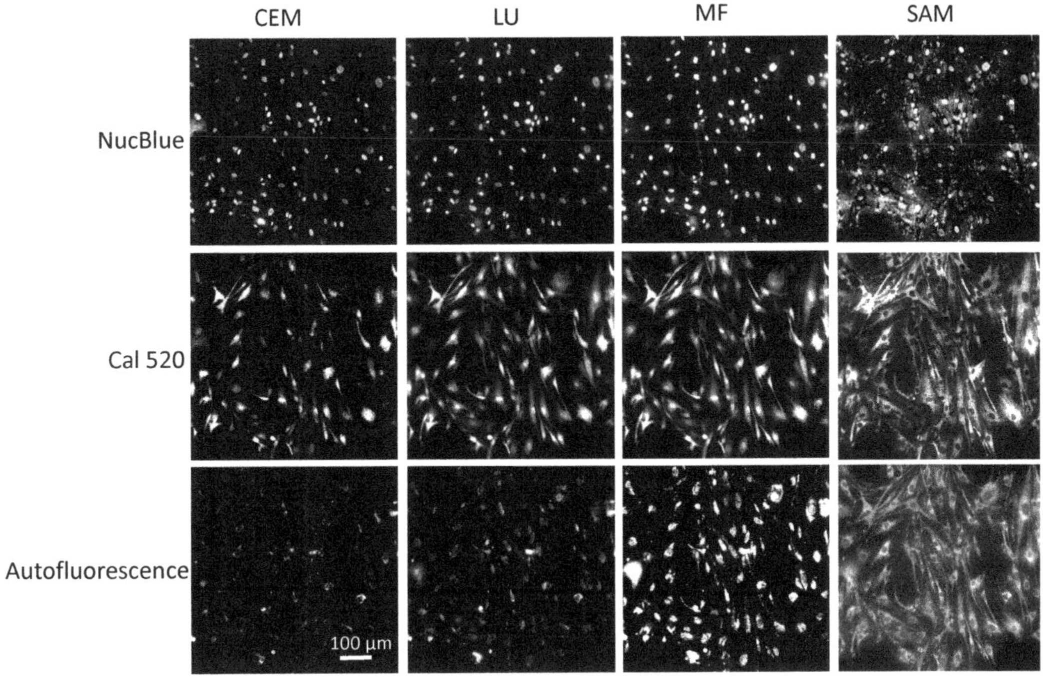

Figure 5. A comparison of 4 common spectral analysis algorithms for identifying endmembers in excitation-scanning spectral image data. Analyzed images were displayed using a greyscale look-up table for each endmember: NucBlue, Cal 520, and autofluorescence (AF). Images were analyzed using constrained energy minimization (CEM), linear unmixing (LU), matched filtering (MF), and spectral angle mapper (SAM). A video showing the full timelapse data set can be viewed in Supplemental Video S1. The same look-up tables were used for both Figures 4 and 5 to permit comparison.

An important consideration when analyzing HSI fluorescence microscopy data, especially from dynamic live cell studies such as this, is that generation of an image data set with a known ground truth is experimentally prohibitive. For cases of fixed cells or fluorescently labeled microspheres, it may be possible to experimentally generate a sample with a known ground truth through very precise fluorescent label titration so as to achieve a specified labeling density. However, for live cell samples, knowledge of the exact labeling density (the number of labels per area or per volume) is prohibitively difficult to ascertain due to an unequal uptake of fluorescent labels by the cell, unequal distributions of cell label binding sites, non-specific binding of the label to other sites, variations in cellular autofluorescence, variations in cell geometry (a typical "fried-egg" shaped cell will be thicker near the nuclear region and very thin near the periphery), and important variations in these parameters over time due to both normal and induced changes in cell physiology. To address these concerns, we have previously utilized a "theoretical sensitivity analysis" framework for probing the sensitivity of a spectral imaging assay, with a corresponding analysis algorithm, to varying theoretical levels of signals that are artificially added post-acquisition to a spectral image dataset [34]. In this study (Sections 3.2 and 3.3), we extend the theoretical sensitivity analysis framework to differentiate between pixels that are above a minimum detectable limit threshold and pixels that are below the detectable limit and may be considered background.

3.2. Theoretical Sensitivity Analysis Results and Discussion

In microscopy, it is often prohibitively difficult to establish a known ground truth for a specific location in a cell signaling experiment. To assess spectral image analysis performance, a theoretical sensitivity framework may be used to estimate detection sensitivity and specificity. This is advantageous as the contributions due to cellular autofluorescence are unknown and variable. Here, we evaluate the performance of four conventional spectral analysis algorithms (LU, SAM, CEM, and MF) for identifying a target endmember in the midst of other endmembers and confounding AF in fluorescence excitation-scanning HSI data (Figure 6).

The TSC was used to visualize the analyzed intensity of each pixel within a region as a function of the amount of target endmember added. LU (Figure 6A) and MF (Figure 6J) analyses provided a linear response (denoted by the black line) to increasing target endmembers, while SAM (Figure 6D) and CEM (Figure 6G) responses were nonlinear. By inspecting the standard deviation (red error bars), it can be seen that the standard deviation for LU and MF were insensitive to the amount of target endmember added, while SAM showed a decreasing standard deviation as a function of increasing endmember added, and CEM provided an increasing standard deviation with each target endmember added.

The threshold positive pixel curve (TPPC) was used to visualize the ability to detect positive pixels in the image when given a specified detection threshold. Specifically, the TPPC provides the ability to visualize both the false-positive level (the baseline number of positive pixels detected when no target endmember signal is added) as well as the slope of the transition from pixels being detected as negative to being detected as positive. LU and MF algorithms (Figure 6B,K) displayed a more vertical slope of the TPPC, indicating the ability to detect positive pixels once the amount of the target endmember was at or above the detection threshold. For SAM (Figure 6E), pixels with angles less than the detection threshold (in this case, 0.98 radians) were defined as positive, as the SAM algorithm calculates the spectral angle in radians where a small spectral angle indicates a high similarity between endmember and unknown pixel spectra. For SAM, 147,267 pixels (29.5% of total pixels) were detected as positive (below the 0.98 radian threshold) prior to adding any Cal 520 signal. This indicates that the AF signature was sufficiently similar to Cal 520 in these pixels to produce a false-positive detection. A 29.5% false-positive detection rate is unacceptable for sensitive signal detection. Hence, if SAM were to be utilized, a more stringent threshold should be selected at the expense of reduced sensitivity for detecting true-positive pixels (see Supplemental Figure S1 for alternative SAM classification threshold that results in decreased false positives, but also inability to detect weak true positive signals).

The ROC curve was used to visualize the overall performance of pixel classification based on results from each analysis algorithm. The area under the ROC curve is typically the most valuable indicator of detection performance, although the shape of the ROC curve may also be informative. All four algorithms produced different ROC curve shapes, with LU (Figure 6C) presenting the best result with an area under the curve (AUC) of 0.963,and SAM, CEM, and MF (Figure 6F,I,L), with areas of 0.570, 0.872, and 0.957, respectively. This signifies that when performing a sensitivity analysis based on the ROC curve, LU and MF provide similar performance that is much higher than SAM or somewhat higher than CEM.

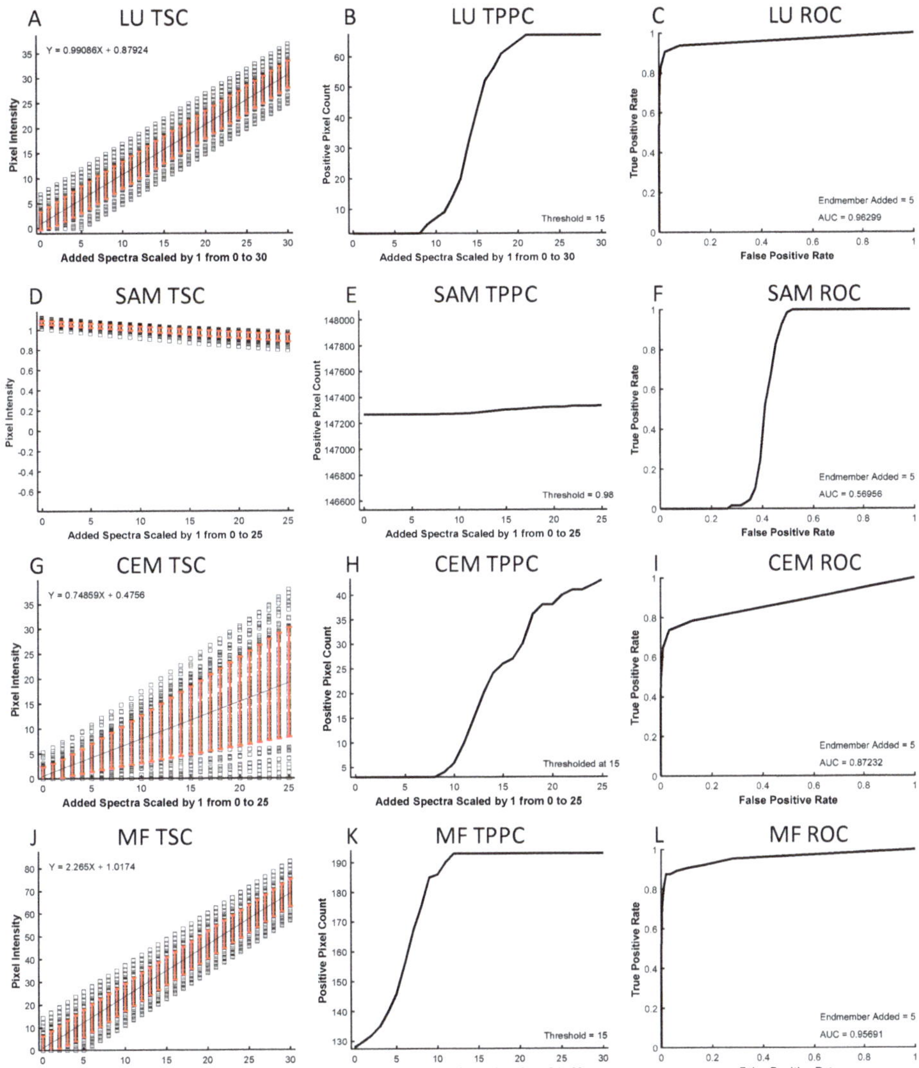

Figure 6. A theoretical sensitivity analysis was applied using 4 spectral analysis algorithms for identifying fluorescence excitation-scanning signals. Three curves were calculated for each analysis algorithm: the Theoretical Sensitivity Curve (TSC), Thresholded Positive Pixel Curve (TPPC), and Receiver Operator Characteristic (ROC) curve. (**A–C**) TSC, TPPC, and ROC for linear unmixing (LU); (**D–F**) TSC, TPPC, and ROC for spectral angle mapper (SAM); (**G–I**) TSC, TPPC, and ROC for constrained energy minimization (CEM); (**J–L**) TSC, TPPC, and ROC for matched filter (MF). TSC and TPPC plots were generated by adding the spectrum of the target endmember, Cal 520, to a specified region of interest (ROI) in set amounts (increments of 1 were 0–30, 0–25, 0–25, and 0–30 for LU, SAM, CEM, and MF, respectively). ROC scale factors were 5 for all algorithms. Threshold values for TPPC were 15 for LU, 0.98 radians for SAM, and 15 for both CEM and MF. ROC threshold values were varied from 0 to 50 in increments of 1 for LU, CEM, and MF. ROC threshold values for SAM varied from 0 to 1.8 radians. ROC performance, as described by area under the curve (AUC), was 0.963 for LU, 0.570 for SAM, 0.872 for CEM, and 0.957 for MF.

3.3. Pixel Threshold Analysis Results and Discussion

Biological samples frequently have higher variability than is encountered in traditional satellite or aerial remote sensing applications. In addition, cellular autofluorescence often confounds measurements of target signatures. We have previously shown that a theoretical sensitivity analysis framework can be used to provide an estimate of the images with varying autofluorescence. A minimum detectable limit (the detection threshold) is evaluated in theoretical sensitivity analysis, and this limit can be further used to identify pixels that have a detectable signal level and those that do not (a signal may be present but is not above the detection threshold). This is advantageous, as traditional cell signaling assays often quantify the signal level for the entire field of view or, at best, specify a region of interest, such as one or several cells. However, when specifying the whole field of view or large regions, it is likely that an error will be present due to non-cell background pixels. In addition, when analyzing data from specific, single-cell regions, there may be pixels within a region that present low or, in practice, unusable levels of fluorescent labels. These weakly labeled pixels may bias extracted time-dependent kinetic measurements.

The Cal 520 signal was filtered using three different threshold levels: 0 (no threshold, Figure 7A), 20 (threshold calculated using the theoretical sensitivity analysis, Figure 7B), and 24 (threshold calculated using Otsu, Figure 7C). The Cal 520 signal levels were quantified for the entire field (Figure 7D). Each pixel filtering threshold level was used to measure traces of the Cal 520 signal for defined single-cell regions (Figure 7E–G). We observed different signal trace kinetics for the entire field (ensemble average response) and for each cell selected, although all cells displayed increased signaling activity after treatment with 50 μM carbachol. These observations can be seen in Supplemental Video S2. The application of pixel filtering through the binary mask resulted in the identification of subtler Cal 520 signal variations for the entire field, as well as for cells 1 and 2. The results of pixel thresholding were least pronounced for cell 3, which appeared to be very well labeled (all pixels above the minimum detectable limit threshold). Hence, a pixel thresholding technique such as this may be helpful in identifying subtle changes in Ca^{2+} of other cell signaling agent dynamics, given the caveat that pixel filtering also may provide an additional analysis step that could introduce signal measurement bias or artifacts if used inappropriately.

To further evaluate the effects of selecting pixels above a minimum detectable limit for measurement of Cal 520 signal dynamics, additional experiments were conducted utilizing HASMCs treated with either 50 μM carbochol (Figure 8) or 50 μM histamine (Figure 9). Hyperspectral images were analyzed using the MF algorithm and thresholded as described above (Figure 7) to retain only pixels that were above a defined minimum detectable limit. Single-cell ROIs were selected, and region-averaged signal traces were extracted for four representative cells. A comparison of time traces before and after pixel thresholding revealed that subtle variations in single-cell Cal 520 signals could be visualized in the pixel-thresholded image sets (compare cell 1 time traces for non-thresholded, Figure 8C, and thresholded, Figure 8F plots). These results indicate that there are likely transient signal variations within cell 1, as well as other cells, that may be masked or altered by pixels below the minimum detectable limit and which become further pronounced with pixel-thresholding techniques.

The effects of pixel thresholding were also investigated using cells treated with an alternative agonist, 50 μM histamine (Figure 9). In this experiment, pixel thresholding resulted in changes in Cal 520 traces for several of the single cells investigated. For example, cell 1 displayed increased fluctuations in Cal 520 signal levels prior to treatment with pixel thresholding (Figure 9F) when compared to non-thresholded images (Figure 9C). These results indicate that there were perhaps time-dependent fluctuations in a subset of pixels within cell 1 but that when averaging all pixels within the cell, these fluctuations were damped by a large number of low-intensity pixels. In contrast, when averaging only pixels above the minimum detectable limit threshold, these fluctuations were more pronounced. In addition, pixel thresholding appeared to bring the baseline value for cells 2–4 to a more consistent value, indicating that perhaps in non-thresholded data, these cells had variable amounts of low-intensity pixels (below the minimum detectable limit) that affected the whole-cell Cal 520 signal level. Taken together, these data indicate that a single-cell ROI may likely contain a mixture of low-intensity pixels that are below a minimum detectable limit for Cal 520 and high-intensity pixels that are above the limit. In addition, changes in intensity may not occur uniformly. In the case of cell 1, there are likely high-intensity pixels that undergo large fluctuations in intensity as well as low-intensity pixels that undergo minimal fluctuations in intensity. Hence, when measuring single-cell Cal 520 time traces from analyzed HSI data, the trend and accuracy of results are likely dependent on both the ability to perform accurate cell segmentation as well as on whether all pixels within a cell ROI are considered or only pixels above a defined minimum detectable limit.

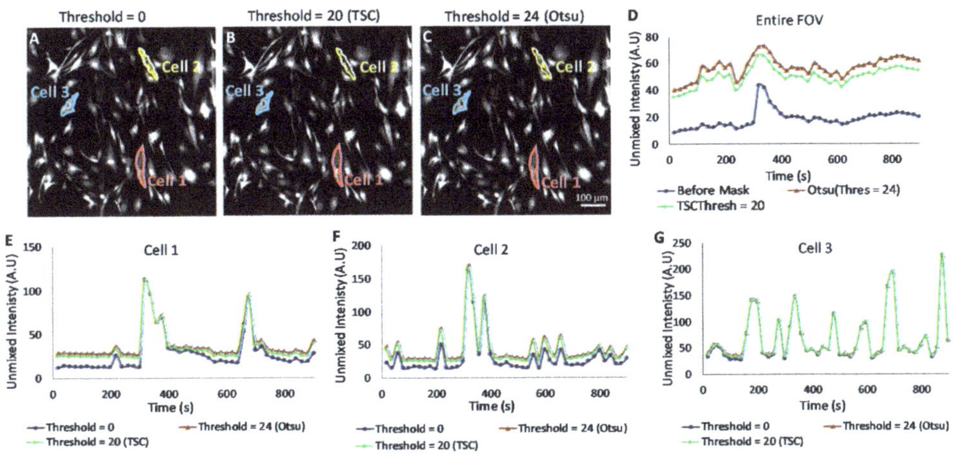

Figure 7. Effects of pixel filtering on time-lapse Ca^{2+} image data acquired using excitation-scanning spectral imaging microscopy and matched filter (MF) analysis. Thresholds for pixel filtering were determined using a theoretical sensitivity analysis and a standard Otsu thresholding algorithm available in ImageJ. (**A**) A representative endmember image of Cal 520 at time point 40 s after treatment obtained using the MF algorithm. The Cal 520 image from panel (**A**) was also processed using pixel filtering to remove low-intensity (below detection limit) pixels using a threshold estimated from the theoretical sensitivity analysis ((**B**), threshold = 20) and from Otsu ((C), threshold = 24). (**D**) The field of view averaged time course for Cal 520 signal as estimated from the original and pixel-filtered image sets. Agonist-induced Ca^{2+} signals in three different cells (cells 1, 2, and 3 shown as red, yellow, and blue regions on image panels (**A**–**C**)) were measured using the original and pixel-filtered image sets ((**E**)—(cell1), (**F**)—(cell2), and (**G**)—(cell3)).

3.4. Considerations for Live-Cell HSI Studies

In this study, fluorescence excitation-scanning HSI microscopy was utilized to study dynamic Ca^{2+} signals in HASMC preparations using time-lapse imaging. Analysis of dynamic signals in live cell samples presents an especially unique set of challenges due to requirements for temporal, spectral, and spatial sampling; effects of photobleaching and cell movement; and confounding factors such as autofluorescence. Detailed descriptions of these requirements for live cell imaging are provided at the beginning of Section 3. The results from this study have indicated that excitation-scanning HSI can provide an effective approach for separating desired fluorescence labels, such as Cal 520, from other labels as well as competing autofluorescence. Of the four spectral analysis algorithms evaluated, LU and MF provided similar results that were deemed to be overall more accurate than SAM or CEM algorithms. MF provided a reduced computational burden and was utilized for further study of time-course measurements from specific cells of interest. The use of pixel-wise thresholding provided an additional ability to improve the reliability of time-course measurements by rejecting pixels below a specified minimum detectable limit.

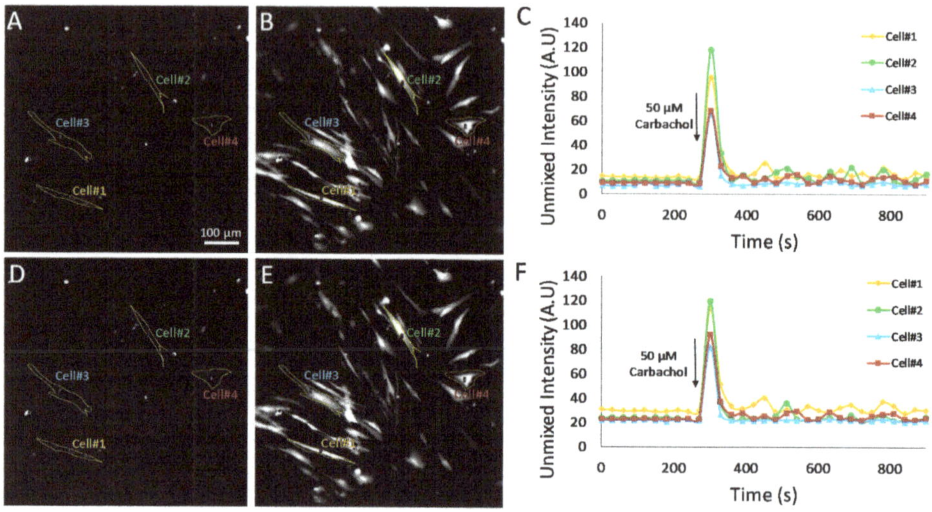

Figure 8. Effects of pixel filtering were further evaluated using time-lapse excitation-scanning HSI data of Cal 520 and NucBlue labeled HASMCs to identify dynamic Ca^{2+} signals. Hyperspectral images were analyzed using a matched filter (MF) algorithm, and thresholds for pixel filtering were set to a value of 20, corresponding to the theoretical sensitivity analysis identified threshold used in Figure 7B. (**A**) The initial time point Cal 520 endmember image (time point 0 s) corresponding to baseline Cal 520 fluorescence. Outlines for single-cell regions of interest (ROIs) are shown in yellow, and the cell number is indicated by colored text. A video showing all time points is provided in Supplemental Video S3. (**B**) The Cal 520 endmember image at a time point of 300 s, corresponding to a time point immediately after addition of 50 µM carbochol. A high-intensity response is seen in most cells. (**C**) The corresponding time trace for Cal 520 intensity for each of the single-cell ROIs. (**D**) The Cal 520 endmember image at time point 0 after applying a pixel threshold of 20. (**E**) The Cal 520 image at time point 300 after applying a pixel threshold of 20. A video showing all time points after pixel thresholding is provided in Supplemental Video S4. (**F**) The corresponding time trace for pixel-thresholded Cal 520 intensity for each of the single-cell ROIs. Periodic oscillations in Cal 520 intensity can be seen in cell 1 in the pixel-thresholded image data that are not visible in the non-thresholded data.

The advantages of this combined excitation-scanning hardware and custom analytical approach include the ability to measure time-dependent fluorescence image data and to discriminate among fluorescent labels and cellular autofluorescence—in general, the approach is effective. However, there are several considerations and limitations that should be noted. First, the current hardware embodiment of the excitation-scanning HSI microscope is limited in terms of the wavelength switch time by the need to mechanically rotate between adjacent TFTFs housed in a filter wheel and the need to tilt the TFTF filter wheel so as to tune the center wavelength band. There is much potential to improve the wavelength tuning speed through the revision of the optomechanical tuning assembly and the potential use of high-speed piezoelectric rotation stages or galvanometers.

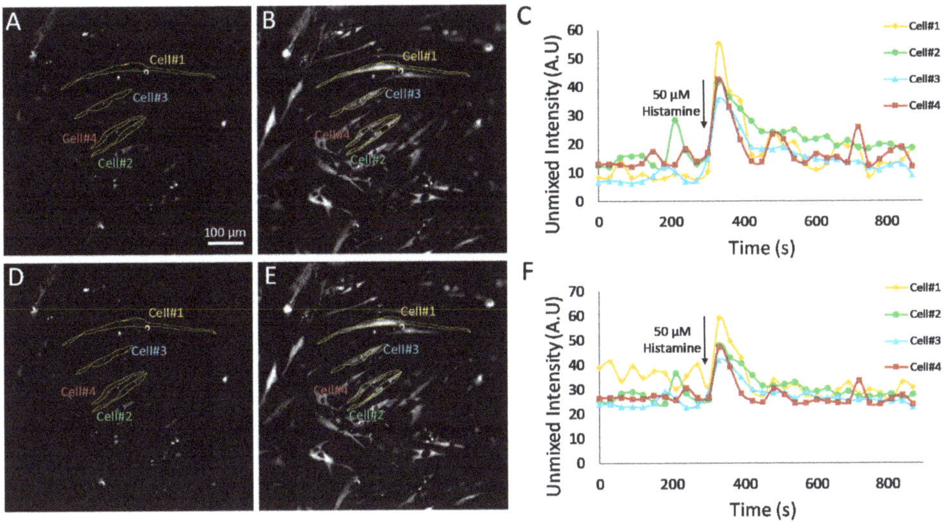

Figure 9. Effects of pixel filtering were also evaluated in hyperspectral images of HASMCs treated with 50 μM histamine. Hyperspectral images were acquired and analyzed identically to images shown in Figure 8. (**A**) The initial time point Cal 520 endmember image (time point 0 s), corresponding to baseline Cal 520 fluorescence. Outlines for single-cell regions of interest (ROIs) are shown in yellow, and the cell number is indicated by colored text. A video showing all time points is provided in Supplemental Video S5. (**B**) The Cal 520 endmember image at a time point of 330 s, corresponding to a time point immediately after addition of 50 μM carbochol. A high-intensity response is seen in most cells. (**C**) The corresponding time trace for Cal 520 intensity for each of the single-cell ROIs. (**D**) The Cal 520 endmember image at time point 0 after applying a pixel threshold of 20. (**E**) The Cal 520 image at time point 330 after applying a pixel threshold of 20. A video showing all time points after pixel thresholding is provided in Supplemental Video S6. (**F**) The corresponding time trace for pixel-thresholded Cal 520 intensity for each of the single-cell ROIs. Fluctuations in Cal 520 signal prior to treatment can be seen in cell 1 in the pixel-thresholded image data that are not present in the non-thresholded data.

Second, when configuring an HSI microscope system for a particular assay, it is important to match the spectral scanning parameters to the fluorescent labels or, alternatively, to select appropriate fluorescent labels that match the spectral scanning capabilities of the microscope system. In this study, the Cal 520 and NucBlue excitation peak wavelengths were separated by ~150 nm (see Figure 3), with a broad cellular autofluorescence covering the entire spectral scan range. While the excitation spectral scan range of 360–480 nm was sufficient to capture the full signature of NucBlue, it was only sufficient to capture part of the spectral signature of Cal 520, with the peak excitation wavelength reported by the manufacturer to occur at ~495 nm. If available, the selection of an alternative Ca^{2+} label with a lower excitation peak—for example, at 425 nm—would have allowed more of the spectral signature to be detected using the current scan range and may have provided an improved ability to separate signals from each fluorescent label. Alternatively, the use of such an alternative fluorescent label could enable a reduced excitation-scanning spectral range, resulting in an increase in overall spectral image acquisition time and enabling either reduced photobleaching or improved temporal sampling.

Third, the approach of fluorescence excitation-scanning HSI carries the potential to generate time-dependent spectral artifacts if high-speed changes in fluorescent label distribution or intensity occur between the onset and the completion of a spectral scan. These artifacts could affect the accuracy of any resulting spectral image analysis algorithms.

However, it should be noted that alternative push–broom or whisk–broom spectral scanning techniques would also present the capacity for a motion-induced artifact, but in this case, the artifact would not affect the spectral domain, but would instead manifest as a spatial artifact, as changes to label distribution or intensity could occur between the scanning of one pixel and an adjacent pixel. Snapshot spectral imaging may present an alternative approach to mitigate these motion artifacts but presents a trade-off in providing reduced spatial resolution/reduced field of view, as well as reduced spectral resolution and the inability to scan continuously across a range of spectral bands. It is important to consider these potential limitations and potential sources of artifacts when selecting a system and settings for HSI fluorescence microscopy.

4. Conclusions

Hyperspectral imaging technologies have undergone rapid development, from initial applications in remote sensing to a wide range of applications, including biomedical imaging and fluorescence microscopy. However, there is still much potential to improve both HSI hardware and software for use in live cell fluorescence microscopy. New approaches are needed that provide improved temporal sampling and reduced photobleaching. The excitation-scanning HSI system utilized in this study has been previously demonstrated to provide improved acquisition sensitivity and speeds in comparison to a similarly configured emission-scanning HSI system. However, these prior studies were performed in non-dynamic experimental conditions. Here, we have demonstrated the use of excitation-scanning HSI for live cell dynamic Ca^{2+} imaging. We have also evaluated four analysis algorithms that have been previously used in the remote sensing field (LU, SAM, CE, and MF) for the ability to detect and discriminate fluorescence excitation signatures. All three subpixel analysis techniques (CEM, LU, and MF) provided a high ability to quantify the Cal 520 signal. As seen in Figure 6, LU and MF provided a linear response to increasing Cal 520 signal and a steep slope of the TPPC, indicating that both are effective algorithms for quantifying Cal 520. At a threshold value of 15, LU resulted in 0 false-positive detections, while MF resulted in 127 (out of a total of 498,436 pixels within the image); hence, LU provided slightly improved false-positive rejection. The classification analysis algorithm (SAM) was able to identify regions of the Cal 520 signal but displayed a nonlinear dependence endmember signal. Hence, SAM could be useful for identifying regions within an image but was not suitable for quantifying Cal 520 signal over time.

To improve the ability to extract dynamic Ca^{2+} signals from small ROIs, the theoretical sensitivity framework was extended to allow the estimation of a minimum detectable limit for use in pixel filtering. The results indicated that pixel filtering could be used to remove bias from unlabeled or weakly labeled pixels, especially when measuring bulk signals from an entire field of view. When pixel filtering was applied to whole-field time-lapse data, the resultant Cal 520 time course included subtle kinetic features that were not present in unfiltered data—features that were likely washed out by averaging with background pixels. This approach could be further extended to a range of dynamic cell signaling experiments and may also be modified to allow for the use of autofluorescence signals as either fiducial or metabolic markers [35]. Hence, the use of excitation-scanning HSI with suitable spectral analysis algorithms and pixel filtering may provide a mechanism for measuring subtle kinetic events in cell signaling and other live cell dynamic assays.

Supplementary Materials: The following supporting information can be downloaded at: https://www.mdpi.com/article/10.3390/bioengineering10060642/s1, Video S1: Supplemental Video S1, Video S2: Supplemental Video S2, Video S3: Supplemental Video S3, Video S4: Supplemental Video S4, Video S5: Supplemental Video S5, Video S6: Supplemental Video S6.

Author Contributions: Conceptualization, S.J.L., M.P., N.S.A. and T.C.R.; methodology, M.P., N.S.A. and S.J.L.; software, M.P., N.S.A. and D.P.; validation, M.P. and N.S.A.; investigation, M.P., N.S.A., Z.I. and J.J.; resources, R.P., D.D., T.C.R. and S.J.L.; writing—original draft preparation, M.P., N.S.A. and S.J.L.; writing—review and editing, M.P., N.S.A., S.J.L., T.C.R., R.P. and D.D.; visualization, M.P.,

N.S.A. and D.P.; supervision, S.J.L. and T.C.R.; project administration, S.J.L. and T.C.R.; funding acquisition, R.P., D.D., T.C.R. and S.J.L. All authors have read and agreed to the published version of the manuscript.

Funding: This research was funded with support from NSF grant 1725937 and NIH grants P01HL066299, R01HL58506, R01HL137030, and R01HL169522.

Institutional Review Board Statement: Not applicable.

Informed Consent Statement: Not applicable.

Data Availability Statement: Data utilized in this manuscript are available at https://www.southalabama.edu/centers/bioimaging/resources.html accessed on 22 April 2023.

Conflicts of Interest: Leavesley and Rich disclose financial interest in a start-up company, SpectraCyte LLC, formed to commercialize spectral imaging technologies.

References

1. Goetz, A.F.H.; Vane, G.; Solomon, J.E.; Rock, B.N. Imaging Spectrometry for Earth Remote Sensing. *Science* **1985**, *228*, 1147–1153. [CrossRef] [PubMed]
2. Harris, A.T. Spectral Mapping Tools from the Earth Sciences Applied to Spectral Microscopy Data. *Cytom. Part A J. Int. Soc. Anal. Cytol.* **2006**, *69A*, 872–879. [CrossRef]
3. Goltz, D.; Attas, M.; Young, G.; Cloutis, E.; Bedynski, M. Assessing Stains on Historical Documents Using Hyperspectral Imaging. *J. Cult. Herit.* **2010**, *11*, 19–26. [CrossRef]
4. Kim, S.J.; Zhuo, S.; Deng, F.; Fu, C.-W.; Brown, M. Interactive Visualization of Hyperspectral Images of Historical Documents. *IEEE Trans. Vis. Comput. Graph.* **2010**, *16*, 1441–1448. [PubMed]
5. Adão, T.; Hruška, J.; Pádua, L.; Bessa, J.; Peres, E.; Morais, R.; Sousa, J.J. Hyperspectral Imaging: A Review on UAV-Based Sensors, Data Processing and Applications for Agriculture and Forestry. *Remote Sens.* **2017**, *9*, 1110. [CrossRef]
6. Guolan, L.; Fei, B. Medical Hyperspectral Imaging: A Review. *J. Biomed. Opt.* **2014**, *19*, 24.
7. Garini, Y.; Young, I.; McNamara, G. Spectral Imaging: Principles and Applications. *Cytom. Part A J. Int. Soc. Anal. Cytol.* **2006**, *69*, 735–747. [CrossRef]
8. Hu, F.; Zeng, C.; Long, R.; Miao, Y.; Wei, L.; Xu, Q.; Min, W. Supermultiplexed Optical Imaging and Barcoding with Engineered Polyynes. *Nat. Methods* **2018**, *15*, 194–200. [CrossRef]
9. Mansfield, J.R.; Hoyt, C.; Levenson, R.M. Visualization of Microscopy-Based Spectral Imaging Data from Multi-Label Tissue Sections. *Curr. Protoc. Mol. Biol.* **2008**, *84*, 14.19.1–14.19.15. [CrossRef]
10. Deal, J.; Britain, A.; Rich, T.; Leavesley, S. Excitation-Scanning Hyperspectral Imaging Microscopy to Efficiently Discriminate Fluorescence Signals. *J. Vis. Exp.* **2019**, *150*, 59448. [CrossRef]
11. Rossetti, B.J.; Wilbert, S.A.; Mark Welch, J.L.; Borisy, G.G.; Nagy, J.G. Semi-Blind Sparse Affine Spectral Unmixing of Autofluorescence-Contaminated Micrographs. *Bioinformatics* **2020**, *36*, 910–917. [CrossRef] [PubMed]
12. Mansfield, J.R.; Gossage, K.W.; Hoyt, C.C.; Levenson, R.M. Autofluorescence Removal, Multiplexing, and Automated Analysis Methods for in-Vivo Fluorescence Imaging. *J. Biomed. Opt.* **2005**, *10*, 41207. [CrossRef] [PubMed]
13. Leavesley, S.J.; Annamdevula, N.; Boni, J.; Stocker, S.; Grant, K.; Troyanovsky, B.; Rich, T.C.; Alvarez, D.F. Hyperspectral Imaging Microscopy for Identification and Quantitative Analysis of Fluorescently-Labeled Cells in Highly Autofluorescent Tissue. *J. Biophotonics* **2012**, *5*, 67–84. [CrossRef] [PubMed]
14. Levy, S.; Wilms, C.D.; Brumer, E.; Kahn, J.; Pnueli, L.; Arava, Y.; Eilers, J.; Gitler, D. SpRET: Highly Sensitive and Reliable Spectral Measurement of Absolute FRET Efficiency. *Microsc. Microanal.* **2011**, *17*, 176–190. [CrossRef] [PubMed]
15. Leavesley, S.J.; Britain, A.; Cichon, L.K.; Nikolaev, V.O.; Rich, T.C. Assessing FRET Using Spectral Techniques. *Cytom. Part A* **2013**, *83*, 898–912. [CrossRef] [PubMed]
16. Annamdevula, N.S.; Sweat, R.; Griswold, J.R.; Trinh, K.; Hoffman, C.; West, S.; Deal, J.; Britain, A.L.; Jalink, K.; Rich, T.C.; et al. Spectral Imaging of FRET-based Sensors Reveals Sustained CAMP Gradients in Three Spatial Dimensions. *Cytom. Part A* **2018**, *93*, 1029–1038. [CrossRef]
17. Rothmann, C.; Bar-Am, I.; Malik, Z. Invited Review: Spectral Imaging for Quantitative Histology and Cytogenetics. *Histol Histopathol* **1998**, *13*, 921–926.
18. Levenson, R.M. Spectral Imaging and Pathology: Seeing More. *Lab. Med.* **2004**, *35*, 244–251. [CrossRef]
19. Levenson, R.M.; Mansfield, J.R. Multispectral Imaging in Biology and Medicine: Slices of Life. *Cytom. Part J. Int. Soc. Anal. Cytol.* **2006**, *69*, 748–758. [CrossRef]
20. Wachman, E.S.; Niu, W.; Farkas, D.L. AOTF Microscope for Imaging with Increased Speed and Spectral Versatility. *Biophys. J.* **1997**, *73*, 1215–1222. [CrossRef]
21. Annamdevula, N.S.; Sweat, B.; Favreau, P.; Lindsey, A.S.; Alvarez, D.F.; Rich, T.C.; Leavesley, S.J. An Approach for Characterizing and Comparing Hyperspectral Microscopy Systems. *Sensors* **2013**, *13*, 9267–9293. [CrossRef] [PubMed]

22. Favreau, P.F.; Hernandez, C.; Lindsey, A.S.; Alvarez, D.F.; Rich, T.C.; Prabhat, P.; Leavesley, S.J. Thin-Film Tunable Filters for Hyperspectral Fluorescence Microscopy. *J. Biomed. Opt.* **2014**, *19*, 011017. [CrossRef] [PubMed]
23. Li, Q.; Zhou, M.; Liu, H.; Wang, Y.; Guo, F. Red Blood Cell Count Automation Using Microscopic Hyperspectral Imaging Technology. *Appl. Spectrosc.* **2015**, *69*, 1372–1380. [CrossRef] [PubMed]
24. Wang, Q.; Sun, L.; Wang, Y.; Zhou, M.; Hu, M.; Chen, J.; Wen, Y.; Li, Q. Identification of Melanoma from Hyperspectral Pathology Image Using 3D Convolutional Networks. *IEEE Trans. Med. Imaging* **2021**, *40*, 218–227. [CrossRef]
25. Sun, L.; Zhou, M.; Li, Q.; Hu, M.; Wen, Y.; Zhang, J.; Lu, Y.; Chu, J. Diagnosis of Cholangiocarcinoma from Microscopic Hyperspectral Pathological Dataset by Deep Convolution Neural Networks. *Methods* **2022**, *202*, 22–30. [CrossRef]
26. Pawley, J.B. *Handbook of Biological Confocal Microscopy*, 3rd ed.; Springer: Berlin/Heidelberg, Germany, 2006; ISBN 978-0-387-25921-5.
27. Liu, Z.; Lavis, L.D.; Betzig, E. Imaging Live-Cell Dynamics and Structure at the Single-Molecule Level. *Mol. Cell* **2015**, *58*, 644–659. [CrossRef]
28. Stehbens, S.; Pemble, H.; Murrow, L.; Wittmann, T. Imaging Intracellular Protein Dynamics by Spinning Disk Confocal Microscopy. *Methods Enzymol.* **2012**, *504*, 293–313.
29. Oreopoulos, J.; Berman, R.; Browne, M. Spinning-Disk Confocal Microscopy: Present Technology and Future Trends. *Methods Cell Biol.* **2014**, *123*, 153–175.
30. Zhang, H.; Tan, C.; Shi, X.; Xu, J. Impacts of Autofluorescence on Fluorescence Based Techniques to Study Microglia. *BMC Neurosci.* **2022**, *23*, 21. [CrossRef]
31. Miyazaki, K.; Lisman, J.E.; Ross, W.N. Improvements in Simultaneous Sodium and Calcium Imaging. *Front. Cell. Neurosci.* **2019**, *12*, 514. [CrossRef]
32. Weigert, M.; Schmidt, U.; Boothe, T.; Müller, A.; Dibrov, A.; Jain, A.; Wilhelm, B.; Schmidt, D.; Broaddus, C.; Culley, S.; et al. Content-Aware Image Restoration: Pushing the Limits of Fluorescence Microscopy. *bioRxiv* **2018**, *15*, 236463. [CrossRef] [PubMed]
33. Favreau, P.F.; Hernandez, C.; Heaster, T.; Alvarez, D.F.; Rich, T.C.; Prabhat, P.; Leavesley, S.J. Excitation-Scanning Hyperspectral Imaging Microscope. *J. Biomed. Opt.* **2014**, *19*, 046010. [CrossRef] [PubMed]
34. Leavesley, S.J.; Sweat, B.; Abbott, C.; Favreau, P.F.; Rich, T.C. A Theoretical-Experimental Methodology for Assessing the Sensitivity of Biomedical Spectral Imaging Platforms, Assays, and Analysis Methods. *J. Biophotonics* **2018**, *11*, e201600227. [CrossRef] [PubMed]
35. Favreau, P.F.; Deal, J.A.; Harris, B.; Weber, D.S.; Rich, T.C.; Leavesley, S.J. Label-Free Spectroscopic Tissue Characterization Using Fluorescence Excitation-Scanning Spectral Imaging. *J. Biophotonics* **2020**, *13*, e201900183. [CrossRef] [PubMed]
36. Larson, J. The Nikon C1si Combines High Spectral Resolution, High Sensitivity, and High Acquisition Speed. *Cytom. Part A J. Int. Soc. Anal. Cytol.* **2006**, *69*, 825–834. [CrossRef] [PubMed]
37. Dicker, D.T.; Lerner, J.; Van Belle, P.; Barth, S.F.; Herlyn, M.; Elder, D.E.; El-Deiry, W.S. Differentiation of Normal Skin and Melanoma Using High Resolution Hyperspectral Imaging. *Cancer Biol. Ther.* **2006**, *5*, 1033–1038. [CrossRef]
38. Horvat, S.J.; Deshpande, D.A.; Yan, H.; Panettieri, R.A.; Codina, J.; DuBose, T.D., Jr.; Xin, W.; Rich, T.C.; Penn, R.B. A-kinase Anchoring Proteins Regulate Compartmentalized CAMP Signaling in Airway Smooth Muscle. *FASEB J.* **2012**, *26*, 3670–3679. [CrossRef]
39. Deshpande, D.A.; Wang, W.C.H.; McIlmoyle, E.L.; Robinett, K.S.; Schillinger, R.M.; An, S.S.; Sham, J.S.K.; Liggett, S.B. Bitter Taste Receptors on Airway Smooth Muscle Bronchodilate by Localized Calcium Signaling and Reverse Obstruction. *Nat. Med.* **2010**, *16*, 1299–1304. [CrossRef]
40. Penn, R.B. Embracing Emerging Paradigms of G Protein-Coupled Receptor Agonism and Signaling to Address Airway Smooth Muscle Pathobiology in Asthma. *Naunyn. Schmiedebergs Arch. Pharmacol.* **2008**, *378*, 149–169. [CrossRef]
41. Agarwal, S.R.; Miyashiro, K.; Latt, H.; Ostrom, R.S.; Harvey, R.D. Compartmentalized CAMP Responses to Prostaglandin EP $_2$ Receptor Activation in Human Airway Smooth Muscle Cells: CAMP Compartmentation in Human Airway Smooth Muscle Cells. *Br. J. Pharmacol.* **2017**, *174*, 2784–2796. [CrossRef]
42. Ostrom, R.S.; Liu, X.; Head, B.P.; Gregorian, C.; Seasholtz, T.M.; Insel, P.A. Localization of Adenylyl Cyclase Isoforms and G Protein-Coupled Receptors in Vascular Smooth Muscle Cells: Expression in Caveolin-Rich and Noncaveolin Domains. *Mol. Pharmacol.* **2002**, *62*, 983–992. [CrossRef] [PubMed]
43. Keshava, N.; Mustard, J.F. Spectral Unmixing. *Signal Process. Mag. IEEE* **2002**, *19*, 44–57. [CrossRef]
44. Keshava, N. A Survey of Spectral Unmixing Algorithms. *Linc. Lab. J.* **2003**, *14*, 55–78.
45. Farrand, W.H.; Harsanyi, J.C. Mapping the Distribution of Mine Tailings in the Coeur d'Alene River Valley, Idaho, through the Use of a Constrained Energy Minimization Technique. *Remote Sens. Environ.* **1997**, *59*, 64–76. [CrossRef]
46. Williams, P.; Hunt, A.; Hunt, E.R., Jr. Estimation of Leafy Spurge Cover from Hyperspectral Imagery Using Mixture Tuned Matched Filtering. *Remote Sens. Environ.* **2002**, *82*, 446–456. [CrossRef]
47. Boardman, J.W.; Kruse, F.A. Analysis of Imaging Spectrometer Data Using-Dimensional Geometry and a Mixture-Tuned Matched Filtering Approach. *IEEE Trans. Geosci. Remote Sens.* **2011**, *49*, 4138–4152. [CrossRef]
48. Kruse, F.; Lefkoff, A.; Boardman, J.; Heidebrecht, K.; Shapiro, A.; Barloon, P.; Goetz, A. The Spectral Image Processing System (SIPS)—Interactive Visualization and Analysis of Imaging Spectrometer Data. *Remote Sens. Environ.* **1993**, *44*, 145–163. [CrossRef]
49. Panettieri, R.A.; Murray, R.K.; DePalo, L.R.; Yadvish, P.A.; Kotlikoff, M.I. A Human Airway Smooth Muscle Cell Line That Retains Physiological Responsiveness. *Am. J. Physiol.-Cell Physiol.* **1989**, *256*, C329–C335. [CrossRef]

50. Schneider, C.A.; Rasband, W.S.; Eliceiri, K.W. NIH Image to ImageJ: 25 Years of Image Analysis. *Nat. Methods* **2012**, *9*, 671–675. [CrossRef]
51. Chang, C.I. *Hyperspectral Imaging: Techniques for Spectral Detection and Classification*; Kluwer Academy, Plenum Publishers: New York, NY, USA, 2003.
52. Du, Q.; Ren, H.; Chang, C.-I. A Comparative Study for Orthogonal Subspace Projection and Constrained Energy Minimization. *IEEE Trans. Geosci. Remote Sens.* **2003**, *41*, 1525–1529.
53. Metz, C.E. *Basic Principles of ROC Analysis*; Elsevier: Amsterdam, The Netherlands, 1978; Volume 8, pp. 283–298.
54. Fox Jr, D.J.; Velde, H.T.; Preza, C.; O'Sullivan, J.A.; Smith, W.H.; Woolsey, T.A. Computational Hyperspectral Interferometry for Studies of Brain Function: Proof of Concept. *Appl. Opt.* **2006**, *45*, 3009–3021. [CrossRef] [PubMed]
55. Hedde, P.N.; Cinco, R.; Malacrida, L.; Kamaid, A.; Gratton, E. Phasor-Based Hyperspectral Snapshot Microscopy Allows Fast Imaging of Live, Three-Dimensional Tissues for Biomedical Applications. *Commun. Biol.* **2021**, *4*, 721. [CrossRef] [PubMed]
56. Browning, C.M.; Browning, C.M.; Browning, C.M.; Mayes, S.; Mayes, S.; Mayes, S.A.; Mayes, S.A.; Rich, T.C.; Rich, T.C.; Leavesley, S.J.; et al. Microscopy Is Better in Color: Development of a Streamlined Spectral Light Path for Real-Time Multiplex Fluorescence Microscopy. *Biomed. Opt. Express* **2022**, *13*, 3751–3772. [CrossRef] [PubMed]
57. Parker, M.; Mayes, S.A.; Browning, C.M.; Deal, J.; Gunn-Mayes, S.; Annamdevula, N.S.; Rich, T.C.; Leavesley, S.J. Multifaceted Mirror Array Illuminator for Fluorescence Excitation-Scanning Spectral Imaging Microscopy. *J. Biomed. Opt.* **2023**, *28*, 026502. [CrossRef]
58. SearchLight Spectra Viewer from Semrock. Available online: https://searchlight.semrock.com/ (accessed on 12 December 2022).

bioengineering

MDPI

Article

CT Imaging of Eustachian Tube Balloon Dilation: Method Development on Cadaver Heads

Selma Cetin-Ferra [1], Miriam S. Teixeira [2], J. Douglas Swarts [3], Tanya J. Rath [4] and Cuneyt M. Alper [3,5,*]

1 Department of Cell Biology, University of Pittsburgh School of Medicine, Pittsburgh, PA 15260, USA
2 Graduate Medical Education Research Division, Arnot Ogden Medical Center, Elmira, NY 14905, USA
3 Department of Otolaryngology, University of Pittsburgh School of Medicine Pittsburgh, PA 15213, USA
4 Department of Radiology, Division of Neuroradiology, Mayo Clinic College of Medicine, Phoenix, AZ 85054, USA
5 Division of Pediatric Otolaryngology, UPMC Children's Hospital of Pittsburgh, 1 Children's Place, 4401 Penn Avenue, Pittsburgh, PA 15224, USA
* Correspondence: cuneyt.alper@chp.edu; Tel.: +1-412-692-7003; Fax: +1-412-692-6074

Abstract: *Objective*: To develop a methodology for the measurement of balloon dilation (BD) effects on Eustachian Tube (ET) structure using Computerized Tomography (CT) images. *Methods*: The BD of the ET was performed on three cadaver heads (five ears) through the nasopharyngeal orifice. The axial CT images of the temporal bones were obtained before dilation, while an inflated balloon was in the lumen of ET, and after balloon removal in each ear. Utilizing Dicom images captured by the ImageJ software 3D volume viewer function, the anatomical landmark coordinates of the ET were matched with their pre- and post-dilation counterparts, and the longitudinal axis of the ET was captured with serial images. The histograms of the regions of interest (ROI) and three different lumen width and length measurements were obtained from captured images. The densities of air, tissue, and bone were determined with histograms as a baseline to determine the BD rate as a function of increased air in the lumen. *Results*: The small ROI box included the area of prominently dilated ET lumen after BD and best represented the visually obvious changes in the lumen, compared to the ROIs that extended the wider areas (longest and longer). Air density was the outcome measure for comparison with each corresponding baseline value. The average increase in air density in the small ROI was 64%, while the longest and long ROI boxes showed 44 and 56% increases, respectively. *Conclusion:* This study describes a method to image the ET and quantify the outcomes of BD of the ET using anatomical landmarks.

Keywords: eustachian tube; eustachian tube dysfunction; balloon dilation of eustachian tube; imaging; CT imaging; cadaver

Citation: Cetin-Ferra, S.; Teixeira, M.S.; Swarts, J.D.; Rath, T.J.; Alper, C.M. CT Imaging of Eustachian Tube Balloon Dilation: Method Development on Cadaver Heads. *Bioengineering* **2023**, *10*, 592. https://doi.org/10.3390/bioengineering10050592

Academic Editor: Maria Francesca Spadea

Received: 5 March 2023
Revised: 5 May 2023
Accepted: 9 May 2023
Published: 14 May 2023

1. Introduction

The Eustachian tube (ET) is an anatomical connector structure between the middle ear (ME) and nasopharynx (NP) [1]. The ET opening is the main means of pressure regulation in the ME and is impaired in patients presenting with otitis media with effusion (OME) [2]. Standard medical therapy for Eustachian tube dysfunction (ETD) is the insertion of ventilation tubes (VTs) to temporarily bypass the ET, equalize ME pressure with the environment, and prevent fluid recurrence. Although widely used, VTs come with potential inconveniences such as otorrhea and rare complications and sequelae such as tympanic membrane perforation, tympanosclerosis, and cholesteatoma, making management of ETD a challenge for otolaryngologists [3]. Balloon dilation (BD) of the ET (BDET) was introduced as an alternative and feasible procedure to treat ETD, with promising outcomes [4–7]. There is still no consensus for patient selection for BDET or for objective outcome measures to evaluate the long-term effect of BDET [8,9]. Our recent clinical trial in patients with ETD aimed to quantify ET function and structural changes using Computerized Tomography

(CT) images that were captured before and after BDET [10]. A pilot study was performed to develop methodology for verification of the anatomical landmarks and the feasibility of quantification of the changes in ET structure utilizing a balloon catheter in the ET lumen for guidance on CT scan X-ray images. The eventual aim was to use this method to evaluate the effect of BDET as a measurable outcome.

2. Materials and Methods

This protocol was approved by the Committee for Oversight of Research and Clinical Training Involving Decedents. Three frozen whole male cadaver heads from the University of Pittsburgh School of Medicine Department of Neurobiology and Office for Oversight of Anatomic Specimens were used.

2.1. Specimen Preparation

Four days before the experiment, the cadaver heads were thawed, first at room temperature and then in a 4 °C refrigerator. On the day of the experiment, an exam of the mouth, nose, and throat was performed, including 0° and 45° rigid nasal endoscopies (Hopkins II, 2.7 mm, 18 cm, Karl Storz Endoscopy, Storz Xenon 300 light source, Germany). The external ear canals were cleaned under microscopy, and myringotomies were performed bilaterally in all tympanic membranes.

2.2. Experimental Protocol

The cadaver head was placed in the supine position, and a plastic ear probe connected to tubing and a three-way valve were placed in the right ear canal to administer contrast medium to the middle ear.

Under the guidance of a 45° rigid endoscope, a 16 mm long balloon catheter (with maximum and minimum diameters of 7 and 3 mm, respectively) was introduced in the left nostril. (Acclarent Balloon Dilation System, Irvine, CA [https://www.jnjmedtech.com/en-US/companies/acclarent, accessed 10 May 2023] . The pharyngeal opening of the ET was identified, and the balloon advanced into the ET lumen. A baseline axial CT scan acquisition was performed by a helical 64 channel multidetector CT scanner (GE LightSpeed VCT; GE Healthcare; Milwaukee, WI, USA). The CT scan protocol parameters were: mAS = 180, kvP = 100, rotation time = 0.6 s, thickness = 0.625 mm, pitch = 0.969, and FOV = 160 mm. The image field of view extended from the tegmen of the temporal bone through the pterygoid processes.

Next, approximately 4–6 mL of 1:3 diluted contrast (Omnipaque [Iohexol] 180 mg/mL, GE Healthcare Inc., Princeton, NJ, USA) was infused through the probe into the right middle ear, and the valve was closed. A balloon was introduced into the left ET lumen and kept uninflated, and a second CT scan acquisition was obtained. A second balloon was then introduced into the right ET lumen, and a third CT scan acquisition of both ears was performed. Both balloons were then inflated to 12 atm for 1 min, and two additional acquisitions were performed: one while both balloons were inflated and another after deflation. Lastly, both balloons were removed, and one final CT scan acquisition was performed.

2.3. Image Analysis

First, axial CT images were formatted into the RadiAnt DICOM viewer (version 1.9.14.7431) to be able to be viewed in the ImageJ program (1.51K National Institutes of Health, Bethesda, MD USA). Images with uninflated balloons on the left side were used as the baseline for the images, while the right side had contrast, which was visible in most axial images although not always in the canal. Secondly, sequential images of each series before, during, and after balloon dilation were viewed in the 3D volume viewer plugin of the ImageJ software (Figure 1). The center of the ET longitudinal axis was saved as a single JPEG image after determining X, Y, and Z coordinates and superior-inferior distances by using the image of the inflated balloon as guidance in each ET. This was followed by 12 sequential images from the superior (cranial) and inferior (caudal) directions of the images recaptured and numbered accordingly to include measurements of the whole lumen dimension.

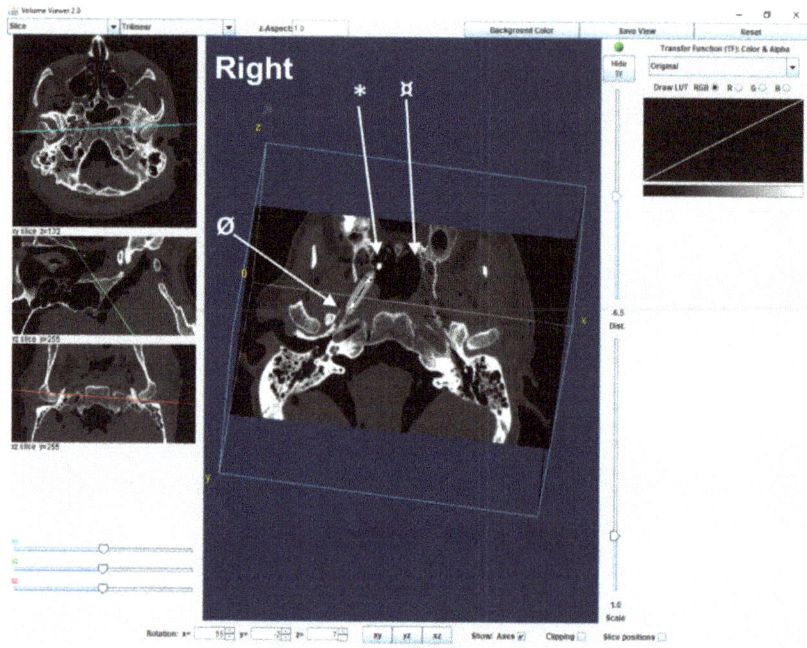

Figure 1. Example of 3D volume viewing of a CT image in ImageJ software under the guidance of anatomical landmarks *(* medial pterygoid plate spine, ¤ opposite medial pterygoid plate spine,* and *Ø sphenoid spine)* and an inflated balloon to determine the ET longitudinal axis coordinates at the center as a starting point of image analysis. The X, Y, and Z coordinates of the image on the right are demonstrated on the smaller images on the left side as blue line (top), green line (middle image), and red line (bottom image).

Third, the corresponding centers of the luminal areas in the pre- and post-dilation conditions were re-imaged by creating a plane in the 3D volume viewer using the medial pterygoid plate, hamulus, and sphenoid spine as landmarks. Twenty-four more total sequential images were captured in similar fashion in the 3D volume viewer for each counterpart.

Finally, vertical rectangles of the regions of interest (ROIs) were created in the ImageJ software at the mid-cartilaginous ET lumen level after rotating all the images to bring lumens in the same direction (Figure 2). Initial small ROIs were chosen in the images with inflated balloons with guidance from the contrast at mid-bead level at the two vertical ends. Then the original rectangle was extended toward the NP orifice until the tip of the torus and at the posterior opening to the tip of the balloon by an equal distance in the number of pixels without including the bony structures (longest box), and the last ROI box was shaped by decreasing 5 pixels from each end of the longest box (long box) (Figure 3). Histograms were constructed in the corresponding ROIs of each pre- and post-dilation series in all ET images (Figure 4), followed by baseline histograms from the nasopharyngeal air, surrounding tissue, and the petrous bone (Figure 5) for the determination of the density range of each structure. The mean gray values of each ROI were also measured with ImageJ software in all series images (Figure 6).

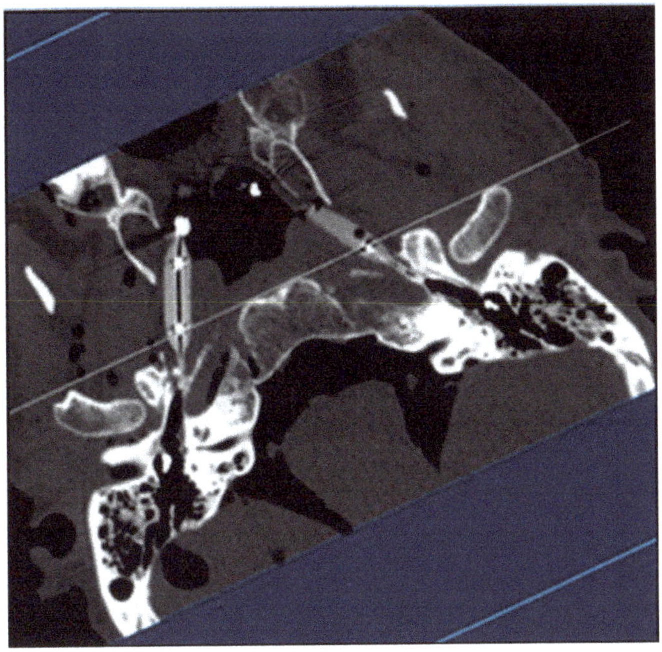

Figure 2. Example of a vertically rotated JPEG image captured at the center of the longitudinal axis from the NP to the isthmus after positioning in ImageJ software (Ear 1—right).

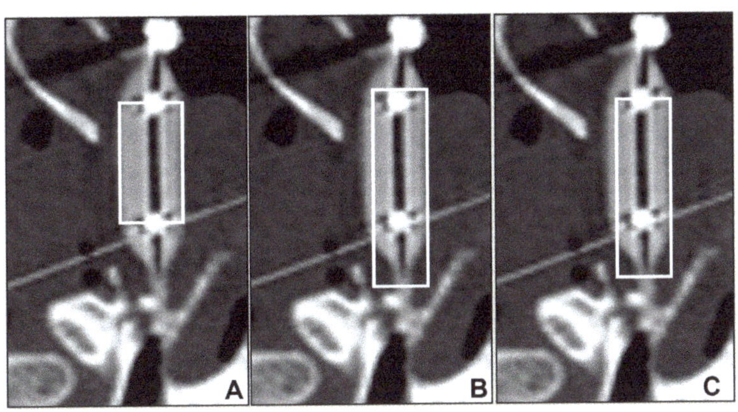

Figure 3. Example of how vertical ROI boxes (white rectangle) (**A**)—small; (**B**)—longest; (**C**)—long) are placed in an inflated balloon-guided ET lumen (Ear 1). Outline of the small ROI boxes (A) was formed to include the main body of the inflated balloons, to include the 6 mm width in the full 16 mm cylindrical segment. Radiopaque markers were used as markers, and short sides of the ROI boxes were set to cross these markers. This segment of the balloon represents the section of the ET that was subjected to the maximum dilation pressure. Proximal and distal ends of the balloons had conical sections which had a relatively less dilation force on the tissues. In order to include a longer section of the ET in the ROI, and to differentiate the effect of these cylindrical and conical sections of the balloons on the tissues, the longest ROI box (B) was extended to the end of the balloon towards the middle ear (bottom of the images) and towards the nasopharyngeal end of the balloon (top of the images) while avoiding inclusion of nasopharyngeal air column. The long ROI (C) was then formed by retracting the box by 5 pixel units from the middle ear and the NP end.

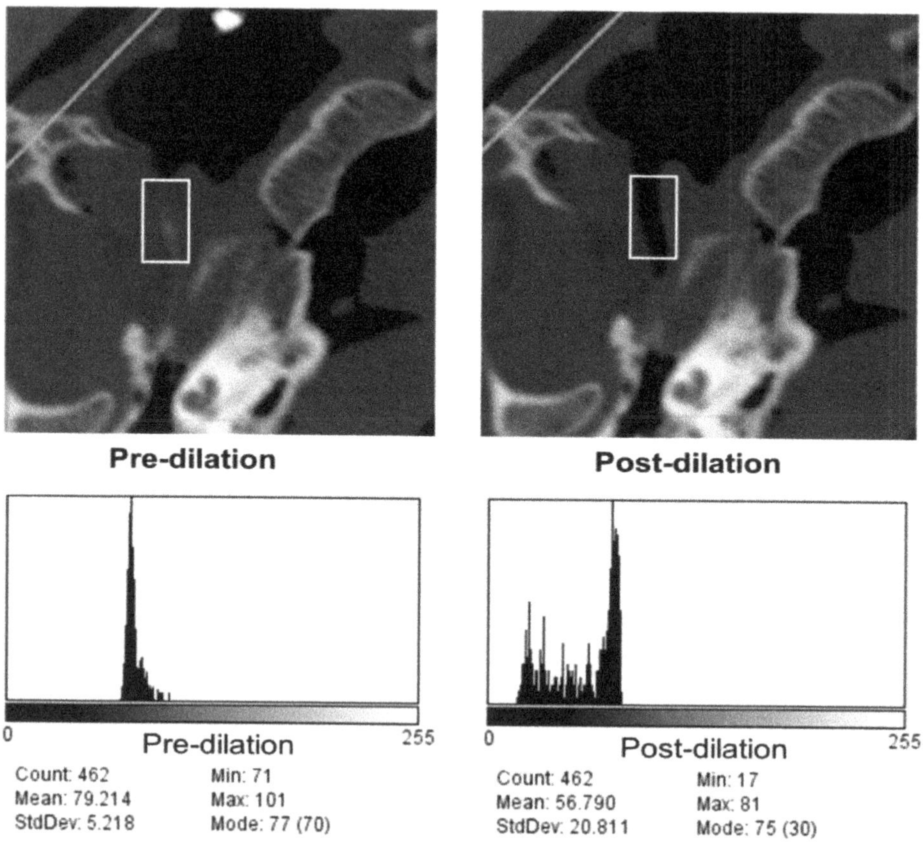

Figure 4. Small ROI placements (white rectangle boxes) in the center of the longitudinal axis of the lumen (**top row**); histograms of the ROIs shown in the top row (**bottom row**).

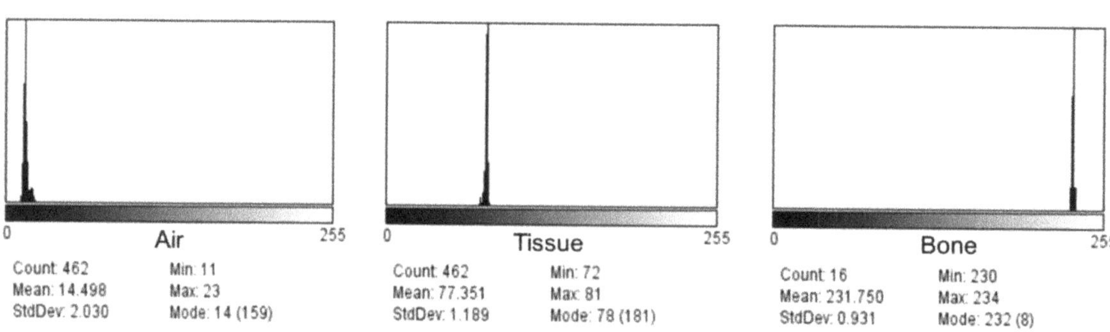

Figure 5. Baseline histograms in nasopharyngeal air, tissue, and bone (Ear 4).

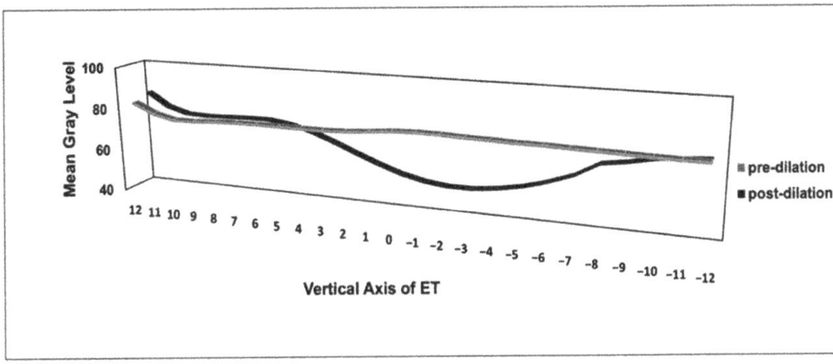

Figure 6. Representative graph for the measurement of mean gray values in small ROIs in pre- and post-dilation images in ET (Ear 4). The graph with mean gray value depicts markedly the visually observed dilation in the serial CT images of the ears pre- versus post dilation image of this example ear.

Additionally, three different measurements were taken for the length of the lumen with the guidance of the inflated balloon. The longest length of the lumen measurement was from the tip of the torus to the posterior end where the tip of the balloon was located (1); the second length measurement was the distance from the torus to the end of the widest part of the balloon (2); and the last length measurement was the distance from the widest part of the balloon to the tip of the balloon (3) (Figure 7). Finally, pre- and post-dilation widths were measured at 3 different levels of the lumen: at the nasopharyngeal opening (4), mid-cartilaginous (5), and at the posterior end (6) (Figure 8), using the longest ROI as a constant locator of the points in all ears.

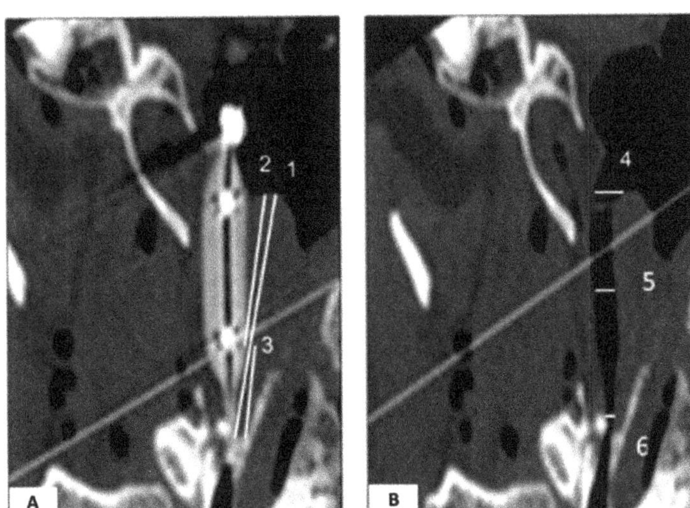

Figure 7. Measurements of ET structure under the guidance of the balloon and ROI. (**A**) Image during dilation at the center of the balloon lumen, representing the center of the longitudinal axis of the ET (Ear 1). The longest luminal length (distance 1), the length from the torus to the end of the widest part of the balloon at the tip (distance 2), and the length from the widest part of the balloon to the tip of the balloon (distance 3). (**B**) Image after dilation at the center of the longitudinal axis of the ET (Ear 1). NP opening at the torus #4 (width 1), the width at the mid-balloon level #5 (width 2), and the width at the isthmus end of the Eustachian lumen #6 (width 3).

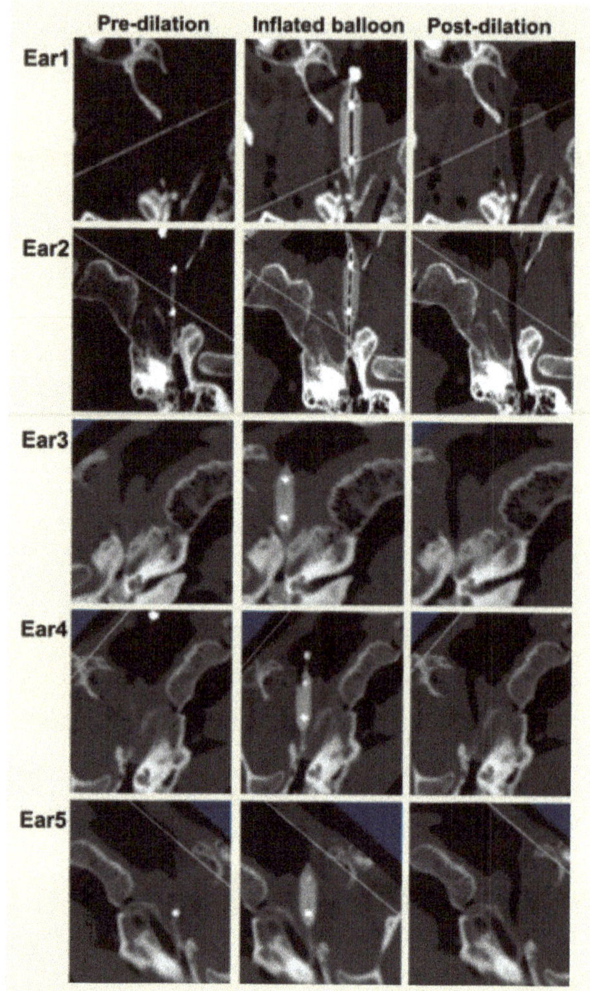

Figure 8. Center of the ET longitudinal axis in images pre-, during, and post-dilation in each ear (Ear 1–Ear 5).

3. Results

One of the balloon catheter insertions created a false passage, which was determined before the image analysis and excluded that ET from the data. Other catheter insertions were adequate from the NP orifice to the isthmic end.

The baseline histograms of air, tissue, and bony structures in the representative mid-section images of each ear illustrated the number of pixels distributed on the images for each level (gray value), from the darkest (0) to the brightest (256), for all ROIs that were saved with the ImageJ software, for each described step of imaging. The pixel values in the air density found 0–16 with the baseline histograms that were taken with locating small ROIs in the nasopharyngeal space and areas out of the frame adjacent to the cadaver head. We also checked the pixel values in the dilated portion of the ET lumen in several sequential images. These measurements increased the air density distribution per pixel up to 31 units in the histograms, suggesting there is a partial volume effect in the lumen. Tissue and bone pixel values were also documented to be within the ranges of 42–88 and 214–255, respectively. We have used two different ranges of pixel values for the air density (0–16 and 0–31) in two different calculations as the total number of pixels in all ROIs in the sequential images of each pre- and post-dilation condition in five ears. Small ROI box measurements depicted the change better than the longest and long box ROIs as pixel distribution in the air density, which were observed with clearly widened lumens in post-dilation images compared to the pre-dilation counterparts (Figure 8). The number of pixels in the air density range pre- and post-dilation for all five ears is demonstrated in Figure 9. This representation with small ROIs was expected since the longest and long boxes included more heterogeneous structures. There was a 20–100% increase in air pixel values with small ROI histograms (average 64%), while the longest and long ROI boxes showed an increase between 6–68 and 8–97% (averages 44 and 56%), respectively (Table 1).

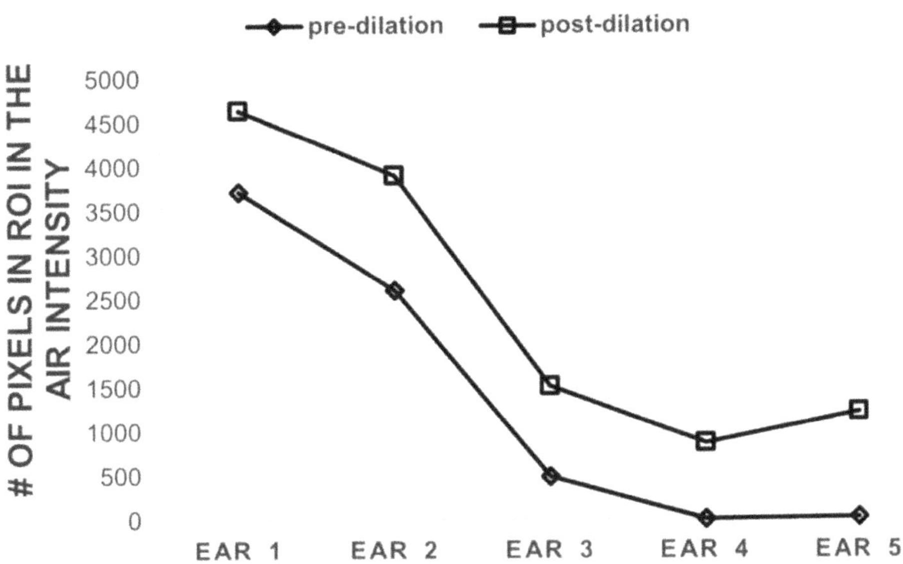

Figure 9. Total pixel values in the ROI (small box) in the air density values (0–31) pre- versus post-dilation in each ear.

Table 1. Percentage increase in air density (pixel unit) pre- versus post-dilation in all three ROIs in each ear. There was a 20–100% increase in air pixel values with small ROI histograms (average 64%), while the longest and long ROI boxes showed an increase between 6–68 and 8–97% (averages 44 and 56%), respectively.

	Small	Longest	Long
Ear 1	20	6	8
Ear 2	33	23	26
Ear 3	68	50	62
Ear 4	100	74	97
Ear 5	98	68	88
% Increase	64	44	56

ET lumen width measurements also showed increases compared to baseline values, especially the width at the center of the balloon (width 2). The width percentage increase was 57.8–93.5% (width 2, average 71.8%, and SD = 16.4), while NP opening and the lumen at the isthmic end dilated between 0 and 25 (width 1, average 8.3, and SD = 9.9) and 0–100 (width 3, average 37.8, and SD = 39.4%, respectively) (Table 2).

Table 2. Percentage increase in lumen width (pre- versus post-dilation) in all ears.

	Width 1	Width 2	Width 3	Average
Ear 1	3.5	57.8	31.6	31
Ear 2	0	93.5	10	34.5
Ear 3	9	75.6	100	61.5
Ear 4	4.2	53.3	47.2	34.9
Ear 5	25	78.9	0	34.6
Average pixel	8.3	71.8	37.8	39.3

The average of the longest lumen in the five ears was 75.42 (SD = 7.4) pixels. The average lengths of the cartilaginous part and the bony part of the ET lumen were 45.26 (SD = 8) and 31.26 (SD = 6.5) pixels, respectively, which are approximated with the distance 2 and 3 measurements as an average in the five ears (Table 3).

Table 3. Distance measurements as pixels in all ETs.

	Distance 1	Distance 2	Distance 3
Ear 1	67.99	50.45	20.82
Ear 2	87.74	56.01	34.41
Ear 3	73.37	44.24	30.16
Ear 4	75.41	37.79	38.09
Ear 5	72.6	37.79	32.82
Average (pixel)	75.42	45.26	31.26

4. Discussion

Balloon Dilation Eustachian Tuboplasty (BDET) is being increasingly used for ETD in adults and now children, first in Europe and then in the US, with the intention of decreasing and/or eliminating the need for long-term ventilation tubes in chronic conditions such as chronic OME [6,7,11–14]. Evaluations of BDET effects are mostly subjective, involving clinical outcome scoring before and after the procedure using a questionnaire such as the Eustachian Tube Dysfunction Questionnaire-7 (ETDQ-7) and/or the Eustachian Tube Score (ETS), in some studies combined with Valsalva and Toynbee maneuvers [15,16]. Tympanometry has been widely used as part of the diagnostic tools for ETD for comparison before and after the procedure [5–7]. The R value measured with tubomanometry while swallowing water ($R \leq 1$ immediate opening, $R > 1$ late opening, non-measurable R closed tube) was recommended as a functional outcome measure for BDET, indicating latency of the ET [17]. In our previous study, we found that the presence of an R value had high sensitivity for detecting an ET opening but low specificity for detecting a non-opening [18]. For this reason, the tubomanometry test by itself may not be an ideal objective measure of the BDET outcome.

The ETDQ-7 is widely used by clinicians, and a recent randomized study indicated that BDET is significantly superior to medical management for persistent ETD [8], but our previous study showed it is only moderately associated with an objective measure of ET function and appears less reliable when applied to people with non-intact tympanic membranes [19]. We think scoring systems should be supported by functional and structural objective measures for the evaluation of BDET effects to confirm its long-term benefits in clinical trials.

Visualization of the ET in imaging studies has always been a challenge due to its normally closed status and small size, unless it is patulous [20]. The first CT studies in a cadaver head and/or live images were mostly performed to determine the length, volume, and angle of the ET lumen for better understanding ET structure and function [10,21–24]. Nasopharyngeal maneuvers such as Valsalva were added to the scans to get a clear opening, but capturing the ET was still complicated by the lack of contrast in the tissues. Three-dimensional (3D) CT imaging studies were very beneficial for the understanding of temporal bone structure and for the diagnosis and follow-up of malignant tumors [25–29].

CT imaging of the ET has been part of the pre-procedure evaluation for BDET to determine if there is dehiscence and abnormality of the carotid artery and to prevent potential lethal complications [23,29,30] CT imaging preoperatively has been part of the protocol for BDET for most surgeons, and sometimes it is obtained after the procedure to see if there is any damage to the structures. A study showed no predictivity of difficulties and complications with preoperative high-resolution CT scanning, and these same authors suggested that the fear of injury to the internal carotid artery during the procedure might be disproportionate since there has not been published data about this complication [23]. Because physicians feel safer visualizing ET structure preoperatively, adding a quantification technique to the image analysis that helps to see the correlation between dilated lumen and symptoms could magnify the benefit of CT exposure pre- and post-operatively. We can also synchronize these imaging studies with the forced response test (FRT), as we established the method in our previous pilot study with five non-intact ears and/or some of the nasopharyngeal maneuvers to capture cartilaginous lumen opening before and after BDET [10]. Measurements would be performed in Hounsfield Units (HU) as in the later study, or histograms can be counted as pixel values after creating the area of interest as detailed in the methods in this paper. The idea is to compare air and tissue intensities in pre- and post-procedure images after determining the baseline values of the known anatomical areas, such as the nasopharyngeal space for air, surrounding tissues, and petrous bone intensities. However, HU is a universal, appropriately standardized metric for the analysis of CT images across scanner models. Even though in the current study, conversion of HU to the non-standard ranges (0, 255) for tissue types has been utilized with ImageJ software, utilizing a 16 or 32 bit image may allow preservation of the previously accepted ranges for various tissue types.

In this cadaver study, we have used contrast instilled through a non-intact tympanic membrane in an attempt to picture the ET before catheter insertion. Although contrast was visible in most images, the time it took to pass the contrast was too short to be able to visualize the entire length of the ET lumen, even with synchronized imaging in the cadaver heads. Using CT images with the inflated balloon served the purposes of determining feasibility and finding anatomical landmarks while viewing the whole anatomy of the head in 3D without a need for contrast locally or systemically. Even though there was some degree of potential mismatch between the pre- and post-dilation CT images, when there were all three landmarks in one plane, it was possible to approximate the counterparts to capture reasonably comparable images at the same plane of anatomy.

While many case series have suggested BDET can successfully restore ET function, systematic reviews have indicated no high-quality evidence exists to support its effectiveness [8,9,31]. Recently, otolaryngologists have been focused on creating a consensus for the eligibility criteria of ETD for BDET and the objective evaluation of its effects with more randomized clinical trials [32].

One study showed CT scanning in the sitting position may visualize ET opening better, especially in patients with a patulous ET [33]. These authors also suggested that vertical CT, which is mainly used in dental clinics, can be used to visualize the ET at a low cost and with less radiation exposure than conventional temporal bone CT imaging [29]. Since we are proposing an additional postoperative CT for BDET, it would be important to decrease the cost and radiation exposure of vertical CT imaging and increase its effectiveness by accommodating the position while scanning.

The proposed image analysis and quantification method with ImageJ is relatively practical for the clinician, and the software is freely available from the NIH. This was our choice of image analysis software due to its availability and our familiarity with it from our previous work. However, there is other, more advanced image analysis software available that may be more user-friendly. Some options may be able to generate ROIs, and most will be able to get various summary data from these ROIs. Even though explicit histograms may or may not be possible with a specific software program, such a feature may not be needed with some study material. Although a time-consuming learning curve is the downside of the ImageJ method, there is the advantage of more than one obtainable measure in matched pre- and post-procedure images of the ET. Additionally, we could depict the increase in width measurements with visually prominent lumen dilation in all ears in the CT images. The length measurements were also comparable to those in the previous literature; the cartilaginous part and the bony part of the ET lumen averaged 45.26 (16.8 mm) and 31.26 (11.3 mm) pixels, respectively, displaying the precision of the technique [28]. The mean gray value was one of the useful measures that can be performed in ImageJ software, and representation as a graph is also available for future studies for the confirmation of dilation rates. We previously presented the ETF test data from the clinical BDET study with eleven adult subjects [14] and are in the process of evaluating the pre- and post-procedure CT images in the same data set with the proposed method. Of course, we did not have the advantage of having the balloon in the canal during inflation on these live images, and dilations were not as prominent as in the cadaver ETs without tissue elasticity, but we found this image analysis method applicable, having three landmarks (medial pterygoid plate where the tissue thickness is the minimum, immediate opposite side medial pterygoid plate hamulus, and sphenoid spine superiorly) in the same plane by the 3D viewing tool. This could be an important part of the quantification of the effect of BDET in clinical research studies and aid physicians in future patient selection and evaluation of BDET outcomes. Future studies should also take into account tissue elasticity, recoil, and elastic fiber remodeling during and after the application of such pressure on the ET with various tissue properties including cartilage, muscle, fat, and possibly inflamed and thickened mucosa. Moreover, even though ET dilation balloons do reach a maximum diameter of 6 mm at pressures of 10 ATM, the biomechanics of the balloons would need to be considered.

5. Study Limitations

1. The main limitation of this study is the small sample size. Even for method development, a larger sample size would have been better;
2. The ImageJ image analysis software may have limitations compared to other newer software that may eliminate some of the labor-intensive steps in standardizing ROIs and generating outcome images and reports;
3. Utilizing a cadaver head in the study has inherent limitations due to tissue elasticity and recoil properties similar to those in live human subjects. On the other hand, it was possible to have six consecutive CT scans as per the protocol on cadaver heads for the purpose of method development, while this would not be justified in live human subjects;
4. The effects of balloon shape and length, as well as different durations of balloon inflation, have not been explored in this study. Future studies incorporating and controlling the biomechanics of the balloons and their effects on tissues may provide a better understanding of the role of this treatment and its long-term outcomes and consequences.

6. Conclusions

Imaging of the ET, particularly comparison of the effect of an intervention, has inherent challenges due to the technique and image analysis. This study is an attempt to describe a method to image the ET and quantify the outcomes of BD of the ET using anatomical landmarks.

Author Contributions: Conceptualization, C.M.A., S.C.-F., J.D.S. and T.J.R.; methodology, C.M.A., S.C.-F., M.S.T., J.D.S. and T.J.R.; software, S.C.-F. and J.D.S.; validation, C.M.A., S.C.-F., M.S.T., J.D.S. and T.J.R.; formal analysis, S.C.-F.; investigation, C.M.A.; resources, J.D.S. and M.S.T.; data curation, S.C.-F. and J.D.S.; writing—original draft preparation, S.C.-F.; writing—review and editing, C.M.A., S.C.-F., M.S.T., J.D.S. and T.J.R.; visualization, S.C.-F.; supervision, C.M.A., M.S.T., J.D.S. and T.J.R.; project administration, C.M.A.; funding acquisition, C.M.A. All authors have read and agreed to the published version of the manuscript.

Funding: NIH Grant DC007667.

Institutional Review Board Statement: Committee for Oversight of Research and Clinical Training Involving Decedents (CORID) of University of Pittsburgh approved the protocol"A modified Computerized Tomography Pilot Study for Anatomic and Functional Evaluation of the Eustachian Tube (CORID No. 435) on 8/14/2013.

Informed Consent Statement: Could not be obtained, due to cadaver of the deceased humans.

Data Availability Statement: Not applicable.

Acknowledgments: The authors acknowledge Ellen M. Mandel for editing the manuscript.

Conflicts of Interest: The authors declare no conflict of interest.

References

1. Bluestone, C.D.; Bluestone, M.B. *Eustachian Tube: Structure, Function, and Role in Middle-Ear Disease*, 2nd ed.; Peoples Medical Publishing House: Raleigh, NC, USA, 2017.
2. Doyle, W.J. A formal description of middle ear pressure-regulation. *Hear Res.* **2017**, *354*, 73–85. [CrossRef] [PubMed]
3. Bluestone, C.D.; Klein, J.O. *Otitis Media in Infant and Children*, 5th ed.; B. C. Decker, Inc.: Hamilton, OT, Canada, 2007.
4. Poe, D.S.; Silvola, J.; Pyykko, I. Balloon dilation of the cartilaginous eustachian tube. *Otolaryngol. Head Neck Surg.* **2011**, *144*, 563–569. [CrossRef] [PubMed]
5. Schroder, S.; Lehmann, M.; Ebmeyer, J.; Upile, T.; Sudhoff, H. Balloon Eustachian tuboplasty: A retrospective cohort study. *Clin. Otolaryngol.* **2015**, *40*, 629–638. [CrossRef] [PubMed]
6. Poe, D.; Anand, V.; Dean, M.; Roberts, W.H.; Stolovitzky, J.P.; Hoffmann, K.; Nachlas, N.E.; Light, J.P.; Widick, M.H.; Sugrue, J.P.; et al. Balloon dilation of the eustachian tube for dilatory dysfunction: A randomized controlled trial. *Laryngoscope* **2018**, *128*, 1200–1206. [CrossRef] [PubMed]
7. Meyer, T.A.; O'Malley, E.M.; Schlosser, R.J.; Soler, Z.M.; Cai, J.; Hoy, M.J.; Slater, P.W.; Cutler, J.L.; Simpson, R.J.; Clark, M.J.; et al. A Randomized Controlled Trial of Balloon Dilation as a Treatment for Persistent Eustachian Tube Dysfunction With 1-Year Follow-Up. *Otol. Neurotol.* **2018**, *39*, 894–902. [CrossRef] [PubMed]
8. Hwang, S.Y.; Kok, S.; Walton, J. Balloon dilation for eustachian tube dysfunction: Systematic review. *J. Laryngol. Otol.* **2016**, *130* (Suppl. 4), S2–S6. [CrossRef] [PubMed]
9. Huisman, J.M.L.; Verdam, F.J.; Stegeman, I.; de Ru, J.A. Treatment of Eustachian tube dysfunction with balloon dilation: A systematic review. *Laryngoscope* **2018**, *128*, 237–247. [CrossRef]
10. Alper, C.M.; Rath, T.J.; Teixeira, M.S.; Swarts, J.D. A Novel Imaging Method for the Cartilaginous Eustachian Tube Lumen: Computerized Tomography During the Forced Response Test. *Ann. Otol. Rhinol. Laryngol.* **2018**, *127*, 13–20. [CrossRef]
11. Catalano, P.J.; Jonnalagadda, S.; Yu, V.M. Balloon catheter dilatation of Eustachian tube: A preliminary study. *Otol. Neurotol.* **2012**, *33*, 1549–1552. [CrossRef]
12. McCoul, E.D.; Anand, V.K. Eustachian tube balloon dilation surgery. *Int. Forum Allergy Rhinol.* **2012**, *2*, 191–198. [CrossRef]
13. Tisch, M.; Maier, H.; Sudhoff, H. Balloon dilation of the Eustachian tube: Clinical experience in the management of 126 children. *Acta Otorhinolaryngol. Ital.* **2017**, *37*, 509–512. [CrossRef] [PubMed]
14. Alper, C.M.; Teixeira, M.S.; Rath, T.J.; Hall-Burton, D.; Swarts, J.D. Change in Eustachian Tube Function With Balloon Dilation in Adults With Ventilation Tubes. *Otol. Neurotol.* **2020**, *41*, 482–488. [CrossRef] [PubMed]
15. McCoul, E.D.; Anand, V.K.; Christos, P.J. Validating the clinical assessment of eustachian tube dysfunction: The Eustachian Tube Dysfunction Questionnaire (ETDQ-7). *Laryngoscope* **2012**, *122*, 1137–1141. [CrossRef]
16. Schroder, S.; Lehmann, M.; Sauzet, O.; Ebmeyer, J.; Sudhoff, H. A novel diagnostic tool for chronic obstructive eustachian tube dysfunction-the eustachian tube score. *Laryngoscope* **2015**, *125*, 703–708. [CrossRef]
17. Schroder, S.; Lehmann, M.; Korbmacher, D.; Sauzet, O.; Sudhoff, H.; Ebmeyer, J. Evaluation of tubomanometry as a routine diagnostic tool for chronic obstructive Eustachian tube dysfunction. *Clin. Otolaryngol.* **2015**, *40*, 691–697. [CrossRef] [PubMed]

18. Alper, C.M.; Teixeira, M.S.; Kim, J.H.; Douglas Swarts, J. Diagnostic accuracy of tubomanometry R value in detecting the Eustachian tube pressure equalizing function. *Eur. Arch. Otorhinolaryngol.* **2017**, *274*, 1865–1872. [CrossRef]
19. Teixeira, M.S.; Swarts, J.D.; Alper, C.M. Accuracy of the ETDQ-7 for Identifying Persons with Eustachian Tube Dysfunction. *Otolaryngol. Head Neck Surg.* **2018**, *158*, 83–89. [CrossRef] [PubMed]
20. Niwa, H.; Takahashi, M.; Yanagita, N.; Naganawa, S. Evaluation of clearance function of the Eustachian tube by sequential contrast CT. *Acta Otolaryngol. Suppl.* **1990**, *471*, 43–50. [CrossRef] [PubMed]
21. Shim, H.J.; Choi, A.Y.; Yoon, S.W.; Kwon, K.H.; Yeo, S.G. The Value of Measuring Eustachian Tube Aeration on Temporal Bone CT in Patients with Chronic Otitis Media. *Clin. Exp. Otorhinolaryngol.* **2010**, *3*, 59–64. [CrossRef]
22. Poe, D.S.; Hanna, B.M. Balloon dilation of the cartilaginous portion of the eustachian tube: Initial safety and feasibility analysis in a cadaver model. Evaluation Studies. *Am. J. Otolaryngol.* **2011**, *32*, 115–123. [CrossRef]
23. Abdel-Aziz, T.; Schroder, S.; Lehmann, M.; Gehl, H.B.; Ebmeyer, J.; Sudhoff, H. Computed tomography before balloon Eustachian tuboplasty—A true necessity? *Otol. Neurotol.* **2014**, *35*, 635–638. [CrossRef] [PubMed]
24. Ikeda, R.; Kobayashi, T.; Yoshida, M.; Yoshida, N.; Kikuchi, T.; Oshima, T.; Kawase, T.; Katori, Y. Patulous Eustachian Tube and Otitis Media With Effusion as Complications After Trigeminal Nerve Injury. *Otol. Neurotol.* **2017**, *38*, 1125–1128. [CrossRef] [PubMed]
25. Buytaert, J.A.; Salih, W.H.; Dierick, M.; Jacobs, P.; Dirckx, J.J. Realistic 3D computer model of the gerbil middle ear, featuring accurate morphology of bone and soft tissue structures. *J. Assoc. Res. Otolaryngol.* **2011**, *12*, 681–696. [CrossRef] [PubMed]
26. Yoshida, H.; Kobayashi, T.; Takasaki, K.; Takahashi, H.; Ishimaru, H.; Morikawa, M.; Hayashi, K. Imaging of the patulous Eustachian tube: High-resolution CT evaluation with multiplanar reconstruction technique. *Acta Otolaryngol.* **2004**, *124*, 918–923. [CrossRef]
27. Sheer, F.J.; Swarts, J.D.; Ghadiali, S.N. Three-dimensional finite element analysis of Eustachian tube function under normal and pathological conditions. *Med. Eng. Phys.* **2012**, *34*, 605–616. [CrossRef]
28. Tarabichi, M.; Najmi, M. Visualization of the eustachian tube lumen with Valsalva computed tomography. *Laryngoscope* **2015**, *125*, 724–729. [CrossRef]
29. Ikeda, R.; Kikuchi, T.; Oshima, H.; Miyazaki, H.; Hidaka, H.; Kawase, T.; Katori, Y.; Kobayashi, T. Computed tomography findings of the bony portion of the Eustachian tube with or without patulous Eustachian tube patients. *Eur. Arch. Otorhinolaryngol.* **2017**, *274*, 781–786. [CrossRef]
30. Smith, M.E.; Scoffings, D.J.; Tysome, J.R. Imaging of the Eustachian tube and its function: A systematic review. *Neuroradiology* **2016**, *58*, 543–556. [CrossRef]
31. Randrup, T.S.; Ovesen, T. Balloon eustachian tuboplasty: A systematic review. *Otolaryngol. Head Neck Surg.* **2015**, *152*, 383–392. [CrossRef]
32. Schilder, A.G.; Bhutta, M.F.; Butler, C.C.; Holy, C.; Levine, L.H.; Kvaerner, K.J.; Norman, G.; Pennings, R.J.; Poe, D.; Silvola, J.T.; et al. Eustachian tube dysfunction: Consensus statement on definition, types, clinical presentation and diagnosis. *Clin. Otolaryngol.* **2015**, *40*, 407–411. [CrossRef]
33. Yoshida, H.; Kobayashi, T.; Morikawa, M.; Hayashi, K.; Tsujii, H.; Sasaki, Y. CT imaging of the patulous eustachian tube–comparison between sitting and recumbent positions. *Auris Nasus Larynx* **2003**, *30*, 135–140. [CrossRef] [PubMed]

bioengineering

MDPI

Article

Reference Standard for Digital Infrared Thermography of the Surface Temperature of the Lower Limbs

Ho Yeol Zhang [1], Seong Son [2,*], Byung Rhae Yoo [2] and Tae-Mi Youk [3,4]

1 Department of Neurosurgery, National Health Insurance Service Ilsan Hospital, Yonsei University College of Medicine, Ilsan 10444, Republic of Korea
2 Department of Neurosurgery, Gil Medical Center, Gachon University College of Medicine, Incheon 21565, Republic of Korea
3 Research Institute, National Health Insurance Service Ilsan Hospital, Yonsei University College of Medicine, Ilsan 10444, Republic of Korea
4 Department of Statistics, Korea University, Seoul 02841, Republic of Korea
* Correspondence: sonseong44@gmail.com; Tel.: +82-32-460-3304

Abstract: Digital infrared thermographic imaging (DITI) is a supplementary diagnostic technique to visualize the surface temperature of the human body. However, there is currently no reference standard for the lower limbs for accurate diagnosis. In this study, we performed DITI on the lower limbs of 905 healthy Korean volunteers (411 males and 494 females aged between 20 and 69 years) to obtain reference standard data. Thermography was conducted on the front, back, lateral sides, and sole area, and 188 regions of interest (ROIs) were analyzed. Additionally, subgroup analysis was conducted according to the proximity of ROIs, sex, and age groups. The mean temperatures of ROIs ranged from 24.60 ± 5.06 to 28.75 ± 5.76 °C and the absolute value of the temperature difference between both sides reached up to 1.06 ± 2.75 °C. According to subgroup analysis, the sole area had a significantly lower temperature than any other areas, men had higher temperatures than women, and the elderly had higher temperatures than the young adults except for the 20s age group ($p < 0.001$, respectively). This result could be used as a foundation for the establishment of a reference standard for DITI. Practical patient DITI can be accurately interpreted using these data, and it can serve as a basis for further scientific research.

Keywords: infrared rays; lower limb; reference standard; skin temperature; thermography

Citation: Zhang, H.Y.; Son, S.; Yoo, B.R.; Youk, T.-M. Reference Standard for Digital Infrared Thermography of the Surface Temperature of the Lower Limbs. *Bioengineering* **2023**, *10*, 283. https://doi.org/10.3390/bioengineering10030283

Academic Editor: Cuneyt M. Alper

Received: 6 February 2023
Revised: 15 February 2023
Accepted: 17 February 2023
Published: 21 February 2023

1. Introduction

Digital infrared thermographic imaging (DITI) is a technique used to display the body's surface temperature using thermography [1,2]. DITI has been used as a complementary diagnostic tool in various clinical fields [3–9]. In several disorders involving the lower limbs, such as lumbar radicular pain, chronic regional pain syndrome, vascular disease, and peripheral nerve entrapment, DITI allows visualization of the affected area as hypo-radiant (hypothermia) or hyper-radiant (hyperthermia) compared to the unaffected area [10–15].

In terms of interpreting DITI, the normal range of skin temperature and criteria for hypo-radiant and hyper-radiant are ambiguous. Empirically, the temperature difference between both sides of the lower limbs is significant when it is more than 0.1–0.3 °C, depending on the location of the surface area [1,16,17]. However, a formally approved consensus on the definition of significant difference between both sides, as well as hypo-radiant or hyper-radiant, is still undetermined due to variations in equipment, environment (room temperature and humidity) of the test room, and the ability of the surveyor [5]. Furthermore, despite the consistent examination conditions, the range of normal surface temperature still varies according to baseline characteristics such as sex, age, body mass index, patient condition, and medical history [17].

To overcome these barriers to the definition of abnormality and to establish a correct standard for DITI, a standardized measurement protocol and a reference standard for DITI are necessary. However, there are no studies about reference standard data for DITI of the lower limbs. In this study, we performed DITI on a large group of healthy volunteers using a standardized protocol and provide detailed reference standard data for DITI of the lower extremities.

2. Materials and Methods

2.1. Trial Design and Ethics

This multi-center, single-arm, open-label trial was conducted for 2 years in accordance with the 1964 Helsinki Declaration and its later amendments. All processes of the study were approved by the Institutional Review Board of three different research centers. Additionally, this research was registered as a clinical trial in the Clinical Research Information Service of the Republic of Korea (number KCT0006880).

2.2. Sample Size

The number of samples was calculated as follows:

$$n = \frac{\theta(1-\theta)^2 z_{\alpha/2}^2}{d^2}$$

The population proportion (θ) for the exam was set to 0.85 and the margin of error (d) was set to 0.025. Using the above formula with a significance level of 5% and a confidence level of 95%, a sample size of 784 was calculated. Considering a dropout rate of 15%, a total of 922 participants were necessary.

2.3. Subjects

Healthy test subjects were voluntarily recruited through public announcements and tests were conducted at three institutions from March 2018 to December 2020.

To minimize bias related to subject selection and external effect, inclusion criteria were as follows: (1) adults between the ages of 20 and 69; (2) no specific medical history including diabetes mellitus, peripheral neuropathy, spinal stenosis, disc herniation, joint disease of the leg, previous surgery history of the spine and lower limbs, or recent trauma; (3) no definite present pain or skin lesion in the lower limbs; (4) no potential risk of test such as claustrophobia, pregnant, or lactating women; (5) who can maintain a stationary posture for the required amount of time during the test; (6) without any other reason for disqualification according to the judgment of the researchers. Participation was granted when the requirements were met based on the inclusion criteria questionnaire.

A total of 922 healthy Korean volunteers were registered with an even distribution based on sex and age group. DITI was conducted following a standardized protocol and informed consent was obtained in advance from all participants. Among them, 905 participants were evaluated after excluding 17 participants due to measurement failure and/or withdrawal of consent.

2.4. Equipment and Examination Protocol

All examinations were conducted in outpatient clinics in three different hospitals. DITI was performed using the Iris-XP Digital infrared imaging system (Medicore, Seoul, Republic of Korea). Volunteers scheduled for DITI were informed about general precautions such as avoiding exposure to cold or hot environments, not smoking, and not consuming caffeine for 1 h before the test [18].

The skin temperature of subjects can be affected by environmental temperature and humidity due to sweating evaporation and vasoconstriction/vasodilation response [19–21]. To maintain consistency, we controlled the air temperature and humidity in the test room. Specifically, the room temperature and humidity were maintained at 20.0–23.0 °C and 30–75%, respectively. After undressing completely, the subjects remained in the room for approximately 20 min to acclimate prior to the examination. They were allowed to stand or sit on a chair with a back, depending on their preference.

The measurement reliability of temperature using the DITI equipment was found to be reasonable. The uncertainty of the thermography equipment ranged from 0.000 °C to 0.369 °C, as specified by the Korean Agency for Technology and Standards.

The test was conducted in the front area, back area including buttocks, both lateral-side areas, and the sole area of both feet. A total of 188 regions of interest (ROIs) were manually divided into 15 × 2 ROIs in the front area, 44 × 2 ROIs in both lateral-side areas, 20 × 2 ROIs in the back area, and 15 × 2 ROIs in the sole area (Figure 1). To ensure the accuracy of ROI division and measurement, objective testing and diagraming of pictures based on 188 ROIs of all subjects were performed by five certified surveyors.

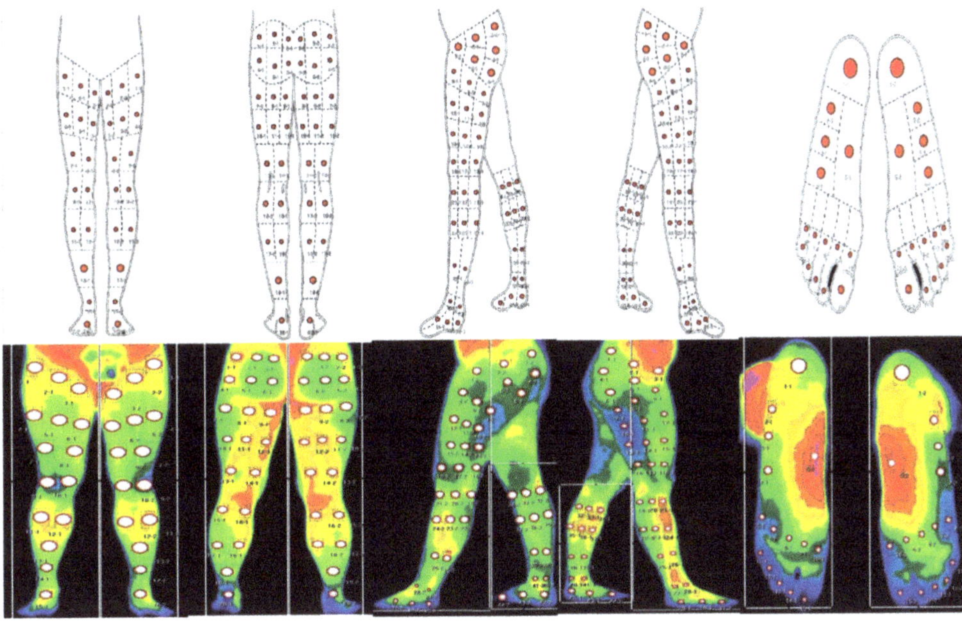

Figure 1. Diagram and practical imaging of regions of interest in the front, lateral-side, back, and sole areas of the leg.

2.5. Statistical Analysis

A quantitative analysis of data was conducted by a specialized doctor and a statistician who was blinded to participant information.

The analysis was performed using SPSS version 27.0 (IBM Corporation, Armonk, NY, USA). The normal distribution of the data was evaluated using the Kolmogorov–Smirnov test, and all data were reported as mean ± standard deviation or mean with 95% confidence intervals (CI). One-way analysis of variance (ANOVA), linear regression analysis, and paired t-tests were performed according to the characteristics of the values. Statistical significance was accepted at $p < 0.05$.

3. Results

3.1. Subjects

The mean age of all participants (n = 905) was 42.86 ± 12.87 years, and 45.4% of the participants were male (n = 411). The demographic distribution of volunteers according to age group was as follows: 183 (97 males and 86 females) in their 20s; 213 (108 males and

105 females) in their 30s; 228 (109 males and 119 females) in their 40s; 177 (65 males and 112 females) in their 50s; 104 (32 males and 72 females) in their 60s.

3.2. Overall Data: The Mean Temperature and Difference between Both Sides (°C)

The mean temperature of the ROIs of each area and the temperature difference between both sides (ΔT, right—left) were as follows: in the front area, the overall mean temperature was 27.69 ± 5.34 (ranged from 26.73 ± 5.12 (extended uncertainty 10.27) to 28.75 ± 5.76 (extended uncertainty 11.54)) and the overall mean difference was 0.03 ± 0.41 (ranged from −0.09 ± 0.33 to 0.24 ± 0.47); in the back area, the overall mean temperature was 27.70 ± 5.38 (ranged from 25.74 ± 5.09 (extended uncertainty 10.21) to 28.48 ± 5.74 (extended uncertainty 11.50)) and the overall mean difference was −0.04 ± 0.42 (ranged from −0.23 ± 0.59 to 0.20 ± 0.42); in the lateral-side area, the overall mean temperature was 27.18 ± 5.49 (ranged from 25.56 ± 5.19 (extended uncertainty 10.41) to 28.53 ± 5.79 (extended uncertainty 11.60)) and the overall mean difference was −0.58 ± 2.72 (ranged from −1.06 ± 2.75 to −0.09 ± 2.37); in the sole area, the overall mean temperature was 25.74 ± 4.98 (ranged from 24.60 ± 5.06 (extended uncertainty 10.1) to 27.69 ± 5.31 (extended uncertainty 10.65)) and the overall mean difference was −0.09 ± 0.80 (ranged from −0.17 ± 0.74 to 0.06 ± 0.71) (Tables 1–4).

Table 1. Overall mean temperature of regions of interest in the front area.

ROI	Mean (°C)	SD	Extended Uncertainty	Difference (ΔT, Right Side—Left Side)				
				Mean (°C)	SD	95% CI Lower	95% CI Upper	Extended Uncertainty
1_1	27.76	5.63	11.28					
1_2	27.85	5.65	11.32	−0.09	0.33	−0.11	−0.07	0.99
2_1	27.92	5.65	11.32					
2_2	27.96	5.67	11.36	−0.04	0.28	−0.06	−0.02	0.93
3_1	28.06	5.67	11.36					
3_2	28.09	5.75	11.52	−0.03	0.43	−0.05	0.00	1.13
4_1	27.68	5.58	11.18					
4_2	27.68	5.54	11.10	0.00	0.30	−0.02	0.02	0.95
5_1	27.68	5.57	11.16					
5_2	27.69	5.58	11.18	−0.01	0.29	−0.03	0.01	0.94
6_1	27.74	5.57	11.16					
6_2	27.72	5.58	11.18	0.02	0.33	0.00	0.05	0.99
7_1	27.19	5.42	10.87					
7_2	27.18	5.40	10.83	0.01	0.36	−0.01	0.03	1.03
8_1	27.61	5.49	11.00					
8_2	27.71	5.58	11.18	−0.10	0.36	−0.12	−0.07	1.03
9_1	27.30	5.33	10.69					
9_2	27.17	5.26	10.55	0.12	0.58	0.09	0.16	1.37
10_1	27.15	5.34	10.71					
10_2	27.17	5.40	10.83	−0.01	0.53	−0.05	0.02	1.29
11_1	28.68	5.71	11.44					
11_2	28.75	5.76	11.54	−0.07	0.38	−0.09	−0.04	1.06
12_1	28.41	5.73	11.48					
12_2	28.17	5.65	11.32	0.24	0.47	0.21	0.27	1.20
13_1	27.89	5.55	11.12					
13_2	27.80	5.53	11.08	0.09	0.40	0.07	0.12	1.09
14_1	27.61	5.48	10.98					
14_2	27.43	5.40	10.83	0.18	0.51	0.15	0.21	1.26
15_1	26.81	5.19	10.41					
15_2	26.73	5.12	10.27	0.07	0.65	0.03	0.11	1.49
Mean	27.69	5.34		0.03	0.41	0.00	0.06	

CI: confidence interval; ROI: region of interest; SD: standard deviation.

Table 2. Overall mean temperature of regions of interest in the back area.

ROI	Mean (°C)	SD	Extended Uncertainty	Difference (ΔT, Right Side—Left Side)				
				Mean (°C)	SD	95% CI Lower	95% CI Upper	Extended Uncertainty
1_1	27.62	5.69	11.40					
1_2	27.69	5.69	11.40	−0.07	0.53	−0.11	−0.04	1.29
2_1	27.52	5.67	11.36					
2_2	27.58	5.67	11.36	−0.06	0.40	−0.09	−0.03	1.09
3_1	27.67	5.64	11.30					
3_2	27.90	5.76	11.54	−0.23	0.59	−0.27	−0.19	1.39
4_1	27.50	5.61	11.24					
4_2	27.60	5.62	11.26	−0.10	0.49	−0.13	−0.06	1.23
5_1	27.52	5.59	11.20					
5_2	27.46	5.56	11.14	0.06	0.36	0.03	0.08	1.03
6_1	27.50	5.57	11.16					
6_2	27.53	5.59	11.20	−0.03	0.39	−0.06	−0.01	1.07
7_1	27.21	5.49	11.00					
7_2	27.34	5.52	11.06	−0.13	0.46	−0.16	−0.10	1.18
8_1	27.87	5.61	11.24					
8_2	27.84	5.60	11.22	0.02	0.36	0.00	0.05	1.03
9_1	28.22	5.66	11.34					
9_2	28.20	5.68	11.38	0.01	0.36	−0.01	0.04	1.03
10_1	27.59	5.54	11.10					
10_2	27.75	5.58	11.18	−0.17	0.41	−0.19	−0.14	1.10
11_1	27.94	5.61	11.24					
11_2	28.00	5.61	11.24	−0.06	0.33	−0.08	−0.03	0.99
12_1	28.14	5.64	11.30					
12_2	28.21	5.67	11.36	−0.06	0.35	−0.09	−0.04	1.02
13_1	27.91	5.61	11.24					
13_2	27.89	5.59	11.20	0.02	0.32	0.00	0.04	0.98
14_1	28.03	5.67	11.36					
14_2	27.82	5.60	11.22	0.20	0.42	0.18	0.23	1.12
15_1	28.05	5.63	11.28					
15_2	28.23	5.68	11.38	−0.19	0.32	−0.21	−0.17	0.98
16_1	28.48	5.74	11.50					
16_2	28.39	5.72	11.46	0.09	0.42	0.07	0.12	1.12
17_1	27.81	5.60	11.22					
17_2	27.72	5.56	11.14	0.09	0.38	0.06	0.11	1.06
18_1	27.93	5.59	11.20					
18_2	28.09	5.63	11.28	−0.17	0.43	−0.20	−0.14	1.13
19_1	27.38	5.50	11.02					
19_2	27.44	5.51	11.04	−0.06	0.42	−0.09	−0.03	1.12
20_1	25.74	5.09	10.21					
20_2	25.75	5.06	10.15	−0.01	0.65	−0.05	0.03	1.49
Mean	27.70	5.38		−0.04	0.42	−0.07	−0.01	

CI: confidence interval; ROI: region of interest; SD: standard deviation.

Table 3. Overall mean temperature of regions of interest in the lateral-side area.

ROI	Mean (°C)	SD	Extended Uncertainty	Difference (ΔT, Right Side—Left Side)				
				Mean (°C)	SD	95% CI Lower	95% CI Upper	Extended Uncertainty
1_1	27.09	5.90	11.82					
1_2	27.54	5.73	11.48	−0.44	2.84	−0.63	−0.26	5.73
2_1	27.40	5.97	11.96					
2_2	28.01	5.85	11.72	−0.60	2.93	−0.79	−0.41	5.91

Table 3. *Cont.*

ROI	Mean (°C)	SD	Extended Uncertainty	Difference (ΔT, Right Side—Left Side)				
				Mean (°C)	SD	95% CI Lower	95% CI Upper	Extended Uncertainty
3_1	27.30	5.92	11.86					
3_2	27.94	5.78	11.58	−0.63	2.95	−0.82	−0.43	5.95
4_1	26.91	5.81	11.64					
4_2	27.46	5.69	11.40	−0.54	2.83	−0.72	−0.35	5.71
5_1	26.84	5.81	11.64					
5_2	27.49	5.69	11.40	−0.64	2.91	−0.83	−0.45	5.87
6_1	26.61	5.65	11.32					
6_2	27.29	5.58	11.18	−0.66	2.75	−0.84	−0.48	5.55
7_1	26.43	5.60	11.22					
7_2	27.05	5.54	11.10	−0.61	2.74	−0.78	−0.43	5.53
8_1	26.79	5.73	11.48					
8_2	27.57	5.70	11.42	−0.77	2.85	−0.95	−0.58	5.75
9_1	27.03	5.81	11.64					
9_2	27.67	5.70	11.42	−0.62	2.80	−0.81	−0.44	5.65
10_1	26.77	5.70	11.42					
10_2	27.48	5.68	11.38	−0.70	2.81	−0.89	−0.52	5.67
11_1	27.22	5.81	11.64					
11_2	27.92	5.74	11.50	−0.69	2.84	−0.88	−0.51	5.73
12_1	27.27	5.86	11.74					
12_2	27.86	5.70	11.42	−0.58	2.85	−0.76	−0.39	5.75
13_1	26.91	5.71	11.44					
13_2	27.54	5.61	11.24	−0.62	2.75	−0.80	−0.44	5.55
14_1	27.19	5.79	11.60					
14_2	27.84	5.68	11.38	−0.64	2.83	−0.83	−0.46	5.71
15_1	27.23	5.78	11.58					
15_2	27.75	5.63	11.28	−0.50	2.83	−0.69	−0.32	5.71
16_1	27.16	5.75	11.52					
16_2	27.78	5.61	11.24	−0.61	2.77	−0.79	−0.42	5.59
17_1	26.88	5.68	11.38					
17_2	27.44	5.56	11.14	−0.55	2.76	−0.73	−0.37	5.57
18_1	26.75	5.63	11.28					
18_2	26.95	5.40	10.83	−0.20	2.71	−0.37	−0.02	5.47
19_1	27.41	5.86	11.74					
19_2	28.00	5.69	11.40	−0.58	2.75	−0.76	−0.40	5.55
20_1	27.21	5.73	11.48					
20_2	27.84	5.60	11.22	−0.62	2.64	−0.79	−0.44	5.33
21_1	27.14	5.71	11.44					
21_2	27.35	5.43	10.89	−0.19	2.73	−0.37	−0.01	5.51
22_1	27.24	5.77	11.56					
22_2	27.96	5.69	11.40	−0.70	2.73	−0.88	−0.52	5.51
23_1	27.71	5.89	11.80					
23_2	28.53	5.79	11.60	−0.80	2.86	−0.99	−0.62	5.77
24_1	27.80	5.92	11.86					
24_2	28.20	5.68	11.38	−0.38	2.91	−0.57	−0.19	5.87
25_1	27.04	5.75	11.52					
25_2	27.86	5.69	11.40	−0.80	2.79	−0.98	−0.62	5.63
26_1	27.26	5.81	11.64					
26_2	27.86	5.67	11.36	−0.59	2.79	−0.77	−0.41	5.63
27_1	26.29	5.45	10.92					
27_2	26.83	5.42	10.87	−0.53	2.44	−0.69	−0.37	4.94
28_1	26.75	5.52	11.06					
28_2	27.25	5.38	10.79	−0.49	2.51	−0.65	−0.33	5.07
29_1	25.56	5.19	10.41					
29_2	25.66	5.09	10.21	−0.09	2.37	−0.24	0.07	4.80
30_1	26.14	5.34	10.71					
30_2	26.52	5.30	10.63	−0.37	2.42	−0.52	−0.21	4.90

Table 3. *Cont.*

ROI	Mean (°C)	SD	Extended Uncertainty	Difference (ΔT, Right Side—Left Side)				
				Mean (°C)	SD	95% CI Lower	95% CI Upper	Extended Uncertainty
31_1	26.20	5.35	10.73					
31_2	26.83	5.35	10.73	−0.61	2.44	−0.77	−0.45	4.94
32_1	26.72	5.59	11.20					
32_2	27.78	5.66	11.34	−1.04	2.79	−1.22	−0.86	5.63
33_1	27.04	5.67	11.36					
33_2	27.64	5.60	11.22	−0.57	2.81	−0.76	−0.39	5.67
34_1	26.74	5.59	11.20					
34_2	26.94	5.40	10.83	−0.18	2.80	−0.37	0.00	5.65
35_1	26.95	5.66	11.34					
35_2	28.04	5.71	11.44	−1.06	2.75	−1.24	−0.88	5.55
36_1	27.47	5.83	11.68					
36_2	28.19	5.72	11.46	−0.69	2.82	−0.88	−0.51	5.69
37_1	27.31	5.78	11.58					
37_2	27.51	5.49	11.00	−0.18	2.89	−0.37	0.01	5.83
38_1	26.65	5.62	11.26					
38_2	27.45	5.59	11.20	−0.78	2.68	−0.96	−0.61	5.41
39_1	26.73	5.64	11.30					
39_2	27.30	5.52	11.06	−0.56	2.70	−0.73	−0.38	5.45
40_1	26.38	5.34	10.71					
40_2	27.31	5.45	10.92	−0.92	2.41	−1.08	−0.76	4.88
41_1	26.64	5.50	11.02					
41_2	27.09	5.40	10.83	−0.44	2.61	−0.61	−0.27	5.27
42_1	26.01	5.31	10.65					
42_2	26.94	5.41	10.85	−0.91	2.44	−1.07	−0.75	4.94
43_1	26.49	5.41	10.85					
43_2	26.87	5.31	10.65	−0.37	2.47	−0.53	−0.21	4.99
44_1	25.74	5.25	10.53					
44_2	26.05	5.13	10.29	−0.30	2.41	−0.46	−0.14	4.88
Mean	27.18	5.49		−0.58	2.72	−0.76	−0.40	

CI: confidence interval; ROI: region of interest; SD: standard deviation.

Table 4. Overall mean temperature of regions of interest in the sole area.

ROI	Mean (°C)	SD	Extended Uncertainty	Difference (ΔT, Right Side—Left Side)				
				Mean (°C)	SD	95% CI Lower	95% CI Upper	Extended Uncertainty
1_1	26.05	5.12	10.27					
1_2	26.07	5.12	10.27	−0.02	0.86	−0.08	0.04	1.87
2_1	26.26	4.97	9.97					
2_2	26.36	5.02	10.07	−0.10	0.75	−0.15	−0.05	1.67
3_1	26.33	5.00	10.03					
3_2	26.42	5.03	10.09	−0.09	0.73	−0.14	−0.04	1.64
4_1	25.96	4.96	9.95					
4_2	26.13	5.00	10.03	−0.17	0.74	−0.22	−0.12	1.65
5_1	27.65	5.29	10.61					
5_2	27.69	5.31	10.65	−0.04	0.61	−0.08	0.00	1.43
6_1	25.69	5.07	10.17					
6_2	25.85	5.07	10.17	−0.15	0.87	−0.21	−0.10	1.89
7_1	25.87	5.12	10.27					
7_2	25.99	5.08	10.19	−0.12	0.76	−0.17	−0.07	1.69
8_1	25.93	5.07	10.17					
8_2	26.02	5.05	10.13	−0.09	0.72	−0.14	−0.04	1.62

Table 4. *Cont.*

ROI	Mean (°C)	SD	Extended Uncertainty	Difference (ΔT, Right Side—Left Side)				
				Mean (°C)	SD	95% CI Lower	95% CI Upper	Extended Uncertainty
9_1	25.99	5.03	10.09					
9_2	26.05	5.04	10.11	−0.06	0.71	−0.11	−0.01	1.60
10_1	25.92	5.04	10.11					
10_2	25.86	5.06	10.15	0.06	0.71	0.02	0.11	1.60
11_1	24.82	5.07	10.17					
11_2	24.95	5.07	10.17	−0.13	0.94	−0.20	−0.07	2.02
12_1	24.60	5.06	10.15					
12_2	24.77	5.06	10.15	−0.16	0.87	−0.22	−0.11	1.89
13_1	24.66	5.13	10.29					
13_2	24.79	5.11	10.25	−0.13	0.88	−0.19	−0.07	1.91
14_1	24.74	5.16	10.35					
14_2	24.84	5.16	10.35	−0.10	0.88	−0.16	−0.04	1.91
15_1	25.19	5.22	10.47					
15_2	25.24	5.24	10.51	−0.05	0.91	−0.11	0.01	1.96
Mean	25.74	4.98		−0.09	0.80	−0.14	−0.04	

CI: confidence interval; ROI: region of interest; SD: standard deviation.

According to the location of the ROIs, the mean temperature of the ROIs was significantly different between the four areas ($p < 0.001$, ANOVA). In particular, the temperature of the sole area was significantly lower than that of any other areas; the difference between the sole and other areas was 1.93 (95% CI, 1.53–2.33), 1.95 (95% CI, 1.57–2.32), and 1.41 (95% CI, 1.08–1.74), respectively ($p < 0.001$, ANOVA post hoc analysis) (Figure 2).

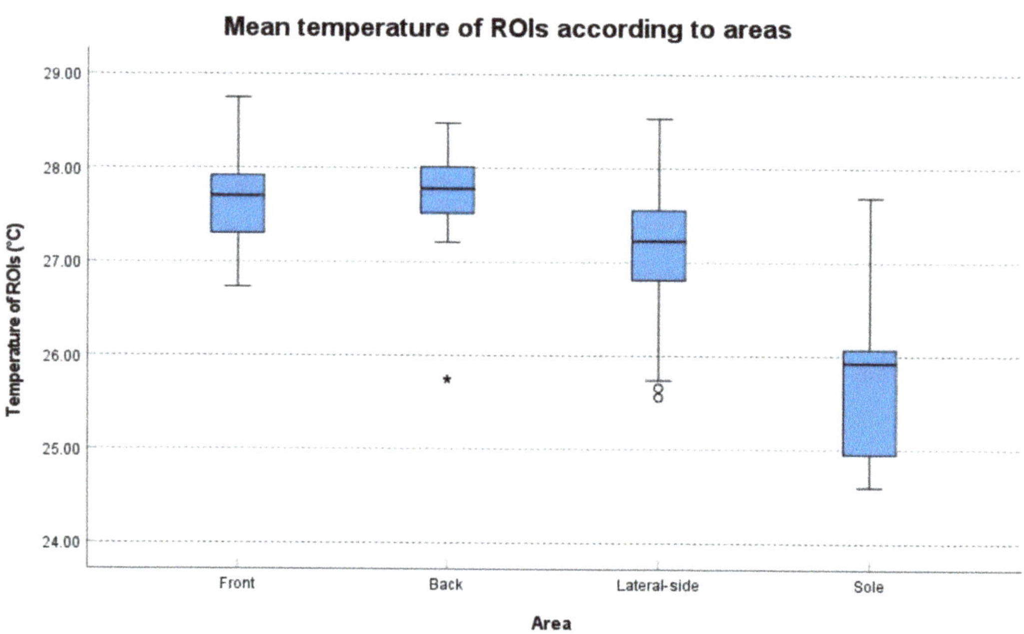

Figure 2. Diagram of the mean temperature according to the four areas. ROIs: region of interests; and * and ○ are labels that are out of range.

Moreover, the absolute value of the temperature difference between both sides (|ΔT|) was also significantly different between the four areas ($p < 0.001$, ANOVA). In particular, the |ΔT| of the lateral-side area was 0.58 ± 2.72 °C (95% CI, 0.40–0.76), which was significantly larger compared to any other area; the mean difference of |ΔT| between lateral-side and other areas was 0.50 (95% CI, 0.38–0.63), 0.48 (95% CI, 0.37–0.60), and 0.48 (95% CI, 0.36–0.60), respectively ($p < 0.001$, ANOVA post hoc analysis) (Figure 3).

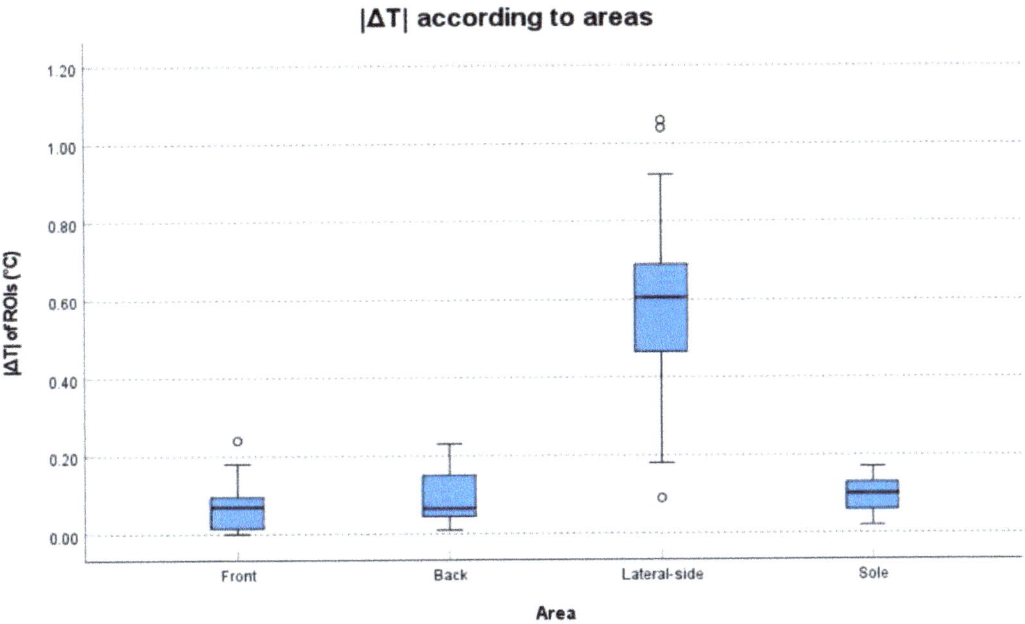

Figure 3. Diagram of the absolute values of difference between both sides (|ΔT|, |right—left|) in each of the four areas. ROIs: region of interests; and ○ are labels that are out of range.

3.3. Subgroup Analysis of the Temperature (°C) According to Proximity of ROIs

The temperature tended to drop from the proximal to the distal part in only the lateral-side area ($p = 0.001$, regression analysis), not in the front or back areas (Figure 4).

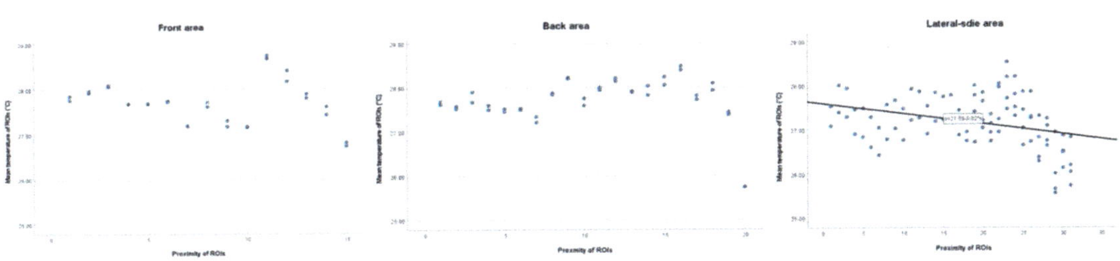

Figure 4. Relation between the mean temperature of the region of interest (ROIs) and the proximity of ROIs in three areas according to regression analysis. Front area temperature = 27.871 − (0.023 × ROIs) ($p = 0.260$, $R^2 = 0.045$); back area temperature = 27.792 − (0.009 × ROIs) ($p = 0.569$, $R^2 = 0.009$); lateral-side area temperature = 27.595 − (0.023 × ROIs) ($p = 0.001$, $R^2 = 0.114$).

In terms of |ΔT| according to the proximity of ROIs, the |ΔT| tended to increase from the proximal to the distal part in only the front area ($p = 0.047$, regression analysis), not in the back or lateral-side areas (Figure 5).

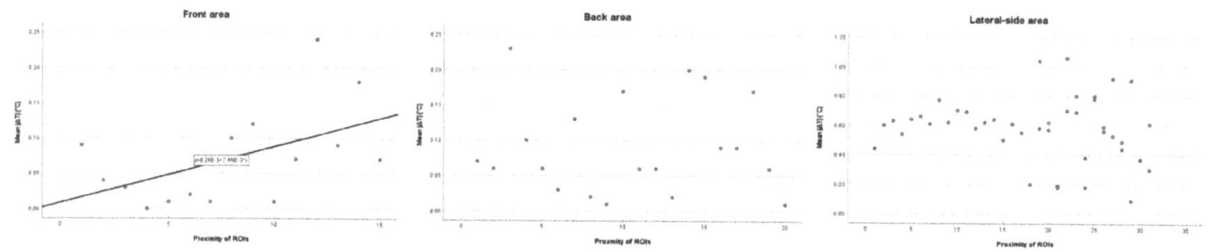

Figure 5. Relation between the absolute values of difference between both sides (|ΔT|, |right—left|) and the proximity of region of interest in each of the three areas according to regression analysis. |ΔT| front area = 0.008 + (0.008 × ROIs) (p = 0.047, R^2 = 0.270); |ΔT| back area = 0.090 + (9.774 × 10^{-5} × ROIs) (p = 0.972, R^2 = 0.000); |ΔT| lateral-side area = 0.645 − (0.004 × ROIs) (p = 0.327, R^2 = 0.023).

3.4. Subgroup Analysis of the Temperature (°C) According to Sex

The mean temperature of each area depending on the sex was as follows: in the front area, the mean temperature was 28.83 ± 5.15 (ranged from 27.10 ± 5.16 to 32.02 ± 4.85) in males and 26.54 ± 5.53 (ranged from 24.03 ± 5.16 to 29.32 ± 5.66) in females; in the back area, the mean temperature was 28.90 ± 5.18 (ranged from 25.99 ± 5.05 to 31.84 ± 4.65) in males and 26.50 ± 5.61 (ranged from 22.86 ± 4.86 to 28.74 ± 5.59) in females; in the lateral-side area, the mean temperature was 27.99 ± 5.45 (ranged from 25.32 ± 5.25 to 32.15 ± 4.71 in males and 26.29 ± 5.55 (ranged from 22.89 ± 4.95 to 29.02 ± 5.58) in females; in the sole area, the mean temperature was 26.59 ± 4.87 (ranged from 25.32 ± 5.08 to 30.36 ± 5.03) in males and 24.85 ± 5.03 (ranged from 22.19 ± 4.69 to 28.21 ± 5.40) in females; and in all areas, the mean temperature was 28.08 ± 5.16 in males and 26.05 ± 5.43 in females (Table 1).

In terms of trends according to sex, the mean surface temperature in the same ROIs was higher in males than in females in all areas of all age groups (p < 0.001, paired t-test), although the mean differences varied depending on the areas and age groups. The males' surface temperatures were at least 0.52 (95% CI, 0.09–0.94) and at most 3.42 (95% CI, 3.19–3.65) higher than that of females (Table 5 and Figure 6).

Table 5. Comparison of mean temperature for the same region of interest according to sex and age group.

Characteristics	Male (°C)	Female (°C)	Mean (°C)	Difference Between Sexes (°C)	p Value
Front area					<0.001 [a]
20s (n = 183)	28.46 ± 5.02	27.50 ± 5.68	28.01 ± 5.33	0.96 (95% CI, 0.70–1.21)	<0.001 [b]
30s (n =213)	27.96 ± 5.51	24.54 ± 5.34	26.27 ± 5.43	3.42 (95% CI, 3.19–3.65)	<0.001 [b]
40s (n = 228)	28.83 ± 5.18	26.74 ± 5.84	27.74 ± 5.52	2.09 (95% CI, 1.80–2.38)	<0.001 [b]
50s (n = 177)	29.82 ± 4.91	27.31 ± 5.16	28.23 ± 5.07	2.51 (95% CI, 2.24–2.78)	<0.001 [b]
60s (n = 104)	30.86 ± 4.71	28.15 ± 5.42	28.98 ± 5.20	2.71 (95% CI, 2.39–3.04)	<0.001 [b]
Sum	28.83 ± 5.15	26.54 ± 5.53	27.69 ± 5.34	2.29 (95% CI, 2.18–2.39)	<0.001 [b]
Back area					<0.001 [a]
20s (n = 183)	28.46 ± 5.05	27.54 ± 5.83	28.03 ± 5.42	0.92 (95% CI, 0.65–1.18)	<0.001 [b]
30s (n =213)	27.96 ± 5.57	24.46 ± 5.49	26.23 ± 5.53	3.50 (95% CI, 3.27–3.74)	<0.001 [b]
40s (n = 228)	28.96 ± 5.27	26.65 ± 5.85	27.75 ± 5.57	2.31 (95% CI, 2.05–2.56)	<0.001 [b]
50s (n = 177)	29.97 ± 4.88	27.30 ± 5.14	28.28 ± 5.04	2.67 (95% CI, 2.43–2.91)	<0.001 [b]
60s (n = 104)	31.06 ± 4.62	28.09 ± 5.44	29.00 ± 5.19	2.97 (95% CI, 2.67–3.26)	<0.001 [b]
Sum	28.90 ± 5.18	26.50 ± 5.61	27.70 ± 5.38	2.40 (95% CI, 2.34–2.46)	

Table 5. *Cont.*

Characteristics	Male (°C)	Female (°C)	Mean (°C)	Difference Between Sexes (°C)	*p* Value
Lateral-side area					<0.001 [a]
20s (n = 183)	28.09 ± 5.06	27.05 ± 5.73	27.60 ± 5.37	1.04 (95% CI, 0.87–1.20)	<0.001 [b]
30s (n =213)	27.25 ± 5.68	24.45 ± 5.40	25.89 ± 5.54	2.79 (95% CI, 2.61–2.98)	<0.001 [b]
40s (n = 228)	27.84 ± 5.56	26.52 ± 5.77	27.15 ± 5.67	1.32 (95% CI, 1.07–1.57)	<0.001 [b]
50s (n = 177)	28.67 ± 5.43	27.16 ± 5.19	27.71 ± 5.28	1.51 (95% CI, 1.23–1.78)	<0.001 [b]
60s (n = 104)	29.34 ± 5.54	27.69 ± 5.54	28.21 ± 5.54	1.65 (95% CI, 1.29–2.00)	<0.001 [b]
Sum	27.99 ± 5.45	26.29 ± 5.55	27.18 ±5.49	1.70 (95% CI, 1.53–1.87)	
Sole area					<0.001 [a]
20s (n = 183)	25.97 ± 4.35	25.46 ± 4.76	25.73 ± 4.54	0.52 (95% CI, 0.09–0.94)	0.019 [b]
30s (n =213)	26.46 ± 5.13	23.30 ± 4.80	24.90 ± 4.97	3.16 (95% CI, 2.79–3.53)	<0.001 [b]
40s (n = 228)	26.47 ± 4.96	24.91 ± 5.37	25.65 ± 5.17	1.57 (95% CI, 1.14–2.00)	<0.001 [b]
50s (n = 177)	27.47 ± 4.93	25.75 ± 5.12	26.38 ± 5.02	1.72 (95% CI, 1.35–2.10)	<0.001 [b]
60s (n = 104)	27.49 ± 5.19	26.22 ± 5.26	26.61 ± 5.24	1.26 (95% CI, 0.76–1.77)	<0.001 [b]
Sum	26.59 ± 4.87	24.85 ± 5.03	25.74 ± 4.98	1.74 (95% CI, 1.70–1.77)	
All areas					<0.001 [a]
20s (n = 183)	27.75 ± 4.87	26.89 ± 5.50	27.34 ± 5.17	0.86 (95% CI, 0.59–1.12)	<0.001 [b]
30s (n =213)	27.41 ± 5.47	24.19 ± 5.26	25.82 ± 5.37	3.22 (95% CI, 2.82–3.62)	<0.001 [b]
40s (n = 228)	28.03 ± 5.24	26.21 ± 5.71	27.07 ± 5.48	1.82 (95% CI, 1.77–1.87)	<0.001 [b]
50s (n = 177)	28.98 ± 5.04	26.88 ± 5.15	27.65 ± 5.11	2.10 (95% CI, 1.83–2.37)	<0.001 [b]
60s (n = 104)	29.70 ± 5.02	27.54 ± 5.42	28.20 ± 5.29	2.16 (95% CI, 1.88–2.44)	<0.001 [b]
Sum	28.08 ± 5.16	26.05 ± 5.43	26.97 ± 5.31	2.03 (95% CI, 1.74–2.32)	<0.001 [b]

CI: confidence interval. [a] ANOVA between five age groups; [b] paired *t*-test between both sexes.

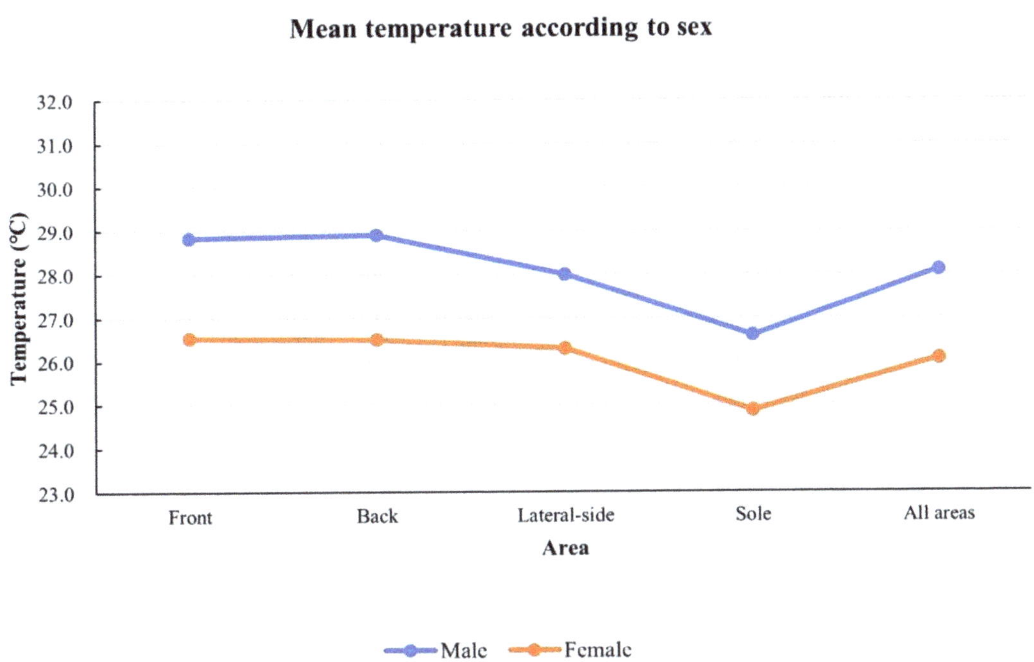

Figure 6. Trend of temperature according to sex in each area.

3.5. Subgroup Analysis of the Temperature (°C) According to Age Group

The mean temperature of each area depending on the age group was as follows: in the 20s age group, the mean temperature was 27.34 ± 5.17 (ranged from 24.14 ± 4.69 to

29.33 ± 5.21); in the 30s age group, the mean temperature was 25.82 ± 5.37 (ranged from 22.19 ± 4.69 to 28.95 ± 5.74); in the 40s age group, the mean temperature was 26.21 ± 5.71 (ranged from 23.74 ± 5.30 to 30.00 ± 5.43); in the 50s age group, the mean temperature was 27.65 ± 5.11 (ranged from 24.67 ± 5.16 to 30.84 ± 4.95); in the 60s age group, the mean temperature was 28.20 ± 5.29 (ranged from 25.06 ± 5.34 to 32.15 ± 4.85) (Table 5).

In terms of trends according to age group, the surface temperature increased as age increased, except for the 20s age group, in all areas ($p < 0.001$, ANOVA). Among all age groups, the 30s age group of both sexes showed the lowest temperature in all areas ($p < 0.001$, ANOVA post hoc analysis) (Table 5 and Figure 7).

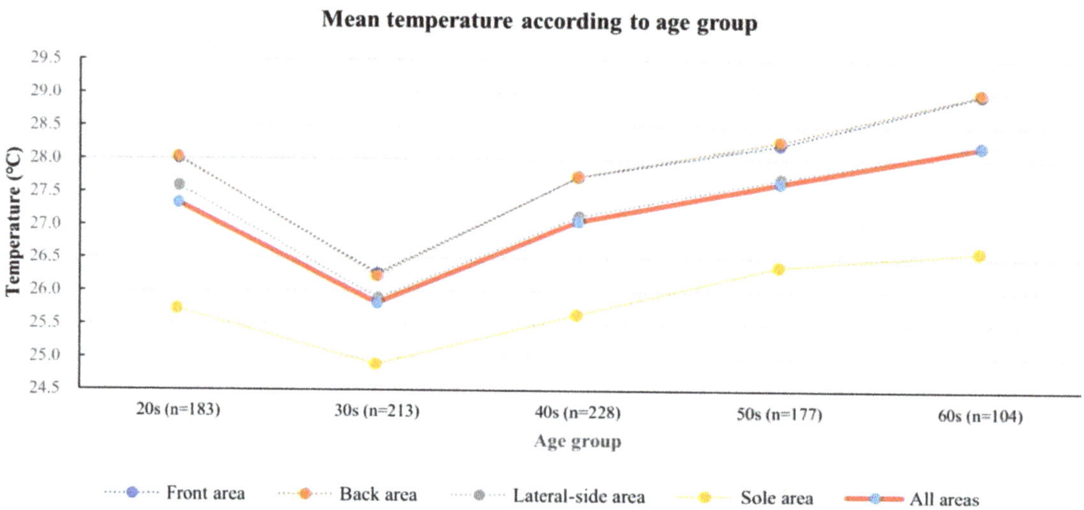

Figure 7. Trend of temperature according to age group in each area.

4. Discussion

4.1. The Mean Temperature and Difference between Both Sides

The surface temperature of the lower limb was difficult to define as a single numerical average because the mean temperature of ROIs was significantly different between the areas and location of ROIs (ranging from 24.60 ± 5.06 °C to 28.75 ± 5.76 °C). Notably, the temperature of the sole area was significantly lower than that of other areas. This finding supports the previous suggestion that distal skin regions, including feet and hands, are hypo-radiant areas because they are further away from the body's main thermal cores, such as the great vessels and viscera [22,23].

In previous clinical studies, the normal range of $|\Delta T|$ was limited to 0.2 °C, although it varied depending on the region [10,15,17]. According to the present study, $|\Delta T|$ was within 0.2 °C, as suggested previously in almost all ROIs except in the lateral-side area. In the lateral-side area, $|\Delta T|$ was higher than 0.2 °C at 0.58 ± 2.72 °C (95% CI, 0.40–0.76). The reason for this variability requires further study, and careful interpretation should be taken when evaluating $|\Delta T|$ in the lateral-side area.

The results of this study, which include mean temperature and mean ΔT of each ROI from a large sample, can serve as a reference standard in DITI. Based on this reference standard, objective hypo-radiant/hyper-radiant and clinical significance of ΔT can be determined by comparing the practical DITI results with this reference data. However, it may be difficult to subdivide the ROIs in practical clinics as was done in this study. In such cases, abbreviations for representative RIOs can be used and analyzed. It is important to compare values in each ROI of each area, rather than relying on simple overall averages. Additionally, because various individual characteristics, such as sex, age, medical condition, body composition, and circadian rhythm, can affect the skin temperature [24–27], normal values for an individual may fall outside the range of reference standards from this study. Therefore, further research is needed to assess the sensitivity and specificity of this reference standard by applying it to actual patients in the future.

4.2. Subgroup Analysis

We investigated subgroup analysis, including the trend of surface temperature according to the proximity of ROIs, sex, and age groups, to verify existing claims and controversies related to surface temperature. A previous suggestion that the surface temperature decreases and $|\Delta T|$ increases from the proximal to the distal part was limited to only certain areas. Additionally, it was confirmed that men have significantly higher surface temperatures than women in all areas and in all age groups. In terms of the trend of surface temperature related to age, a complex phenomenon that the temperature increased with age except for the 20s age group concluded several existing controversies.

This study's findings that the surface temperature in the lateral-side area decreases from the proximal to the distal part and that the temperature of the sole is the lowest among the four areas are consistent with existing hypotheses. Previous studies suggested that the body temperature would drop from the proximal part to the distal part of the body [6]. We speculate that the reason for this phenomenon is that the bloodstream temperature of the proximal part is warmer than that of the distal part. However, in the front and back areas, this hypothesis does not apply and is, therefore, still controversial.

Based on the results of this study, $|\Delta T|$ increased toward the distal part in the front area only, but not in other areas. Furthermore, $|\Delta T|$ was within 0.2 °C in the sole area and it was measured as higher than 0.2 in the lateral-side area. In contrast, some authors have suggested that $|\Delta T|$ increases toward the distal part of the body and that the normal range of $|\Delta T|$ can be larger than 0.2 °C in the periphery [10,17]. As a result, previous findings are controversial, and it is necessary to be careful about the interpretation of the significance of $|\Delta T|$ depending on the area.

According to this study, there was a significant trend that the surface temperature of females was lower than that of males in all areas of lower limbs. The influence of sex on surface temperature has been a subject of controversy. Some studies have reported that sex differences were significant for only certain ROIs, while others have suggested that women have lower skin temperatures [24]. This study concludes the previous controversy in this regard. A possible explanation is that the higher level of subcutaneous adipose tissue in women is associated with decreased surface temperature [28]. Greater subcutaneous fat in females may provide a significant insulating layer that blunts heat transfer from the core of the body or major blood vessels [29].

In terms of the relationship between surface temperature and aging, there has been controversy. Several studies have identified lower surface temperatures in the elderly and suggested that elderly individuals may have a lower skin temperature due to the association of aging with the loss of muscle mass, which reduces metabolism and limits heat generation [30–32]. In contrast, other studies have found that the elderly tend to have similar or higher skin temperature than young people and insisted that the blood flow of the human skin is controlled by the sympathetic nervous system innervation, which is mainly affected by the aging process [33–35].

According to the present study, the trend of surface temperature across age groups revealed a complicated pattern. Specifically, although the surface temperature tended to increase with age, the surface temperature in the 20s age group was paradoxically higher than that in the 30s age group. Consequently, individuals in their 30s showed the lowest temperature compared to all other age groups in both sexes. This trend may suggest that body surface temperature can vary according to age, even within the same individual.

Aging is related to the alteration of the cutaneous vasodilation and vasoconstriction reflexes that modify peripheral circulation [3]. In other words, older adults have a higher skin temperature due to increased blood flow in the skin caused by a deficit in the venous return [36,37]. On the other hand, the higher metabolic rate of the muscle tissue and active reproductive organs with higher blood supply could have a positive influence on the higher surface temperature of young adults, especially the 20s age group [30,31]. We presume that the 30s age group has a paradoxical heat loss (heat redistribution) in the subcutaneous or skin layer, although they also have a high core temperature. Abundant subcutaneous fat in the 30s age group compared to the 20s age group could lower skin temperature by blocking heat transfer from the core of the body or major blood vessels [29]. In summary, we assume that the trend of body surface temperature according to age reveals a complicated pattern due to complex physiological phenomena related to heat generation and heat redistribution according to the aging process.

4.3. Limitations

This study has some limitations regarding various intrinsic or extrinsic factors that can influence surface temperature [38,39]. Firstly, the environment of the subjects was not entirely controlled during the DITI exam. All physiological and psychological factors, such as time of the test, circadian rhythm, menstrual cycle, menopause, emotional stress, and season, that can affect autonomic function, were neither controlled nor collected. Secondly, the control range of the test room temperature and humidity was too wide. We were unable to strictly control these factors due to practical limitations in the test room environment and seasonal variations. Thirdly, other demographic factors, such as ethnic variability, body mass index, and body fat percentage, were not considered.

However, we tried to minimize errors by limiting the inclusion criteria to healthy subjects with no specific medical history or symptoms, although healthy volunteers did not rule out the possibility of varicose veins, knee osteoarthritis, scars on the lower limbs, limb shortening, postural deviations, or extra hair on the legs. Additionally, a large number of participants who underwent consistent measurement techniques could strengthen the reliability of our examination and analysis. To the best of our knowledge, this study is the first report to investigate the reference standard data of DITI.

5. Conclusions

This large-scale study of DITI of the lower limb could serve as a basis for establishing a reference standard for DITI. Additionally, we were able to analyze the trend of the surface temperature of lower limbs in terms of the proximity of ROIs, sex, and age. Using these reference standards and trends, DITI results in various diseases can be assessed through comparative analysis. Furthermore, the results of the present study can be used in the development of diagnosis using deep learning-based artificial intelligence in the future.

Author Contributions: Conceptualization: H.Y.Z. and S.S.; Methodology: H.Y.Z. and S.S.; Software: T.-M.Y. and S.S.; Validation: T.-M.Y. and S.S.; Formal analysis: T.-M.Y. and S.S.; Investigation: H.Y.Z.; Resources: H.Y.Z.; Data curation: T.-M.Y. and S.S.; Writing—original draft preparation: S.S.; Writing—review and editing: S.S. and B.R.Y.; Visualization: S.S.; Supervision: H.Y.Z.; Project administration: H.Y.Z.; Funding acquisition: H.Y.Z. All authors have read and agreed to the published version of the manuscript.

Funding: This research was supported by the Korea Evaluation Institute of Industrial Technology (Grant No. 20016383) and the National Research Foundation of Korea (NRF) funded by the Korean government (MIST) (Grant No. 2021M3I2A1077405).

Institutional Review Board Statement: The study was conducted according to the guidelines of the Declaration of Helsinki and approved by the Institutional Review Board of Ilsan Hospital (IRB No. NHIMC 2018–01-018, 20 January 2018), Gangnam Severance Hospital (IRB No. 3-2018-0021, 5 January 2018), and Ajou University Hospital (IRB No. AJIRB-MED-MDB-18-032, 2 February 2018).

Informed Consent Statement: Informed consent was obtained from all subjects involved in the study.

Data Availability Statement: The data presented in this study are available on request from the corresponding author. The data are not publicly available due to the privacy of participants.

Acknowledgments: We would like to express our gratitude to the surveyors, Bo Eun Lee, Ko Woon Yoon, Ho Gun Lee, Ji Eun Lee, and Ji Hye Lee, for their hard work in conducting the DITI examination for this study.

Conflicts of Interest: The authors declare no conflict of interest.

References

1. Uematsu, S.; Edwin, D.H.; Jankel, W.R.; Kozikowski, J.; Trattner, M. Quantification of thermal asymmetry. Part 1: Normal values and reproducibility. *J. Neurosurg.* **1988**, *69*, 552–555. [CrossRef] [PubMed]
2. Tattersall, G.J. Infrared thermography: A non-invasive window into thermal physiology. *Comp. Biochem. Physiol. Part A Mol. Integr. Physiol.* **2016**, *202*, 78–98. [CrossRef] [PubMed]
3. Ring, F. Thermal imaging today and its relevance to diabetes. *J. Diabetes Sci. Technol.* **2010**, *4*, 857–862. [CrossRef] [PubMed]
4. Brelsford, K.L.; Uematsu, S. Thermographic presentation of cutaneous sensory and vasomotor activity in the injured peripheral nerve. *J. Neurosurg.* **1985**, *62*, 711–715. [CrossRef]
5. Edeiken, J.; Shaber, G. Thermography: A reevaluation. *Skelet. Radiol.* **1986**, *15*, 545–548. [CrossRef]
6. Abe, N.; Kodama, H. Distal-proximal skin temperature gradient prior to sleep onset in infants for clinical use. *Pediatr. Int. Off. J. Jpn. Pediatr. Soc.* **2015**, *57*, 227–233. [CrossRef]
7. Balbinot, L.F.; Robinson, C.C.; Achaval, M.; Zaro, M.A.; Brioschi, M.L. Repeatability of infrared plantar thermography in diabetes patients: A pilot study. *J. Diabetes Sci. Technol.* **2013**, *7*, 1130–1137. [CrossRef]
8. Resende, P.R.; Brioschi, M.L.; De Meneck, F.; Neves, E.B.; Teixeira, M.J. Predicting lipoabdominoplasty complications with infrared thermography: A delta-R analysis. *Arch. Plast. Surg.* **2021**, *48*, 553–558. [CrossRef]
9. Haddad, D.S.; Brioschi, M.L.; Arita, E.S. Thermographic and clinical correlation of myofascial trigger points in the masticatory muscles. *Dento Maxillo Facial Radiol.* **2012**, *41*, 621–629. [CrossRef]
10. Park, T.Y.; Son, S.; Lim, T.G.; Jeong, T. Hyperthermia associated with spinal radiculopathy as determined by digital infrared thermographic imaging. *Medicine* **2020**, *99*, e19483. [CrossRef]
11. Tuzgen, S.; Dursun, S.; Abuzayed, B. Electrical skin resistance and thermal findings in patients with lumbar disc herniation. *J. Clin. Neurophysiol.* **2010**, *27*, 303–307. [CrossRef]
12. Wakisaka, S.; Kajander, K.C.; Bennett, G.J. Abnormal skin temperature and abnormal sympathetic vasomotor innervation in an experimental painful peripheral neuropathy. *Pain* **1991**, *46*, 299–313. [CrossRef]
13. Krumova, E.K.; Frettlöh, J.; Klauenberg, S.; Richter, H.; Wasner, G.; Maier, C. Long-term skin temperature measurements—A practical diagnostic tool in complex regional pain syndrome. *Pain* **2008**, *140*, 8–22. [CrossRef]
14. Park, E.S.; Park, C.I.; Jung, K.I.; Chun, S. Comparison of sympathetic skin response and digital infrared thermographic imaging in peripheral neuropathy. *Yonsei Med. J.* **1994**, *35*, 429–437. [CrossRef]
15. Ra, J.Y.; An, S.; Lee, G.H.; Kim, T.U.; Lee, S.J.; Hyun, J.K. Skin temperature changes in patients with unilateral lumbosacral radiculopathy. *Ann. Rehabil. Med.* **2013**, *37*, 355–363. [CrossRef]
16. Uematsu, S.; Jankel, W.R.; Edwin, D.H.; Kim, W.; Kozikowski, J.; Rosenbaum, A.; Long, D.M. Quantification of thermal asymmetry. Part 2: Application in low-back pain and sciatica. *J. Neurosurg.* **1988**, *69*, 556–561. [CrossRef]
17. Zhang, H.Y.; Kim, Y.S.; Cho, Y.E. Thermatomal changes in cervical disc herniations. *Yonsei Med. J.* **1999**, *40*, 401–412. [CrossRef]
18. Naulleau, C.; Jeker, D.; Pancrate, T.; Claveau, P.; Deshayes, T.A.; Burke, L.M.; Goulet, E.D.B. Effect of Pre-Exercise Caffeine Intake on Endurance Performance and Core Temperature Regulation During Exercise in the Heat: A Systematic Review with Meta-Analysis. *Sports Med.* **2022**, *52*, 2431–2445. [CrossRef]
19. Xu, X.; Karis, A.J.; Buller, M.J.; Santee, W.R. Relationship between core temperature, skin temperature, and heat flux during exercise in heat. *Eur. J. Appl. Physiol.* **2013**, *113*, 2381–2389. [CrossRef]
20. Okamoto-Mizuno, K.; Tsuzuki, K.; Mizuno, K. Effects of humid heat exposure in later sleep segments on sleep stages and body temperature in humans. *Int. J. Biometeorol.* **2005**, *49*, 232–237. [CrossRef]
21. Petrofsky, J.; Berk, L.; Alshammari, F.; Lee, H.; Hamdan, A.; Yim, J.E.; Patel, D.; Kodawala, Y.; Shetye, G.; Chen, W.T.; et al. The effect of moist air on skin blood flow and temperature in subjects with and without diabetes. *Diabetes Technol. Ther.* **2012**, *14*, 105–116. [CrossRef] [PubMed]
22. Johnson, J.M.; Kellogg, D.L., Jr. Thermoregulatory and thermal control in the human cutaneous circulation. *Front. Biosci.* **2010**, *2*, 825–853. [CrossRef]
23. Braverman, I.M. The cutaneous microcirculation: Ultrastructure and microanatomical organization. *Microcirculation* **1997**, *4*, 329–340. [CrossRef] [PubMed]
24. Neves, E.B.; Salamunes, A.C.C.; de Oliveira, R.M.; Stadnik, A.M.W. Effect of body fat and gender on body temperature distribution. *J. Therm. Biol.* **2017**, *70*, 1–8. [CrossRef]
25. Chudecka, M.; Lubkowska, A. Thermal Imaging of Body Surface Temperature Distribution in Women with Anorexia Nervosa. *Eur. Eat. Disord. Rev. J. Eat. Disord. Assoc.* **2016**, *24*, 57–61. [CrossRef]
26. Te Lindert, B.H.W.; Van Someren, E.J.W. Skin temperature, sleep, and vigilance. *Handb. Clin. Neurol.* **2018**, *156*, 353–365. [CrossRef]
27. Van Someren, E.J.; Raymann, R.J.; Scherder, E.J.; Daanen, H.A.; Swaab, D.F. Circadian and age-related modulation of thermoreception and temperature regulation: Mechanisms and functional implications. *Ageing Res. Rev.* **2002**, *1*, 721–778. [CrossRef]

28. Salamunes, A.C.C.; Stadnik, A.M.W.; Neves, E.B. The effect of body fat percentage and body fat distribution on skin surface temperature with infrared thermography. *J. Therm. Biol.* **2017**, *66*, 1–9. [CrossRef]
29. Savastano, D.M.; Gorbach, A.M.; Eden, H.S.; Brady, S.M.; Reynolds, J.C.; Yanovski, J.A. Adiposity and human regional body temperature. *Am. J. Clin. Nutr.* **2009**, *90*, 1124–1131. [CrossRef]
30. Marion, G.S.; McGann, K.P.; Camp, D.L. Core body temperature in the elderly and factors which influence its measurement. *Gerontology* **1991**, *37*, 225–232. [CrossRef]
31. Fox, R.H.; Woodward, P.M.; Exton-Smith, A.N.; Green, M.F.; Donnison, D.V.; Wicks, M.H. Body temperatures in the elderly: A national study of physiological, social, and environmental conditions. *Br. Med. J.* **1973**, *1*, 200–206. [CrossRef]
32. Niu, H.H.; Lui, P.W.; Hu, J.S.; Ting, C.K.; Yin, Y.C.; Lo, Y.L.; Liu, L.; Lee, T.Y. Thermal symmetry of skin temperature: Normative data of normal subjects in Taiwan. *Zhonghua Yi Xue Za Zhi Chin. Med. J. Free China Ed* **2001**, *64*, 459–468.
33. Kenney, W.L.; Munce, T.A. Invited review: Aging and human temperature regulation. *J. Appl. Physiol.* **2003**, *95*, 2598–2603. [CrossRef]
34. Kenney, W.L.; Armstrong, C.G. Reflex peripheral vasoconstriction is diminished in older men. *J. Appl. Physiol.* **1996**, *80*, 512–515. [CrossRef]
35. Borisov, V.V.; Lin, D.C. Temperature fluctuations in the lower limbs of young and elderly individuals during activities of daily living. *Exp. Gerontol.* **2014**, *57*, 243–249. [CrossRef]
36. Holowatz, L.A.; Thompson-Torgerson, C.; Kenney, W.L. Aging and the control of human skin blood flow. *Front. Biosci.* **2010**, *15*, 718–739. [CrossRef]
37. Holowatz, L.A.; Kenney, W.L. Peripheral mechanisms of thermoregulatory control of skin blood flow in aged humans. *J. Appl. Physiol.* **2010**, *109*, 1538–1544. [CrossRef]
38. Sivanandam, S.; Anburajan, M.; Venkatraman, B.; Menaka, M.; Sharath, D. Medical thermography: A diagnostic approach for type 2 diabetes based on non-contact infrared thermal imaging. *Endocrine* **2012**, *42*, 343–351. [CrossRef]
39. Morrissey, M.C.; Wu, Y.; Zuk, E.F.; Livingston, J.; Casa, D.J.; Pescatello, L.S. The impact of body fat on thermoregulation during exercise in the heat: A systematic review and meta-analysis. *J. Sci. Med. Sport* **2021**, *24*, 843–850. [CrossRef]

bioengineering MDPI

Systematic Review

Artificial Intelligence Advances in Transplant Pathology

Md Arafatur Rahman [1,2], Ibrahim Yilmaz [1,3], Sam T. Albadri [1], Fadi E. Salem [1], Bryan J. Dangott [1,3], C. Burcin Taner [4], Aziza Nassar [1] and Zeynettin Akkus [1,3,*]

1 Department of Laboratory Medicine and Pathology, Mayo Clinic, Jacksonville, FL 32224, USA
2 Department of Mathematics, Florida State University, Tallahassee, FL 32306, USA
3 Computational Pathology and Artificial Intelligence, Mayo Clinic, Jacksonville, FL 32224, USA
4 Department of Transplantation Surgery, Mayo Clinic, Jacksonville, FL 32224, USA
* Correspondence: akkus.zeynettin@mayo.edu

Abstract: Transplant pathology plays a critical role in ensuring that transplanted organs function properly and the immune systems of the recipients do not reject them. To improve outcomes for transplant recipients, accurate diagnosis and timely treatment are essential. Recent advances in artificial intelligence (AI)-empowered digital pathology could help monitor allograft rejection and weaning of immunosuppressive drugs. To explore the role of AI in transplant pathology, we conducted a systematic search of electronic databases from January 2010 to April 2023. The PRISMA checklist was used as a guide for screening article titles, abstracts, and full texts, and we selected articles that met our inclusion criteria. Through this search, we identified 68 articles from multiple databases. After careful screening, only 14 articles were included based on title and abstract. Our review focuses on the AI approaches applied to four transplant organs: heart, lungs, liver, and kidneys. Specifically, we found that several deep learning-based AI models have been developed to analyze digital pathology slides of biopsy specimens from transplant organs. The use of AI models could improve clinicians' decision-making capabilities and reduce diagnostic variability. In conclusion, our review highlights the advancements and limitations of AI in transplant pathology. We believe that these AI technologies have the potential to significantly improve transplant outcomes and pave the way for future advancements in this field.

Keywords: transplant pathology; artificial intelligence; kidney transplant; heart transplant; liver transplant; lung transplant; digital pathology

Citation: Rahman, M.A.; Yilmaz, I.; Albadri, S.T.; Salem, F.E.; Dangott, B.J.; Taner, C.B.; Nassar, A.; Akkus, Z. Artificial Intelligence Advances in Transplant Pathology. *Bioengineering* **2023**, *10*, 1041. https://doi.org/10.3390/bioengineering10091041

Academic Editors: Cuneyt M. Alper and Alan Wang

Received: 28 July 2023
Revised: 15 August 2023
Accepted: 31 August 2023
Published: 4 September 2023

1. Introduction

According to the Centers for Disease Control and Prevention (CDC), the most commonly transplanted organs in the USA are the kidneys, liver, heart, lungs, pancreas, and intestines. Although there are approximately 100,000 people waiting for organ transplants every day, organ supplies remain limited. An estimated 14,000 deceased organ donors are available, each providing an average of 3.5 organs, while living donors supply only 6000 organs each year [1]. In the standard of care for organ transplantation, both donors and recipients undergo a pre-transplant assessment of histocompatibility, pathology, and clinical case histories. When a matched pair is found, the organ transplantation will proceed and then subsequently be monitored. Post-transplant monitoring includes electronic medical record (EMR) review, blood and body fluid assessment for organ function and formation of donor-specific antibodies, and protocol biopsies if there is a suspicion of rejection.

Recent advancements in deep learning-based AI approaches have revolutionized the field of digital pathology by enabling the development of AI-empowered diagnostic models for analyzing digitized glass slides of biopsy specimens. AI-empowered digital pathology can be extremely helpful in transplant pathology, as it could reduce inter-reader variability between pathologists, allow teleconsultation for pre and post transplantation, and provide a second opinion as well as assessing several morphological parameters and their spatial relationships. Over the years, several AI models have been developed for assessing transplant-related heart, lung, kidney, and liver pathologies [2–5]. Transplant pathology is a highly specialized field in which AI-empowered digital pathology tools could aid pathologists in making better decisions and eliminating/reducing the diagnostic variability among them. In addition to AI advances in transplant pathology, AI has also been used in organ allocation and donor-recipient matching [6–9], transplant oncology [10], and immunosuppressive therapy [11].

In spite of the impressive success of AI models in integrating information from multi-modality data sources (e.g., histopathological reports, lab test results, radiological features, and patient demographics), our focus in this review is on the AI-empowered analysis of whole-slide images (WSIs), which is the cornerstone of diagnosis and prognosis in transplant pathology. Within the scope of this review, we aim to present the advancements of AI in the diagnosis and prognosis of transplant-related heart, lung, liver, and kidney pathologies. Furthermore, we offer insights into the future trajectory of AI-assisted approaches in the domain of transplant pathology. A flowchart of AI-assisted transplant pathology is depicted in Figure 1. An example of a web-based framework that could be used for transplant pathology is suggested in Akkus et al. [12]

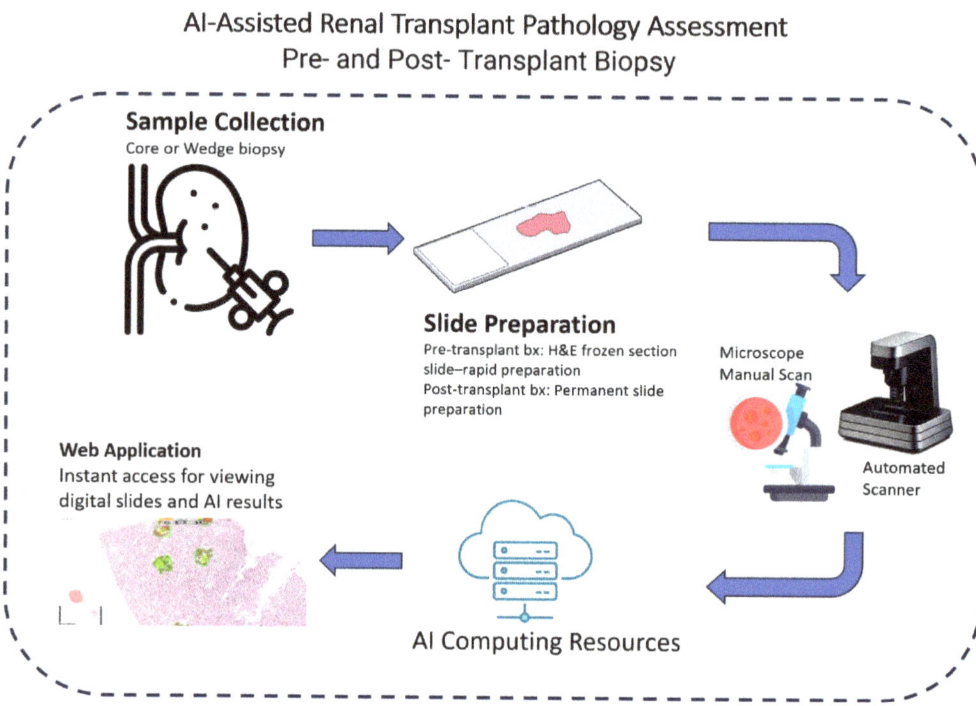

Figure 1. A depiction of an AI-assisted renal transplant pathology workflow.

2. Methods

2.1. Literature Search Strategy

We performed a thorough analysis of the literature using the Web Of Science and PubMed search engines. We included peer-reviewed journal publications and conference proceedings addressing the application of deep learning-based AI to transplant pathology before 30 April 2023. This systematic review was conducted using the Preferred Reporting Items for Systematic Reviews and Meta-Analysis (PRISMA) checklist [13]. The details of the search and our systematic review protocol are registered in the OSF public registries of systematic reviews: https://doi.org/10.17605/OSF.IO/XFGZN, (accessed on 1 May 2023).

2.2. Study Inclusion and Exclusion Criteria

We included all full-text articles focusing on deep learning-based artificial intelligence in transplant pathology. We excluded studies using classical machine learning algorithms that rely on hand-crafted feature extraction. Furthermore, we also searched reference lists of retrieved articles and reviewed articles in the field to identify eligible studies that met inclusion and exclusion criteria. MAR and ZA reviewed titles and abstracts. To facilitate a structured analysis, the gathered reports were categorized into four distinct groups based on the transplanted organs: heart, lungs, kidneys, and liver. (See Figure 2 for a detailed flowchart used for identification, screening, eligibility, and inclusion).

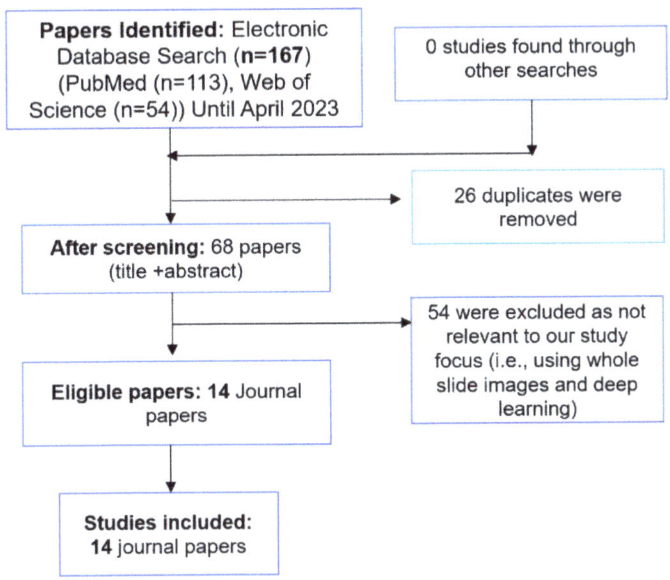

Figure 2. An overview of the systematic review process.

3. Results

The initial systematic literature search yielded a total of 167 relevant articles. After removing duplicates, a rigorous selection process resulted in the inclusion of 14 journal papers that aligned with the defined scope and objectives of this review.

3.1. AI Advances in Transplant Heart Pathology

Cardiac allograft rejection is a critical concern in heart transplantation, primarily due to the limited availability of donor organs. Although endomyocardial biopsy with histopathology grading is the standard of care for diagnosing cardiac allograft rejection, substantial inter- and intra- observer variability among pathologists may result in inappropriate treatment with immunosuppressive drugs, redundant follow-up biopsies, and deteriorated transplant outcomes.

To date, there were four studies that discussed the use of artificial intelligence to assess cardiac allograft rejection and survival prediction. Giuste et al. [14] studied synthetic image generation to improve the risk assessment of rare pediatric heart transplant rejection. They trained progressive and inspirational Generative Adversarial Networks (GANs) to generate high resolution synthetic images with rejection signs that helped improve the performance of their allograft rejection classifier model. Despite their limited dataset consisting of only 12 non-rejection and 12 rejection slides, their approach significantly improved the performance of their allograft rejection classifier model, which achieved an area under receiver operating curve (AUC) of 98.84% for image tile-based rejection detection and 95.56% for biopsy rejection prediction at the WSI level. Peyster et al. [2] presented an automated WSI analysis pipeline based on handcrafted feature extraction and selection for histological grading of cardiac allograft rejection. Their study cohort included 2472 biopsy slides from three major US transplant centers. Their model performance was comparable to pathologists' agreement (65.9% vs. 60.7%). Although this study did not meet our inclusion criteria, we refer readers to this paper for comparison to deep learning-based AI models. A limitation of this study is that the ground truth diagnosis label was obtained based on scoring only one section of tissue digitally while pathologists examine multiple slide sections under the microscope to reach the consensus of a final grade. Later, Lipkova et al. [15] published a study based on deep learning for the assessment of cardiac allograft rejection from the WSIs of endomyocardial biopsies. Their model demonstrated allograft rejection with a notable AUC of 0.962, which is a significant improvement compared to the study of Peyster et al. [2] that used hand-crafted features and a classical machine learning approach. Additionally, the AUC for differentiating between low and high-grade rejection was reported as 0.83. In their pipeline, they used a pre-trained CNN model to extract features from image patches and fine-tuned on three fully connected layers and a separate classifier was trained to estimate the rejection grade. The model was trained on 80% of 1690 internal WSI image datasets and validated on two external datasets: 1717 WSI slides of 585 patients from Turkey and 123 WSI slides of 123 patients from Switzerland. Lastly, Glass et al. [16] fine-tuned a pre-trained VGG model to predict myocyte damage in cardiac transplant acute cellular rejection (ACR). The authors annotated 19,617 regions including 10,855 regions of ACR, 5002 of healing injury, and 3760 of normal from 200 H&E slides and reported 94% for validation accuracy. A summary of these studies is provided in Table 1.

Table 1. A summary of previous AI studies investigating transplant heart pathology.

Author, Year	Objective	AI Model	Dataset	Performance
Giuste et al., 2023 [14]	Enhancing risk assessment of rare pediatric heart transplant rejection through the generation of synthetic images	Progressive and inspirational GAN [17,18]	12 non rejection and 12 rejection slides	95.56% AUROC for biopsy level rejection detection with 83.33% sensitivity and 66.67% specificity
Lipkova et al., 2022 [15]	Assessment of cardiac allograft rejection from endomyocardial biopsies	Pre-trained deep residual CNN [19]	Training: 1352 WSI slides; Validation: 1840 WSI slides	Allograft reject detection with an AUC of 0.962
Peyster et al., 2021 [2]	Histological grading of cardiac allograft rejection	Computer-Assisted Cardiac Histologic Evaluation (CACHE) grader pipeline	2472 endomyocardial biopsy slides	Differentiate low- and high-grade rejection with an AUC of 0.83
Glass et al., 2020 [16]	Determine myocyte damage in cardiac transplant acute cellular rejection	Pre-trained VGG16 [20]	19,617 annotations (10,855 regions of ACR; 5002 healing injury; 3760 normal)	Detection of myocyte damage (Grade 1R2) from non-myocyte damage (Grade 1R1A) with 94% validation accuracy

3.2. *AI Advances in Transplant Lung Pathology*

Despite advances in the immunosuppressive therapies and immunosuppressive drugs used, one-third of lung transplant recipients experience at least one episode of treated acute rejection in the first year after transplantation according to the report of the registry of the International Society of Heart and Lung Transplantation [21]. However, contrasting findings from the Organ Procurement and Transplantation Network/Scientific Registry of Transplant Recipients indicate a lower incidence of less than one-fifth in the first year [22].

Gholamzadeh et al. [23] published an in-depth systematic review about classical machine learning-based techniques to improve lung transplantation outcomes and complications. However, we exclusively focused on deep learning-based AI studies in our review and refer readers to their paper for further information. In one notable study, Davis et al. [3] investigated detecting acute cellular rejection (ACR) in lung transplant biopsies using AI. Board-certified lung transplant pathologists annotated a total of 3349 annotations (2580 regions of normal, 769 lesions of A1/A2 rejection). They included 614 A1/A2 lesions and 2064 regions of normal for the training set. On the other hand, the validation set included 155 A1/A2 lesions and 156 regions of normal to evaluate their AI model performance. Remarkably, their AI model distinguished the vascular component of ACR from normal alveolated lung tissue with 95% validation accuracy. Throughout our search, we identified only one study about lung transplant pathology using AI and one study about donor-recipient matching using AI [24]. Davis et al. presented promising results in identifying ACR, which is related to chronic lung allograft rejection, in lung transplant patients. The main limitation of their research is the lack of multi-institutional validation testing.

3.3. AI Advances in Transplant Kidney Pathology

Kidney transplantation is the most frequently performed solid organ transplantation worldwide. According to the United Network of Organ Sharing (UNOS), more than 25,000 people in the USA had kidney transplants performed in 2022, which was 3.4% more than in 2021 [25]. Despite the growing demand for kidney transplantation, the field of pathology is facing a decline in the available workforce. As per the Organ Procurement and Transplantation Network (OPTN), there are 88629 patients in the US currently on the waiting list for a kidney transplant as of July 23, 2023 [26]. Based on transplants conducted between 2008 and 2015, the 5 years post-transplant survival rates for males and females are 85.85% and 88.2%, respectively [26]. Although leveraging AI is new in this field, it could be useful in many ways such as finding a better matchmaking process between donors and patients, assessing histopathology of kidney biopsies, and guiding the treatment and management of transplant patients. The most common kidney biopsy scoring systems are Remuzzi [27], Banff [28], Leuven [29], and the Maryland Aggregate Pathology Index (MAPI) [30,31]. Table 2 provides a comparative overview of these models. The application of AI in transplant kidney pathology has been examined in a number of studies [4,32–34].

Table 2. Comparison of kidney biopsy scoring system models [30,35–37].

	Remuzzi	Banff	Leuven	MAPI
Scoring System	Scoring range between 0–12	3 grades (Mild, Moderate, and Severe) and 6 rejection categories	Score > 60 or Score < 60	Scoring range between 0–15
Selection Criteria	Glomerulosclerosis, tubular atrophy, interstitial fibrosis, and arterial narrowing	Vascular and other histologic abnormalities	Glomerulosclerosis, donor age, interstitial fibrosis, and tubular atrophy	Histologic parameters (glomerular sclerosis, arteriolar hyalinosis, cortical scar, and periglomerular fibrosis)

Hermsen et al. [32] employed a UNet [38] architectural CNN model for multiclass segmentation of digitized kidney biopsy tissue sections with periodic acid-Schiff (PAS) staining. The model was trained on 40 WSIs and validated 10 WSIs from their home institution and 10 WSIs from an external institution. In another study, Hermsen at al. [33] trained a UNet model to quantify inflammatory and chronic features in kidney transplant biopsies. Kers et al. [34] investigated the performance of multiple CNN architectures to predict transplant rejection. The study encompassed retrospective multicentered data, including 5844 digital whole-slide images of kidney allograft biopsies obtained from 1948 patients. A 3-fold cross-validation approach was employed, and an external dataset consisting of 101 WSIs was used for evaluation. The limitations of this study encompass its concentration on Western European institutions, lack of ethnicity data due to legal restrictions, absence of certain baseline characteristics, and restricted staining variability. Smith et al. [4] used a binary thresholding approach and trained a UNet model for glomeruli segmentation to assess interstitial inflammation from CD45-stained digital slides. They included a total of 60 biopsies from 53 patients in their study and observed a strong correlation between their automated inflammation scoring and Banff scoring. This study's limitations include focusing on pixel counting rather than cell counting, potential complexity in implementing a deep learning approach, and a retrospective design with limited statistical power. Wilbur et al. [39] trained a modified version of the AlexNet CNN model to identify glomeruli on renal biopsies containing four stains (H&E, trichrome, silver, and PAS) from multiple institutions. The study encompassed 71 biopsies that were split into training/validation (n = 52) and testing (n = 19). The authors emphasized the importance of diverse datasets for developing generalizable AI models. The sensitivity of their model ranged from 90% to 93% for the intra-institutional dataset vs. 77% for the inter-institutional dataset. A summary of these studies is shown in Table 3. Furthermore, several review papers discussing AI in kidney transplantation and the assessment of renal transplant prognosis using classical machine learning approaches are available for interested readers in the additional resources [27,40–45].

Table 3. Summary of previous AI studies on transplant kidney pathology.

Author, Year	Objective	AI Model	Dataset	Performance
Hermsen et al., 2019 [32]	Multiclass segmentation of digitized kidney biopsy tissue sections	UNet [38]	Training: 40 WSIs; Validation: 20 WSIs (Home: 10; External Institution: 10)	Detected 92.7% of all glomeruli in nephrectomy samples with 10.4% false positives
Hermsen et al., 2022 [33]	Quantifying the chronic and inflammatory lesions in kidney transplant biopsies	UNet [38]	125 WSI pairs of periodic acid-schiff- and CD3-stained slides	The tissue class glomeruli was segmented with precision, recall, and dice scores of 0.96, 0.94, 0.95, respectively
Kers et al., 2022 [34]	Classifying histology of kidney allograft biopsies	Single CNN (InceptionV3) [46], Serial CNN	5844 WSIs from 1948 patients	AUROC (Single CNN) of 0.86, 0.78, and 0.70 for the normal and rejection disease classes, respectively
Smith et al., 2023 [4]	Quantifying the amount of non-glomerular inflammation within the cortex	UNet [38]	60 biopsies from 53 patients	Precision, recall, and dice scores for glomeruli identification were 0.888, 0.830, and 0.858, respectively
Wilbur et al., 2021 [39]	Identifying glomeruli on renal biopsy containing four stains from multiple institutions	Modified version of AlexNet	71 biopsies (Training: 52; Testing: 19)	Sensitivity of 90–93% for intra-institutional and 77% for inter-institutional dataset

3.4. *AI Advances in Transplant Liver Pathology*

Like other organ transplants, Liver Transplantation (LT) is a crucial treatment for patients with end-stage liver diseases. Recent advancements in surgical techniques, improved management of immunosuppressive drugs, and enhanced understanding of post-transplant morbidities have led to a significant increase in LT procedures. This surge in demand, however, has also led to a shortage of organ donors, resulting in a substantial waiting list for LT candidates [47]. Despite the disparity between organ supply and demand, over a third of donor livers are being rejected due to the risk of early allograft dysfunction (EAD), based on histopathologic findings [48]. The management of LT is complex, and the current approaches are not sufficient in clinical decision making. So, a data-driven LT could be useful in both pre- and post- LT settings [48,49]. Although this research area is quite new, several studies have already been conducted. Narayan et al. [50] published research regarding the use of AI for predicting donor liver allograft steatosis and early post-transplantation graft failure. They developed a Computer Vision AI platform (CVAI) to score donor liver steatosis and compared its capability for predicting EAD against pathologist steatosis scores. The study included liver biopsy slides data from 2014 to 2019 consisting of 25,494 images from 90 liver biopsies. The results indicated that the CVAI platform demonstrated slightly better calibration scores than pathologist steatosis scores. Their study was chiefly limited by a small sample of donor liver from a single institution and the presence of selection bias in the test for association with EAD. Yu et al. [51] designed a Multiple Up-sampling and Spatial Attention guided UNet model (MUSA-UNet) to segment liver portal tract regions in liver WSI that correlates with the stage of liver fibrosis. The dataset consisted of 53 WSIs, 30 of which were used for training and 23 for testing. They obtained an average of 0.94 precision, 0.85 recall, 0.89 F1 score, 0.89 (accuracy), and 0.80 (Jaccard Index) for their model MUSA-UNet. The major limitation of their study is the need for a more diverse training dataset with stain variations and annotations from multiple pathologists at various institutions.

Lu et al. [5] proposed an improved deep learning classifier (MobileNetV2_HCC_class) that could predict hepatocellular carcinoma (HCC) recurrence after liver transplantation. Their study was conducted on 1118 patients where 642 patients were used for training, 144 for testing, and 302 for validation. The hazard ratio obtained from the classifier in the LT set was 3.44 (95% CI 2.01–5.87, $p < 0.001$) and 2.55 (95% CI 1.64–3.99, $p < 0.001$) when known prognostic factors were adjusted.

Sun et al. [52] developed a deep learning model constructed from the pretrained VGG16 architecture [20] to estimate the percent steatosis in donor liver biopsy frozen sections. Their model generated a probability map from an input WSI to the output percent steatosis. Their research dataset comprised 96 WSIs, with 30 slides allocated for training and 66 for testing purposes. During the testing phase, their AI model demonstrated a notable correlation coefficient (r) of 0.85 and intraclass correlation coefficient (ICC) of 0.85, both of which surpassed the on-service pathologist's performance (r = 0.74 and ICC = 0.72). A summary of these studies can be found in Table 4.

Table 4. Summary of previous AI studies on transplant liver pathology.

Author, Year	Objective	AI Model	Dataset	Performance
Narayan et al., 2022 [50]	Prediction of donor liver allograft steatosis and early post-transplantation graft failure	CVAI model consisting of Fully Convolutional Networks (FCN) [53] and UNet	25,494 images from 90 liver biopsies	CVAI peak mean IU 0.80; steatosis score median (CVAI 3% vs. pathologist 20%)
Yu et al., 2022 [51]	Segmentation of portal tract regions from whole-slide images of liver tissue biopsies	MUSA-UNet, FCN, UNet, DeepLab V3 [54]	53 WSIs (Training and Validation: 30; Testing: 20)	The precision, recall, and F1 score of MUSA-UNet were 0.94, 0.85, and 0.89, respectively
Liu et al., 2022 [5]	Predict HCC recurrence after liver transplantation	UNet, MobileNetV2_HCC_class	1118 patients (Training: 642; Testing: 144; Validation: 302)	Hazard ratio of 3.44 for LT set and 2.55 with adjusted known prognostic factors
Sun et al., 2020 [52]	Quantify percent steatosis in donor liver biopsy frozen sections	Pre-trained VGG16 (truncated at bottleneck layer)	96 WSIs (Training: 30; Testing: 66)	r = 0.85, ICC = 0.85 on testing samples

4. Discussion

In this comprehensive review, we have presented the latest advancements in AI-assisted transplant pathology, with a specific focus on the utilization of WSI. As we embark upon the emerging era of digital pathology coupled with the promising potential of AI technologies, therein lies a significant opportunity to revolutionize various aspects of solid organ transplantation. These advancements hold the capacity to positively impact organ procurement processes, optimize the dynamic adaptation of immunosuppressive drugs, and enhance graft and patient survival through effective post-transplant monitoring. Notably, numerous deep learning-based AI applications have emerged, catering to different transplant organs for the precise assessment of histopathology in tissue biopsies. The majority of these proposed AI models, however, have been trained primarily on relatively modest datasets and lack external validation. In order to ensure robust and reliable outcomes, it is imperative to have well-trained and rigorously validated AI models that assess the histopathology of transplant organ biopsies. With such AI models, the field will be able to gain valuable insights into the decision-making process surrounding rejection or acceptance of organs, possibly predict poor outcomes in advance, and reduce postoperative complications significantly. By enhancing patient care and outcomes in the area of solid organ transplantation, this development promises to elevate transplant pathology to new heights.

The histopathological evaluation of organ biopsies could play a crucial role in the decision-making process of accepting or discarding the organs of deceased donors for transplantation and in post-transplant monitoring if there is a suspicion of rejection. The major issue in the current practice of pre-transplant histopathology assessment is the low agreement between the pathologists [55]. The presence of wide variability and subjectivity among pathologists raises a big concern in biopsy scoring that leads to the suboptimal usage of organs. The issue of variability in pre-transplant biopsies can be attributed to the rapid embedding and fixation methods used for examination, resulting in frozen artifacts and generating lower quality slides. Conversely, permanent slides with more elaborate preparations are employed for post-transplant biopsy examinations. Despite the fact that histopathology assessment of transplant organ biopsy requires specialized training and expert consultancy, pre-transplant biopsies are frequently interpreted by on-call general pathologists on most occasions. We, therefore, believe that adaptation of digital pathology and exploiting AI and informatics tools could eliminate the wide variability and subjectivity among pathologists, provide a second virtual expert opinion, and improve clinical workflow efficiency. Embracing these advancements could shape a new era in solid organ transplantation, improving diagnostic accuracy, and helping patients in need of life-saving surgery.

The procurement of transplant organs primarily relies on donations from deceased or unrelated individuals. Therefore, immunosuppressive drugs are given to patients to increase tolerance to transplanted organs in the long term and prevent rejections. Monitoring significant histopathologic changes of allografts over time is essential in assessing transplant rejection and the refinement of immunosuppressive regimens to improve outcomes of transplant patients. In this context, the integration of AI-empowered digital pathology and EMR data proves valuable, as it can support informed decision making for tailoring individualized immunosuppression treatment regimens, thereby optimizing the management of transplant patients.

High-throughput digital pathology slide scanners have recently become available and are being adapted in clinical workflows. Concurrently, developments in computing hardware, cloud resources, AI tools, data storage, and network speed have facilitated more efficient processing of vast quantities of WSI data. Moreover, the availability of portable single slide scanners allows for rapid onsite evaluation of pathology slides which may prove beneficial in organ procurement scenarios. In the realm of pathology departments, digital pathology, automated image analysis of WSI with AI, and web/cloud-based applications have emerged which enables instant access and sharing of WSI and diagnostic reports.

These technologies facilitate ready access to second opinions. The adoption of AI-assisted digital pathology workflows holds immense potential for improving patient outcomes and optimizing organ utilization in transplant pathology. Digital workflows, however, pose challenges due to their cost-intensive nature. Additionally, ensuring reproducibility and generalizability of AI models requires the development of datasets from multiple institutions and scanners. In addition, it is difficult to integrate existing AI tools effectively into clinical workflows at scale while maintaining efficiency. It will be essential to address these challenges and integrate AI-assisted digital pathology into transplant pathology to enhance patient care and organ preservation.

5. Conclusions

AI-empowered digital pathology is currently showing promise in facilitating the histopathological evaluation of organ biopsies, with the potential to mitigate the issues of variability and subjectivity encountered among pathologists. While AI applications in solid organ transplantation remain a nascent field, ongoing research is being conducted by investigators to explore its full potential. It is crucial to acknowledge that certain limitations, as previously discussed, warrant careful consideration and resolution before the widespread implementation of these AI tools in clinical workflows. Referring to the global aspect, cross-border organ transplantation presents significantly greater complexities due to a range of intricate challenges such as the illicit trafficking of human organs, variation of legal and regulatory frameworks, cultural and ethical differences, and financial and insurance matters, as well as travel and visa restrictions, etc. All of these factors need to be carefully addressed to ensure the safety, success, and ethical integrity of the procedures.

Author Contributions: Conceptualization, Z.A., I.Y. and M.A.R.; methodology, Z.A., I.Y. and M.A.R.; investigation, Z.A., I.Y. and M.A.R.; resources, Z.A., I.Y. and M.A.R.; data curation Z.A., I.Y. and M.A.R.; writing—original draft preparation, Z.A., I.Y. and M.A.R.; writing—review and editing, Z.A., I.Y., M.A.R., F.E.S., S.T.A., B.J.D., C.B.T. and A.N.; visualization: Z.A., I.Y. and M.A.R. All authors have read and agreed to the published version of the manuscript.

Funding: This research received no external funding.

Institutional Review Board Statement: Not applicable.

Informed Consent Statement: Not applicable.

Conflicts of Interest: The authors declare no conflict of interest.

References

1. Key Facts. Available online: https://www.cdc.gov/transplantsafety/overview/key-facts.html (accessed on 1 May 2023).
2. Peyster, E.G.; Arabyarmohammadi, S.; Janowczyk, A.; Azarianpour-Esfahani, S.; Sekulic, M.; Cassol, C.; Blower, L.; Parwani, A.; Lal, P.; Feldman, M.D.; et al. An Automated Computational Image Analysis Pipeline for Histological Grading of Cardiac Allograft Rejection. *Eur. Heart J.* **2021**, *42*, 2356–2369. [CrossRef]
3. Davis, H.; Glass, C.; Davis, R.C.; Glass, M.; Pavlisko, E.N. Detecting Acute Cellular Rejection in Lung Transplant Biopsies by Artificial Intelligence: A Novel Deep Learning Approach. *J. Heart Lung Transplant.* **2020**, *39*, S501–S502. [CrossRef]
4. Smith, B.; Grande, J.; Ryan, M.; Smith, M.; Denic, A.; Hermsen, M.; Park, W.; Kremers, W.; Stegall, M. Automated Scoring of Total Inflammation in Renal Allograft Biopsies. *Clin. Transplant.* **2023**, *37*, e14837. [CrossRef]
5. Liu, Z.; Liu, Y.; Zhang, W.; Hong, Y.; Meng, J.; Wang, J.; Zheng, S.; Xu, X. Deep Learning for Prediction of Hepatocellular Carcinoma Recurrence after Resection or Liver Transplantation: A Discovery and Validation Study. *Hepatol. Int.* **2022**, *16*, 577–589. [CrossRef]
6. Peloso, A.; Moeckli, B.; Delaune, V.; Oldani, G.; Andres, A.; Compagnon, P. Artificial Intelligence: Present and Future Potential for Solid Organ Transplantation. *Transpl. Int.* **2022**, *35*, 10640. [CrossRef]
7. Gorgen, A.; Muaddi, H.; Zhang, W.; McGilvray, I.; Gallinger, S.; Sapisochin, G. The New Era of Transplant Oncology: Liver Transplantation for Nonresectable Colorectal Cancer Liver Metastases. *Can. J. Gastroenterol. Hepatol.* **2018**, *2018*, 9531925. [CrossRef]
8. Briceño, J. Artificial Intelligence and Organ Transplantation: Challenges and Expectations. *Curr. Opin. Organ Transplant.* **2020**, *25*, 393–398. [CrossRef]
9. Cruz-Ramírez, M.; Hervás-Martínez, C.; Fernández, J.C.; Briceño, J.; de la Mata, M. Predicting Patient Survival after Liver Transplantation Using Evolutionary Multi-Objective Artificial Neural Networks. *Artif. Intell. Med.* **2013**, *58*, 37–49. [CrossRef]

10. Ivanics, T.; Patel, M.S.; Erdman, L.; Sapisochin, G. Artificial Intelligence in Transplantation (machine-Learning Classifiers and Transplant Oncology). *Curr. Opin. Organ Transplant.* **2020**, *25*, 426–434. [CrossRef]
11. Tang, J.; Liu, R.; Zhang, Y.L.; Liu, M.Z.; Hu, Y.F.; Shao, M.J.; Zhu, L.J.; Xin, H.W.; Feng, G.W.; Shang, W.J.; et al. Application of Machine-Learning Models to Predict Tacrolimus Stable Dose in Renal Transplant Recipients. *Sci. Rep.* **2017**, *7*, 42192. [CrossRef]
12. Akkus, Z.; Dangott, B.; Nassar, A. A Web/Cloud Based Digital Pathology Platform Framework for AI Development and Deployment. *bioRxiv* **2022**, arXiv:2022.11.04.514741.
13. PRISMA. Available online: http://prisma-statement.org/PRISMAstatement/checklist.aspx (accessed on 1 May 2023).
14. Giuste, F.O.; Sequeira, R.; Keerthipati, V.; Lais, P.; Mirzazadeh, A.; Mohseni, A.; Zhu, Y.; Shi, W.; Marteau, B.; Zhong, Y.; et al. Explainable Synthetic Image Generation to Improve Risk Assessment of Rare Pediatric Heart Transplant Rejection. *J. Biomed. Inform.* **2023**, *139*, 104303. [CrossRef]
15. Lipkova, J.; Chen, T.Y.; Lu, M.Y.; Chen, R.J.; Shady, M.; Williams, M.; Wang, J.; Noor, Z.; Mitchell, R.N.; Turan, M.; et al. Deep Learning-Enabled Assessment of Cardiac Allograft Rejection from Endomyocardial Biopsies. *Nat. Med.* **2022**, *28*, 575–582. [CrossRef]
16. Glass, C.; Davis, R.; Xiong, B.; Dov, D.; Glass, M. The Use of Artificial Intelligence (AI) Machine Learning to Determine Myocyte Damage in Cardiac Transplant Acute Cellular Rejection. *J. Heart Lung Transplant.* **2020**, *39*, S59. [CrossRef]
17. Karras, T.; Aila, T.; Laine, S.; Lehtinen, J. Progressive Growing of GANs for Improved Quality, Stability, and Variation. *arXiv* **2017**, arXiv:1710.10196.
18. Rozière, B.; Riviere, M.; Teytaud, O.; Rapin, J.; LeCun, Y.; Couprie, C. Inspirational Adversarial Image Generation. Available online: https://ieeexplore.ieee.org/document/9381602 (accessed on 19 July 2023).
19. He, K.; Zhang, X.; Ren, S.; Sun, J. Deep Residual Learning for Image Recognition. In Proceedings of the IEEE Conference on Computer Vision and Pattern Recognition (CVPR), Boston, MA, USA, 7–12 June 2015.
20. Simonyan, K.; Zisserman, A. Very Deep Convolutional Networks for Large-Scale Image Recognition. *arXiv* **2014**, arXiv:1409.1556.
21. Chambers, D.C.; Yusen, R.D.; Cherikh, W.S.; Goldfarb, S.B.; Kucheryavaya, A.Y.; Khusch, K.; Levvey, B.J.; Lund, L.H.; Meiser, B.; Rossano, J.W.; et al. The Registry of the International Society for Heart and Lung Transplantation: Thirty-Fourth Adult Lung and Heart-Lung Transplantation Report-2017; Focus Theme: Allograft Ischemic Time. *J. Heart Lung Transplant.* **2017**, *36*, 1047–1059. [CrossRef]
22. Valapour, M.; Lehr, C.J.; Skeans, M.A.; Smith, J.M.; Uccellini, K.; Goff, R.; Foutz, J.; Israni, A.K.; Snyder, J.J.; Kasiske, B.L. OPTN/SRTR 2018 Annual Data Report: Lung. *Am. J. Transplant* **2020**, *20*, 427–508. [CrossRef]
23. Gholamzadeh, M.; Abtahi, H.; Safdari, R. Machine Learning-Based Techniques to Improve Lung Transplantation Outcomes and Complications: A Systematic Review. *BMC Med. Res. Methodol.* **2022**, *22*, 331. [CrossRef]
24. Dueñas-Jurado, J.M.; Gutiérrez, P.A.; Casado-Adam, A.; Santos-Luna, F.; Salvatierra-Velázquez, A.; Cárcel, S.; Robles-Arista, C.J.C.; Hervás-Martínez, C. New Models for Donor-Recipient Matching in Lung Transplantations. *PLoS ONE* **2021**, *16*, e0252148. [CrossRef]
25. Organ Transplant Trends. Available online: https://unos.org/data/transplant-trends/ (accessed on 5 June 2023).
26. National Data-OPTN. Available online: https://optn.transplant.hrsa.gov/data/view-data-reports/national-data/ (accessed on 28 July 2023).
27. Neri, F.; Eccher, A.; Rigotti, P.; Girolami, I.; Zaza, G.; Gambaro, G.; Mastrosimini, M.; Bencini, G.; Bella, C.D.; Mescoli, C.; et al. Advantages of Using a Web-Based Digital Platform for Kidney Preimplantation Biopsies. *J. Pathol. Inform.* **2021**, *12*, 41. [CrossRef]
28. Liapis, H.; Gaut, J.P.; Klein, C.; Bagnasco, S.; Kraus, E.; Farris, A.B., 3rd; Honsova, E.; Perkowska-Ptasinska, A.; David, D.; Goldberg, J.; et al. Banff Histopathological Consensus Criteria for Preimplantation Kidney Biopsies. *Am. J. Transpl.* **2017**, *17*, 140–150. [CrossRef]
29. Hall, I.E.; Parikh, C.R.; Schröppel, B.; Weng, F.L.; Jia, Y.; Thiessen-Philbrook, H.; Reese, P.P.; Doshi, M.D. Procurement Biopsy Findings Versus Kidney Donor Risk Index for Predicting Renal Allograft Survival. *Transpl. Direct* **2018**, *4*, e373. [CrossRef] [PubMed]
30. Munivenkatappa, R.B.; Schweitzer, E.J.; Papadimitriou, J.C.; Drachenberg, C.B.; Thom, K.A.; Perencevich, E.N.; Haririan, A.; Rasetto, F.; Cooper, M.; Campos, L.; et al. The Maryland Aggregate Pathology Index: A Deceased Donor Kidney Biopsy Scoring System for Predicting Graft Failure. *Am. J. Transpl.* **2008**, *8*, 2316–2324. [CrossRef]
31. Chen, K.; Guntur, G.; Stalam, T.; Okonkwo, N.; Drachenberg, C.; Goussous, N.; Bruno, D.; Mas, V.; Bromberg, J.; Haririan, A. Deceased-Donor Kidney Biopsy Scoring Systems for Predicting Future Graft Function: A Comparative Study. *Transplant. Proc.* **2021**, *53*, 906–912. [CrossRef]
32. Hermsen, M.; de Bel, T.; den Boer, M.; Steenbergen, E.J.; Kers, J.; Florquin, S.; Roelofs, J.J.T.H.; Stegall, M.D.; Alexander, M.P.; Smith, B.H.; et al. Deep Learning-Based Histopathologic Assessment of Kidney Tissue. *J. Am. Soc. Nephrol.* **2019**, *30*, 1968–1979. [CrossRef]
33. Hermsen, M.; Ciompi, F.; Adefidipe, A.; Denic, A.; Dendooven, A.; Smith, B.H.; van Midden, D.; Bräsen, J.H.; Kers, J.; Stegall, M.D.; et al. Convolutional Neural Networks for the Evaluation of Chronic and Inflammatory Lesions in Kidney Transplant Biopsies. *Am. J. Pathol.* **2022**, *192*, 1418–1432. [CrossRef] [PubMed]
34. Kers, J.; Bülow, R.D.; Klinkhammer, B.M.; Breimer, G.E.; Fontana, F.; Abiola, A.A.; Hofstraat, R.; Corthals, G.L.; Peters-Sengers, H.; Djudjaj, S.; et al. Deep Learning-Based Classification of Kidney Transplant Pathology: A Retrospective, Multicentre, Proof-of-Concept Study. *Lancet Digit. Health* **2022**, *4*, e18–e26. [CrossRef] [PubMed]

35. Roufosse, C.; Simmonds, N.; Groningen, M.C.; Haas, M.; Henriksen, K.J.; Horsfield, C.; Loupy, A.; Mengel, M.; Perkowska-Ptasińska, A.; Rabant, M.; et al. A 2018 Reference Guide to the Banff Classification of Renal Allograft Pathology. *Transplantation* **2018**, *102*, 1795. [CrossRef] [PubMed]
36. Remuzzi, G.; Grinyò, J.; Ruggenenti, P.; Beatini, M.; Cole, E.H.; Milford, E.L.; Brenner, B.M.; The Double Kidney Transplant Group (DKG). Early Experience with Dual Kidney Transplantation in Adults Using Expanded Donor Criteria. *J. Am. Soc. Nephrol.* **1999**, *10*, 2591. [CrossRef]
37. Jeong, H.J. Diagnosis of Renal Transplant Rejection: Banff Classification and beyond. *Kidney Res. Clin. Pract.* **2020**, *39*, 17. [CrossRef]
38. Ronneberger, O.; Fischer, P.; Brox, T. U-Net: Convolutional Networks for Biomedical Image Segmentation. In *Medical Image Computing and Computer-Assisted Intervention–MICCAI 2015*; Lecture Notes in Computer Science; Navab, N., Hornegger, J., Wells, W.M., Frangi, A.F., Eds.; Springer International Publishing: Cham, Switzerland, 2015; Volume 9351, pp. 234–241. ISBN 9783319245737.
39. Wilbur, D.C.; Smith, M.L.; Cornell, L.D.; Andryushkin, A.; Pettus, J.R. Automated Identification of Glomeruli and Synchronised Review of Special Stains in Renal Biopsies by Machine Learning and Slide Registration: A Cross-Institutional Study. *Histopathology* **2021**, *79*, 499–508. [CrossRef]
40. Schwantes, I.R.; Axelrod, D.A. Technology-Enabled Care and Artificial Intelligence in Kidney Transplantation. *Curr. Transplant. Rep.* **2021**, *8*, 235–240. [CrossRef]
41. Ayorinde, J.O.O.; Citterio, F.; Landrò, M.; Peruzzo, E.; Islam, T.; Tilley, S.; Taylor, G.; Bardsley, V.; Liò, P.; Samoshkin, A.; et al. Artificial Intelligence You Can Trust: What Matters Beyond Performance When Applying Artificial Intelligence to Renal Histopathology? *J. Am. Soc. Nephrol.* **2022**, *33*, 2133–2140. [CrossRef]
42. Becker, J.U.; Mayerich, D.; Padmanabhan, M.; Barratt, J.; Ernst, A.; Boor, P.; Cicalese, P.A.; Mohan, C.; Nguyen, H.V.; Roysam, B. Artificial Intelligence and Machine Learning in Nephropathology. *Kidney Int.* **2020**, *98*, 65–75. [CrossRef]
43. Girolami, I.; Pantanowitz, L.; Marletta, S.; Hermsen, M.; van der Laak, J.; Munari, E.; Furian, L.; Vistoli, F.; Zaza, G.; Cardillo, M.; et al. Artificial Intelligence Applications for Pre-Implantation Kidney Biopsy Pathology Practice: A Systematic Review. *J. Nephrol.* **2022**, *35*, 1801–1808. [CrossRef]
44. Paquette, F.-X.; Ghassemi, A.; Bukhtiyarova, O.; Cisse, M.; Gagnon, N.; Della Vecchia, A.; Rabearivelo, H.A.; Loudiyi, Y. Machine Learning Support for Decision-Making in Kidney Transplantation: Step-by-Step Development of a Technological Solution. *JMIR Med. Inform.* **2022**, *10*, e34554. [CrossRef]
45. Raynaud, M.; Aubert, O.; Divard, G.; Reese, P.P.; Kamar, N.; Yoo, D.; Chin, C.-S.; Bailly, É.; Buchler, M.; Ladrière, M.; et al. Dynamic Prediction of Renal Survival among Deeply Phenotyped Kidney Transplant Recipients Using Artificial Intelligence: An Observational, International, Multicohort Study. *Lancet Digit. Health* **2021**, *3*, e795–e805. [CrossRef] [PubMed]
46. Szegedy, C.; Vanhoucke, V.; Ioffe, S.; Shlens, J.; Wojna, Z. Rethinking the Inception Architecture for Computer Vision. In Proceedings of the IEEE Conference on Computer Vision and Pattern Recognition, Boston, MA, USA, 7–12 June 2015.
47. Briceño, J.; Calleja, R.; Hervás, C. Artificial Intelligence and Liver Transplantation: Looking for the Best Donor-Recipient Pairing. *Hepatobiliary Pancreat. Dis. Int* **2022**, *21*, 347–353. [CrossRef] [PubMed]
48. Chu, M.J.J.; Dare, A.J.; Phillips, A.R.J.; Bartlett, A.S.J.R. Donor Hepatic Steatosis and Outcome after Liver Transplantation: A Systematic Review. *J. Gastrointest. Surg.* **2015**, *19*, 1713–1724. [CrossRef] [PubMed]
49. Bhat, M.; Rabindranath, M.; Chara, B.S.; Simonetto, D.A. Artificial Intelligence, Machine Learning, and Deep Learning in Liver Transplantation. *J. Hepatol.* **2023**, *78*, 1216–1233. [CrossRef] [PubMed]
50. Narayan, R.R.; Abadilla, N.; Yang, L.; Chen, S.B.; Klinkachorn, M.; Eddington, H.S.; Trickey, A.W.; Higgins, J.P.; Melcher, M.L. Artificial Intelligence for Prediction of Donor Liver Allograft Steatosis and Early Post-Transplantation Graft Failure. *HPB* **2022**, *24*, 764–771. [CrossRef] [PubMed]
51. Yu, H.; Sharifai, N.; Jiang, K.; Wang, F.; Teodoro, G.; Farris, A.B.; Kong, J. Artificial Intelligence Based Liver Portal Tract Region Identification and Quantification with Transplant Biopsy Whole-Slide Images. *Comput. Biol. Med.* **2022**, *150*, 106089. [CrossRef] [PubMed]
52. Sun, L.; Marsh, J.N.; Matlock, M.K.; Chen, L.; Gaut, J.P.; Brunt, E.M.; Swamidass, S.J.; Liu, T.-C. Deep Learning Quantification of Percent Steatosis in Donor Liver Biopsy Frozen Sections. *EBioMedicine* **2020**, *60*, 103029. [CrossRef]
53. Long, J.; Shelhamer, E.; Darrell, T. Fully Convolutional Networks for Semantic Segmentation 2014. *arXiv* **2014**, arXiv:1411.4038.
54. Chen, L.-C.; Papandreou, G.; Schroff, F.; Adam, H. Rethinking Atrous Convolution for Semantic Image Segmentation. *arXiv* **2017**, arXiv:1706.05587.
55. Husain, S.A.; Shah, V.; Alvarado Verduzco, H.; King, K.L.; Brennan, C.; Batal, I.; Coley, S.M.; Hall, I.E.; Stokes, M.B.; Dube, G.K.; et al. Impact of Deceased Donor Kidney Procurement Biopsy Technique on Histologic Accuracy. *Kidney Int. Rep.* **2020**, *5*, 1906–1913. [CrossRef]

MDPI
St. Alban-Anlage 66
4052 Basel
Switzerland
www.mdpi.com

Bioengineering Editorial Office
E-mail: bioengineering@mdpi.com
www.mdpi.com/journal/bioengineering

www.ingramcontent.com/pod-product-compliance
Lightning Source LLC
Chambersburg PA
CBHW061559210926
51850CB00009B/2151
9783725801312